2nd International Symposium on Sustainable Landscape Development 2016

IOP Conference Series: Earth and Environmental Science
Volume 91

Bogor, Indonesia
9 - 10 November 2016

Editor:

Regan Leonardus Kaswanto

ISBN: 978-1-5108-5195-5
ISSN: 1755-1307

Printed from e-media with permission by:

Curran Associates, Inc.
57 Morehouse Lane
Red Hook, NY 12571

Some format issues inherent in the e-media version may also appear in this print version.

Printed by Curran Associates, Inc. (2017)

Institute of Physics
Dirac House, Temple Back
Bristol BS1 6BE UK

Phone: 44 1 17 929 7481
Fax: 44 1 17 920 0979

techtracking@iop.org

Additional copies of this publication are available from:

Curran Associates, Inc.
57 Morehouse Lane
Red Hook, NY 12571 USA
Phone: 845-758-0400
Fax: 845-758-2633
Email: curran@proceedings.com
Web: www.proceedings.com

Table of contents

Volume 91

2nd International Symposium for Sustainable Landscape Development

9–10 November 2016, Bogor, Indonesia

Accepted papers received: 16 October 2017Published online: 8 November 2017

Preface

Green Landscape

Socio-Economic Cultural Landscape

IOP Conf. Series: Earth and Environmental Science **91** (2017) 011001 doi:10.1088/1755-1315/91/1/011001

FOREWORD

Landscape development is facing very high acceleration changes, considering the growth and development of a region. As a result of human activities, various negative impacts arose for the landscape development, such as the loss of greenery open space (GOS) and blue open space (BOS), decreased quality of comfort-related micro-climate, the omission of the potential landscape services as well as a variety of others environmental problems.

Awareness of the importance of sustainable landscape development has been proclaimed by the United Nations (UN) in the Sustainable Development Goals (SDGs) program at Goal 11th, namely *to make cities and human settlements inclusive, safe, resilient and sustainable* (source: https://sustainabledevelopment.un.org/sdg11). A way to ensure the achievement entire SDGs (2016-2030) are to involve communities in the implementation and attainment in accordance with **nawacita** Indonesia. Therefore, it is needed to bridge and facilitate the role of the community in order to realize the SDGs with **nawacita** that can be obtained in the form of scientific activities such as a symposium. The 2nd International Symposium for Sustainable Landscape Development (**The 2nd ISSLD**) is a continuation of the annually symposium, the sustainability of this activity is very important to be managed on a regular basis. The topic for this annually symposium has been adapted to specific conditions for each year.

The 2nd ISSLD was held on November 9-10, 2016 at the IPB International Convention Centre (IICC), Bogor, Indonesia. The 2nd ISSLD is jointly organized by the Department of Landscape Architecture, Faculty of Agriculture, Bogor Agriculture University (IPB), and the Directorate of Research and Innovation IPB (DRI IPB). This symposium produced 40 (forty) articles which are published in this issue of IOP Conference Series: Earth and Environmental Science (EES). Those articles were come from subtopic about green landscape, socio-cultural landscape, and landscape planning and design.

The 2nd ISSLD aims to assess the landscape issue through a variety of case studies related to the effort for achieving sustainable landscape development. Various ideas towards a sustainable landscape are currently an intense research topic in the international level such as research on urban resilience, ecocity, sustainable city, until the low carbon landscapes. At the 2nd ISSLD, the researchers, students, businessman, practitioners, stakeholders, and governments can meet each other and exchange ideas to jointly formulate effort to create sustainable and comfortable tropical landscape. Around 150 participants have joint the 2nd ISSLD that come from various academicians (A), businessman (B), government (G), community (C) and mass media (M) to reach an ABGCM penta helix collaboration for sustainable landscape development.

We are thankful to all authors who have contributed to yield a high scientific standard to this proceeding of the 2nd ISSLD. We are also grateful to IOP Publishing, especially Anete Ashton (Publisher, IOP Conference Series), for allowing us to guide some of the symposium results for being poublished in the IOP Conference Series: Earth and Environmental Science (EES). We are trully thank to all dedicated editorial and scientific committes who have reviewed and encouraged the authors to revised their papers. We also would like to thank Ministry of Research, Technology and Higher Education (*Kementerian Riset, Teknologi dan Perguruan Tinggi* - RISTEKDIKTI) for the opportunity to join the World Class University (WCU) agenda.

Regan Leonardus Kaswanto
Editor-in-Chief
Proceeding the 2nd ISSLD 2016

IOP Conf. Series: Earth and Environmental Science **91** (2017) 011001 doi:10.1088/1755-1315/91/1/011001

Editorial Committees

Regan Leonardus Kaswanto, *Department of Landscape Architecture, Faculty of Agriculture, Bogor Agricultural University IPB, Jl. Meranti Kampus IPB Darmaga 16680, Bogor, West Java, Indonesia.*

Hadi Susilo Arifin, *Department of Landscape Architecture, Faculty of Agriculture, Bogor Agricultural University IPB, Jl. Meranti Kampus IPB Darmaga 16680, Bogor, West Java, Indonesia.*

Wahju Qamara Mugnisjah, *Department of Landscape Architecture, Faculty of Agriculture, Bogor Agricultural University IPB, Jl. Meranti Kampus IPB Darmaga 16680, Bogor, West Java, Indonesia..*

Bambang Sulistyantara, *Department of Landscape Architecture, Faculty of Agriculture, Bogor Agricultural University IPB, Jl. Meranti Kampus IPB Darmaga 16680, Bogor, West Java, Indonesia.*

Tatsuaki Kobayashi, *Departement of Environmental Science and Landscape Architecture, Graduate School of Horticulture, Faculty of Horticulture, Chiba University, Chiba 271-8510, Japan.*

Nor Atiah Ismail, *Department of Landscape Architecture, Faculty of Design and Architecture, Universiti Putra Malaysia, UPM Serdang, Selangor Darul Ehsan, Malaysia.*

Scientific Committees

Afra DN Makalew, *Bogor Agricultural University, Indonesia*
Andi Gunawan, *Bogor Agricultural University, Indonesia*
Aris Munandar, *Bogor Agricultural University, Indonesia*
Damian Tang, *International Federation of Landscape Architects – Asia Pasific Region (IFLA-APR), Singapore*
Hadi Susilo Arifin, *Bogor Agricultural University, Indonesia*
Indung Sitti Fatimah, *Bogor Agricultural University, Indonesia*
Nizar Nasrullah, *Bogor Agricultural University, Indonesia*
Nor Atiah Ismail, *Universiti Putra Malaysia, Malaysia*
Nurhayati HS Arifin, *Bogor Agricultural University, Indonesia*
Prita Indah Pratiwi, *Bogor Agricultural University, Indonesia*
Qodarian Pramukanto, *Bogor Agricultural University, Indonesia*
Regan Leonardus Kaswanto, *Bogor Agricultural University, Indonesia*
Siti Nurisjah, *Indonesian Society of Landscape Architects (ISLA / IALI), Indonesia*
Saiful Arif Abdullah, *the National University of Malaysia, Malaysia*
Syartinilia, *Bogor Agricultural University, Indonesia*
Tati Budiarti, *Bogor Agricultural University, Indonesia*
Tatsuaki Kobayashi, *Chiba University, Japan*
Wahju Qamara Mugnisjah, *Bogor Agricultural University, Indonesia*

Assistant Editors

Rosyidamayanti Twinsari Manningtyas, Yulius Budi Prastyo, Nike Dyah Permata
Department of Landscape Architecture, Faculty of Agriculture, Bogor Agricultural University IPB, Jl. Meranti Kampus IPB Darmaga 16680, Bogor, West Java, Indonesia.

IOP Conf. Series: Earth and Environmental Science **91** (2017) 011001 doi:10.1088/1755-1315/91/1/011001

The 2nd International Symposium for Landscape Sustainable (The 2nd ISSLD)

The Opening Ceremony

All participants take a photo session together

IOP Conf. Series: Earth and Environmental Science **91** (2017) 011001 doi:10.1088/1755-1315/91/1/011001

All keynote and invited speakers are discussing their ideas.

Mayor of Bogor City is giving a keynote speech.

Bogor City towards Sustainable Landscape Development.

IOP Conf. Series: Earth and Environmental Science **91** (2017) 011001 doi:10.1088/1755-1315/91/1/011001

Invited speakers are giving their presentation.

Giving appreciation to invited speaker for their presentation.

Excursion activity in Bogor Regency.

IOP Conf. Series: Earth and Environmental Science **91** (2017) 011002 doi:10.1088/1755-1315/91/1/011002

Peer review statement

All papers published in this volume of *IOP Conference Series: Earth and Environmental Science* have been peer reviewed through processes administered by the proceedings Editors. Reviews were conducted by expert referees to the professional and scientific standards expected of a proceedings journal published by IOP Publishing.

IOP Conf. Series: Earth and Environmental Science **91** (2017) 012001 doi:10.1088/1755-1315/91/1/012001

Land Ecological on Public Transport Infrastructure Development In Indonesia

N Sari
Land Transport Academy, Ministry of Transportation
Jl. Raya setu No.89 CibitungBekasi 17001

E-mail: miss_vita20@yahoo.com

Abstract. The development of public transport infrastructure in Indonesia has been growing rapidly since the last five years. The utilization of area as public transport infrastructure, for example bus depot, bus Station and terminal requires wide area and influences many elements, such as land ecological quality, water supplies, power supplies, and environmental balance. However the development of public transport infrastructure now days is less considering on environmental approach, especially for green and catchment area for water conservation (water balance).This paper aims to propose the concept of Public Transport Infrastructure using green concept. The green design concept is using GBCI (Green Building Council Indonesia) standard, which contains seven categories: land ecological enhancement, movement and connectivity, water management and conservation, solid waste and material, community wellbeing strategy, building and energy, and also innovation and future development. The result is, by using the GBCI standard for the green design of Public Transport Infrastructure, the land ecological impact could be decreased. The effective areas that required are at least 5000 m^2, from which the green areas for public increase 36% and 76% of areas could be used as catchment area for water conservation.

1. Introduction

Urban transportation is a complex system on land use planning and urban design. The provision of transportation systems has a large influence on the form of the built environment and people's quality of life. Sustainable transport development concepts include some point such as vehicle design, vehicle operational, transport system sewage, quality of land use, and reduce capacity (volume). This concept design is to support human ecosystem activity, flora and fauna in the future. Sustainable transport is a transportation that does not cause harmful effects on public health or ecosystems and could support the existing mobility consistently, and is considering to: (a) the use of renewable energy resources at a lower rate from the regeneration rate; and (b) the use of non-renewable resource at a lower level from the development of alternative renewable resources level. (Organization for Economic Co-Operation and Development (OECD), 1994). Concerning on sustainability, the public transport development has given environmental impacts that can impose economic, social and ecological costs. Global air pollution, the durable effects of manufactured toxins, degrades natural resources such as fresh water and fisheries, and the cross-border nature of many environmental problems which all highlight the need to view human impacts from a broad perspective.

Floods are one of the types of disasters more frequently occurring, compared to earthquakes, volcanic eruption, drought and landslides. Floods are even occurring more frequently in the present days. Floods are not only caused by rain that consequently causes the overflowing of surface (run-off) water, but also by the change of land usage, for example forest/land clearing and development in the water absorption area. A number of studies has even shown that human is the main cause of flood, compared to nature itself.

On early 2014, there has been flood disaster in Indonesia. Floods occurred in six provinces, i.e. DKI Jakarta (capital city of Indonesia), Banten, West Java, Yogyakarta, Central Java, and South Sulawesi. Based on United Nations Office for the Coordination on Humanitarian Affairs (OCHA) report, for the last two weeks, Jakarta and outlying areas have experienced continued rains causing river overflows and inundation since 12 January 2014. Thousands of houses, buildings, roads have been flooded. The Provincial Agency for Disaster Management (BPBD) DKI Jakarta reported that as of 21 January 2014, approximately 134,662 persons or 38,672 households in 100 urban villages are directly affected by floods. Twelve casualties have been recorded. At least 62,819 persons are displaced who are sheltered in 253 displacement centers. Floodwaters had blocked some major roads. Flood water levels were vary from 0.2 m up to 1.2 m and around 2 m at the river bank.

The development of public transport infrastructure in Indonesia has been growing rapidly since the last five years. The utilization of area as public transport Infrastructure, for example bus depot, bus Station and terminal requires wide area and influences many elements, such as land ecological quality, water supplies, power supplies, and environmental balance. However the development of public transport infrastructure now days is less considering on environmental approach, especially for green and catchment area for water conservation (water balance).Area terminals and bus depot are usually using flexible pavement or rigid pavement, which this pavement can't absorb rainwater, and the drainage using open drainage system. It can cause ecological impacts from urban land use. The aim of this research is to propose the concept of Public Transport Infrastructure using green concept.

Limitation from this research focuses on public transport infrastructure development for terminals and bus depot. Sample of terminals and bus depot is only in Jabodetabek area, and the analysis is using Green Building Council Indonesia (GBCI) standard.

2. Literature Review

Two major environmental concepts influence contemporary thinking about sustainable developmen. The traditional concept is on of "environmental services", those beneficial functions (such as maintenance of water-flow patterns and recycling of waste) that natural area perform, the full range of genetic diversity (plant and animal species and population) and ecosystem (in which plants and animals exist).

Environmental service are the economically valuable benefits to society that natural area provide. These include creation and protection of soil, stabilization of water-flow patterns, amelioration of climate, breakdown of pollutants, recycling of waste, provision of fish nurseries, and protection against weather damage.

The major sources of negative environmental impacts from urban land use are: (1) displacement or damage of natural areas by intruding development; and (2) pollution of environmental media (air, water and land) by such urban residuals as stormwater runoff or automobile and industrial emissions [7].

Terminals or terminal areas are generally made at the ends or beginning of bus lines, the major transfer point and, generally, areas where buses stop for longer periods of time while still in service. As such, they differ significantly from simple bus stops in that they cover larger areas and accommodate many buses and greater numbers of passengers. They are relative simple, mostly in the open-air structures comprising loading berths, shelters and sidewalks for pedestrians. They are also generally smaller than sophisticated bus stations which usually covered within buildings including shops, restaurants and other facilities. The layout and size bus terminal, depends on the number of bus movements that it will accommodate, the number of passengers, the availability of space, the general effects on the surroundings etc. the critical elements in the design of a bus terminal are therefore : the determination of capacity and the number of loading berths, the geometric characteristics and horizontal layout of these berths, the determination of capacity and the geometric characteristics of the walkways for the movement of pedestrians [5]. Based on Minister of Transportation regulation in Indonesia, classification of bus terminals based on the service isdivided into there catagories : terminal type A, terminal type B and terminal type C.

Bus depot is a garage where buses are stored / parked at night, which is also used to perform maintenance or repair the bus. This depot became the starting point of the bus trip in the morning and the end of the trip at night [4]. The facilities on bus depot include: bus barking area, bus refueling for natural gas and diesel fuels, bus washing area, bus servicing and maintenance facilities, and management building. The activities of a bus depot and workshop can have a significant impact on the environment. Some steps to improve the quality on bus depot area are: rain water collected and will be used to wash down buses and flush toilets, recycling waste oil, using porous well, and recycling bus wash water.

Porous well is the infrastructure to collect and absorb rainwater (surface run-off) into the ground (SNI 03-2453-2002). One of effective method for food mitigation is porous wells. Porous wells is wells or holes made to collect rain water or surface run-off into ground water. By using the porous well could increasing groundwater levels and reducing surface run-off (rain water) [8].

Green Building Council Indonesia (GBCI) is an independent organization established in 2009 by professionals in design and construction industry who concern about green building practices. The main focus of GBC Indonesia is to pursue the socialization and transformation of sustainable green principles, particularly in building construction industry in Indonesia. Rating tools in green buildings is tools for assessing building level that achieve sustainable concept. For the green building benchmarks in Indonesia, GBC Indonesia issued a rating tool titled GREENSHIP. GREENSHIP is developed by Green Building Council Indonesia (GBC Indonesia) by considering the conditions, natural character and regulations as well standards that apply in Indonesia. GBCI (Green Building Council Indonesia) for neighborhood, contains seven categories: Land Ecological Enhancement, Movement and connectivity, water management and conservation, solid waste and material, community wellbeing strategy, building and energy, and also innovation and future development.

3. Research Methodology

The research method used is as follows:

3.1 Collecting data

This research collects data from secondary data sources. The data are collected by using significant source such as books, journals, internet sources, Indonesia government standard, government database and also from developing theoretical reviews and case study. The data that used in this research are land use map, sample of terminals and bus depots area, and rain intensity.

3.2 Method of Analysis

The steps of this research are : literature study and the research method uses GBCI (Green Building Council Indonesia) standard, which contains seven categories: Land Ecological Enhancement, Movement and connectivity, water management and conservation, solid waste and material, community wellbeing strategy, building and energy, and also innovation and future development.

4. Research Analysis

Research area is Jabotabek area, with the area consist of: DKI Jakarta, Bogor, Tangerang, and Bekasi. Figure 1 presents the map of Jabotabek area for public transport network.

Figure 1. Public Transport Network Masterplan for Jabotabek Area 2014-2030
Source : *Minister of Transportation Regulation* [11]

The number of terminal in Jabotabek area is shown in the Table 1 below:

Table 1. Terminals in Jabotabek

Type of Terminal	Quantity
Terminal Type A	15
Terminal Type B	16
Terminal Type C	1

Source : BPTJ, 2016

Bus depot sample, inventory survay result, shown in the Tabel 2 below :

Table 2. Bus depot in Jabotabek area

Name of Bus Depot	Area (m2)	Name of Bus Depot	Area (m2)
PPD		Transjakarta	each ± 7.000
1. Ciputat	67.875	1. Cawang	
2. Depok	7.000	2. Kramat jati	
3. Pulogadung	7.344	3. Pinang Ranti	
4. Bekasi	17.860	4. Cakung	
5. Tangerang	3.371	5. Kampung Rambutan	
6. Cakung	12.000	6. Perintis Kemerdekaan	
7. Cawang	9.251	7. Pesing	
Mayasari Bhakti		Sinar Jaya (Cibitung)	
1. Pasar rebo			
2. Cibitung			

Jabotabek regional development masterplan, includes : monitoring the distribution of population growth, restriction of the development in catchment area in the South, especially in Bogor area, recommendation for linear development along the East-West track (Bekasi-Tangerang), and the development priorities in the financial sector, trade and Jakarta tourism. The existing distribution of bus depots and terminals in Jabotabek based on development zoning is shown in the Figure 2.

As shown in the Figure 2, the distribution of bus depots and terminals are located on urban high density residential area, it means, it can give the environmental impacts from urban land use such as : (1) displacement or damage of natural areas by intruding development; and (2) pollution of environmental media (air, water and land). The analysis for Greenship neighborhood categories as follow:

4.1 Land Ecological Enhancement

Some points on land ecological enhancement are: basic green area, green area for public, habitat conservation, land revitalization, and micro climate. The goals are to protect the harmony and the ecological balance of the environment and to improve the quality of healthy environment (public health), to encourage interaction with providing green area, to minimize the impact of development of balance and biodiversity natural species, to avoid the development in green areas and opening new land, and to improve the quality of the microclimate around the area of the region and reduce the Urban Heat Island (UHI). Trinanda [9] using Greenship from GBC Indonesia proposed the bus depot could be design with area 149.121 m^2 (the bus depot capacity is 500 buses), from which 66% is effective areas and 34% is green area. The GBC Indonesia criteria for minimum area is 5000 m^2 and green area for public 25% from land area, it means the effective land and green area are fulfill requirement, even green areas for public increase 36%.

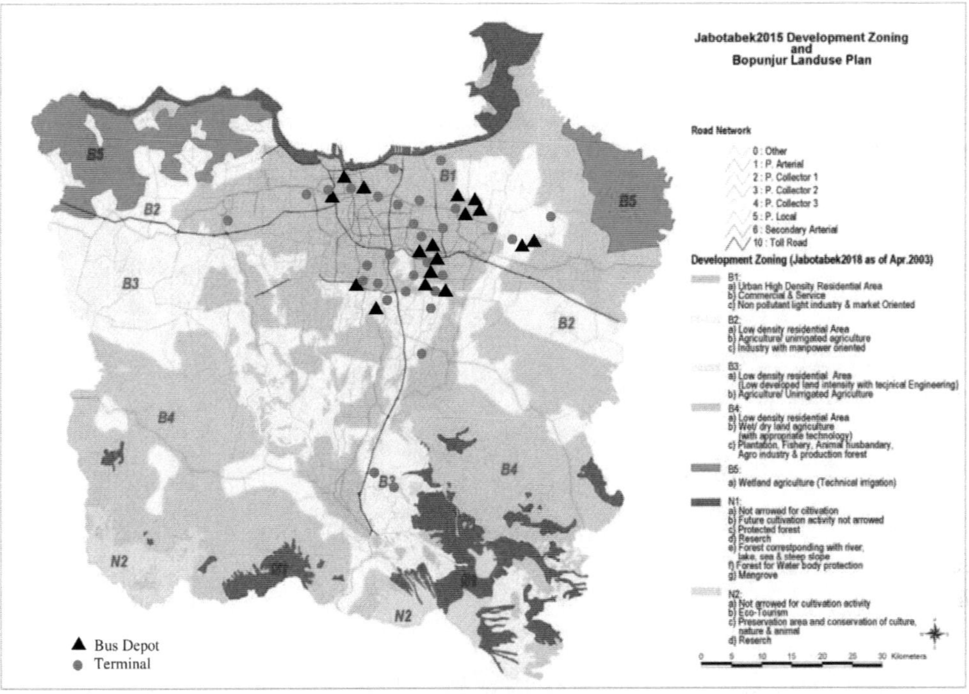

Figure 2. Bus depots and terminals in Jabotabek
Source: Jabotabek Transportation Masterplan with author analytical drawing added, 2016

4.2 Movement and connectivity
Based on movement and connectivity, the accessibility is necessary, it starts from pedestrian network and facilities, people and goods movement analysis, connected area and walkway strategy, public transportation, public utilities and facilities, universal accessibility, bicycle network and storage, and public car park area. The goals are to ensure their accessibility planning, for the movement of people, goods and vehicle, to make the pedestrian become the priority, to open access from the inside area to the outside area, to apply the principle of connectivity, accessibility, safety, comfort and attractive on a pedestrian path. Jehosua, Santosa and Novita [6] proposed the concept of movement and connectivity bus depot shown in the Figure 3 below:

4.3 Water management and conservation
On Bus depot and terminal, water consumption becomes very important, so it should concern on water consumption at the area, to use of alternative water (besides ground water and PDAM water), to design the rainfall management system, and the waste water management. Trinanda [9] propose for Depok area, bus depot with area 149.12 m^2 required 28 units porous wells. Each porous well could collect 14,5m^3 rain water, it means 406 rain water could be collected and 76% of areas could be used as catchment area for water conservation.

4.4 Solid waste and material
In this case, there are several things should be concerned : reduce the negative impact on the environment through solid waste management (garbage), extend the life cycle and add value to the benefits of waste through processing environmental friendly waste, reduce waste taken to landfills

(TPA), using environmental friendly materials, and recycled and reuse materials for road infrastructure.

The main potential problems at terminals and bus depots are traffic congestion caused by buses entering and leaving the area, pollution from exhaust fumes and excessive noise from the vehicles themselves and from other workshop activities. Less visible, but often more serious on bus depots, is environmental damage caused by waste oil or spilled fuel entering the drainage system or polluting nearby rivers. A vehicle workshop generates a considerable quantity of waste oil and if this is not disposed of properly it can cause serious pollution. These environmental problems can be minimized with good design of the facilities, proper maintenance, and good discipline and housekeeping.

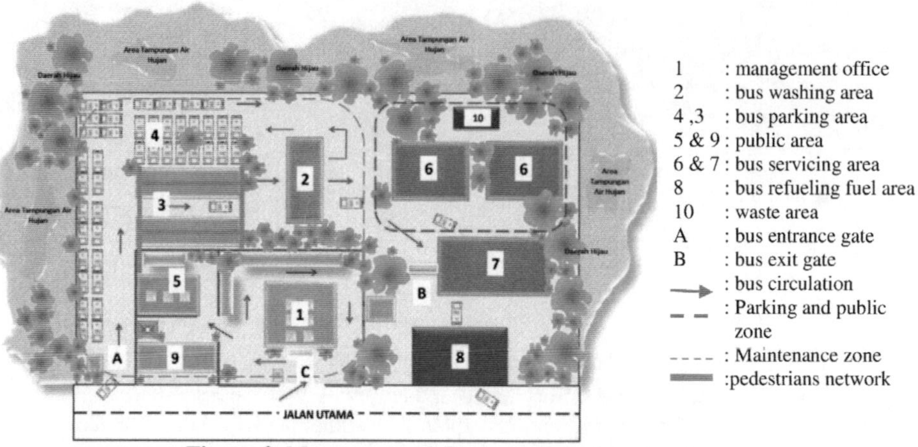

Figure 3. Movement and Connectivity at bus depot area
Source: Jehosua , Santosa and Novita [6]

4.5 Community wellbeing strategy
Provide public facilities for public activity, make the society became priority, increase awareness, knowledge and participation of the society about the concept sustainability, create an area (neighborhood) safe and comfortable.

4.6 Building and energy
Darmanto and Wiguna (2013) suggested that there are some criteria of green building: alternative water resource, thermal comfort, visual comfort (SNI 16-7062-2004), natural lighting , water use reduction (SNI 03-7065-2005), environmental tobacco smoke control and energy measure (SNI 03-6389-2000).

4.7 Innovation and future development
In this case, sustain the implementation of sustainability concept in the area, create the innovation for the environmental sustainability.

5. Conclusion
The result of this reseachare, by using the GBCI standard which contains seven categories: Land Ecological Enhancement, Movement and connectivity, water management and conservation, solid waste and material, community wellbeing strategy, building and energy, and also innovation and future development. The green design concept of Public Transport Infrastructure for bus depot and terminal, the land ecological impact could be decreased. The effective areas that are required at least

5000 m^2. The green areas for public increase 36% and 76% of areas could be used as catchment area for water conservation.

References

[1] _____ 2015 *Greenship Neighborhood versi 1.0,* Green Building council Indonesia (GBCI), Jakarta
[2] Dedy D and I Putu A W 2013 Penilaian Kriteria Green Building pada Gedung Rektorat ITS *JurnalTeknik POMITS Vol.2 No.2* (Surabaya)
[3] Eka AI and Setyawan P 2013 *Rancangan Sumur Resapan Air Hujan sebagai Salah Satu Usaha Konservasi Air Tanah di Perumahan Dayu Baru Kabupaten Sleman Daerah Istimewa Yogyakarta* (Yogyakarta: Universitas Gadjah Mada)
[4] Iskandar A 2011 *Parkir: Pengantar Perencanaan dan Penyelenggaraan Fasilitas Parkir* (Jakarta: Trasindo Gestama Media)
[5] Giannopoulus G A 1989 Bus *Planning and Operation in Urban Area: A Practical Guide* (Avebury, England)
[6] Jehosua J, Santosa P and Novita S 2014 Konsep Desain Eco-Depo Bus *Jurnal Sekolah Tinggi Transportasi Darat*
[7] Keiser E J, Godschaclk D R, and Chapin F S 1995 *Urban Land Use Planning* (Chicago: University of Illinois Press)
[8] Koosdaryani 2009 Penggunaan Modifikasi Desain Sumur Resapan sebagai Pengisian Kembali Air Tanah dan Pengendalian Banjir di Kelurahan Sewu Surakarta *Media Teknik Sipil Vol IX*
[9] Trinanda A 2016 *Konsep Desain Depo Bus Berwawasan Lingkungan* (Jakarta: Fakultas Teknik Sipil Dan Perencanaan Universitas Jayabaya)
[10] SNI: 03-2453-2002 Tata Cara Perencanaan SumurResapan Air Hujan untuk Lahan Perkarangan
[11] PM 54 Tahun 2013 Tentang Rencana Umum Jaringan Angkutan Massal pada Kawasan Perkotaan Jakarta, Bogor, Depok, Tangerang dan Bekasi (JABOTABEK)

IOP Conf. Series: Earth and Environmental Science **91** (2017) 012002 doi:10.1088/1755-1315/91/1/012002

Land Use Cover Changes and Run Off Potention of Cipunten Agung Watershed Banten

A Karima[1] and R L Kaswanto[2]

[1]Research Scholar, Department of Landscape Architecture, Bogor Agricultural University, West Java, Indonesia.
[2]Lecturer, Department of Landscape Architecture, Bogor Agricultural University,

E-mail: karimaafifa@gmail.com

Abstract. The changes of landscape form such as Land Use Cover Changes (LUCC) of Cipunten Agung watershed could be identified periodically in 1995, 2005, and 2015. In general, land utilization in Cipunten Agung classified into protected region and cultivated region. In 2011, total of protected area is 885.80 ha or 22.54% of watershed area. Those conditions affected both positively to the community development and negatively to the water quantity condition in Cipunten Agung such as flooding, run off, and erosion. Therefore, the purpose of this research is to analyze LUCC impacts to run off potential in Cipunten Agung watershed. Supervised classification method and Soil Conservation Services (Qscs) approach were correlated to determine the figure out an optimal solution to reduce the rate of LUCC. Cipunten Agung watershed imagery was classified into five classes, namely water bodies, forest, cultivated tree, settlement and paddy field. The result shows that area of cultivation tree and paddy fields are larger than others in midstream, and settlement is denser in downstream, particularly at riparian landscapes. The LUCC into paddy field often occur at two period 1995 to 2005 and 2005 to 2015 with several area are 530.92 ha and 388.17 ha. The Qscs method calculation result for 1995 until 2015 was affected by land use cover composition in each year and it was defined by Curve Number (CN). High rainfall in 1995 was generating high run off potential volume. Nevertheless, curve number value was increase get near to 100, which indicate the potential of run off volume increases along with LUCC in each year, those are 70.95; 72.47; and 72.81.

1. Introduction

Landscape mozaic changes could occur by the addition of new landscape form and affect its function [1, 2]. The change of landscape form such as Land Use Cover Changes (LUCC) of Cipunten Agung watershed could identify periodically in 1995, 2005, and 2015. In general, land utilization in Cipunten Agung classified into protected region and cultivated region. In 2011, total of protected area is 885.80 ha or 22.54 % of watershed area. The conversion of cultivated tree into paddy field area could decreasing infiltration area, increasing run off, and land erosion. LUCC from forest to developed area such as settlement could increasing run off from high rainfall and flooding in down stream [3]. Thus, when there is LUCC at Cipunten Agung, the function in water flowing also could be changed. The changes of run off volume, could identifed by Curve Number in hydrology analysis. Therefore, the objective of this study was to analyze the effect of LUCC against potential run off volume at Cipunten Agung watershed. This study result could be as consideration in the sustainable and flood free of watershed planingg and management.

2. Methods

2.1 Location, Tool, Materials

This study was located at Cipunten Agung watershed, Pandeglang Regency, Banten Province with coordinate 6°17'30" to 6°22'30" Latitude and 105°50'0" to 105°55'0" Longitude (Figure 1**Error! Reference source not found.**). Cipunten main river length is 23,44 km. The tools used in this study consist of hardware, such as digital camera, GPS, and several geography information system (GIS) softwares. Materials used in this study are Landsat imagery of year 1995, 2005, 2015. Sub-district administration map, watershed boundary map, precipitation map, digital elevation model (DEM), soil type map, and series of Daily rainfall from BMKG (Figure 2).

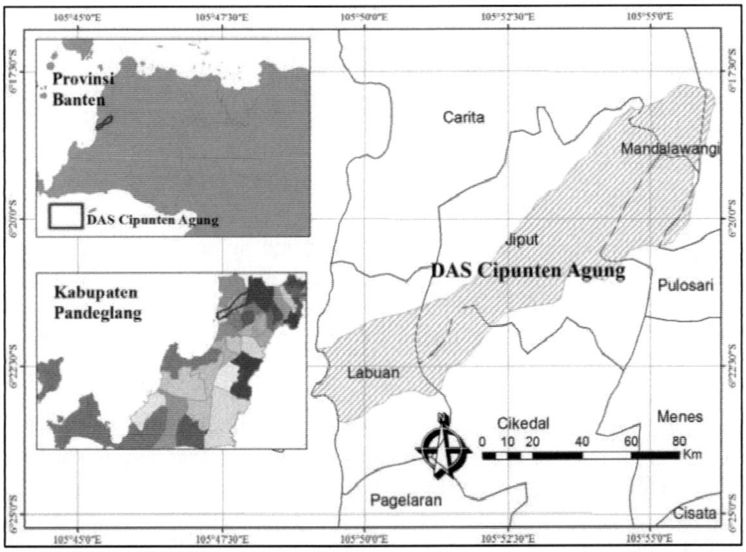

Figure 1. Study area of Cipunten Agung watershed

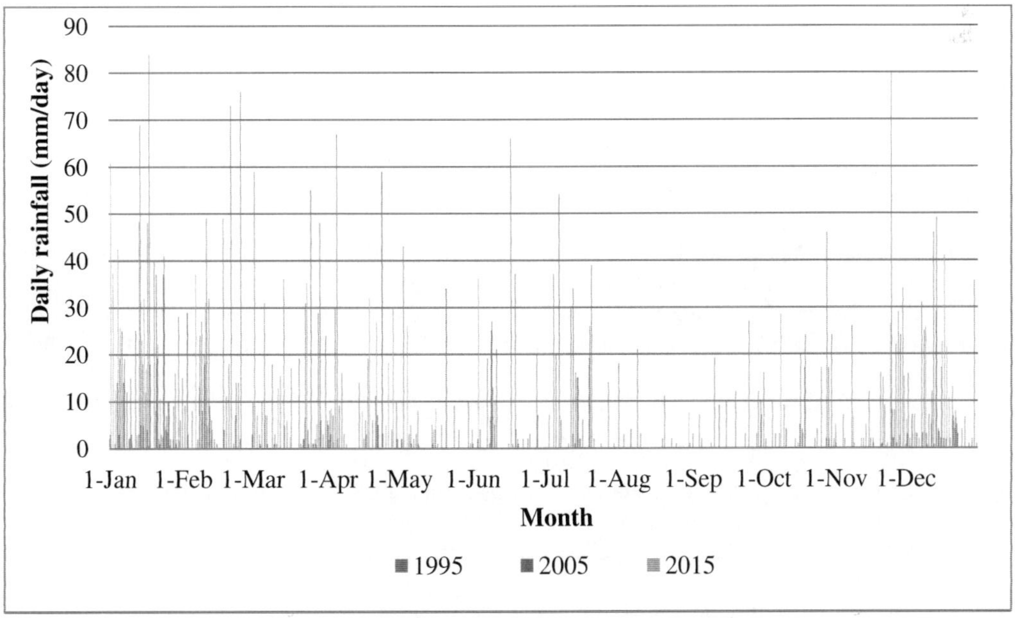

Figure 2. Daily rainfall (Source: BMKG)

2.2 Method

The main method that used in this study was supervised classification for LUCC analyze and Soil Conservation Services for run off volume calculation. Land use and cover was classified into five classes, namely water bodies, forest, cultivated tree, settlement and paddy field. Curve Number (CN) value in Soil Conversion Services (SCS) method is determine by the combination of land use, soil type, and *Antecedent Moisture Categories* (AMC). The estimation of the value of CN can be initiated by determine the Hydrology Soil Group. In this study HSG were identified based on the soil type character. The equation used to determine the CN DAS value as follows :

$$CN\ DAS = \frac{\sum CNi \times Ai}{\sum Ai}$$

Methods CN-SCS has been widely applied to calculate the surface runoff in some watersheds in Java, for example Citarum Watershed in West Java and Ciliwung Watershed in Jakarta [4]. The equation used in determining the amount of runoff volume with SCS method is as follows :

$$S = \frac{25400}{CN} - 254\ and\ Qscs = \frac{(P-0{,}2S)^2}{P+0{,}8S}$$

Qscs is daily surface runoff (mm), S is the maximum potential retention after the start of the run-off (mm), CNI is Curve Number land use, and Ai is the area of land use (ha). CN value is also influenced by Antecedent Moisture Conditions (AMC) or the value of the previous humidity. Value AMC affect the value of the volume and rate of runoff. AMC classification based on:

Condition I : soil in the dry state but not to the wilting poiny, once planted with satisfactory result

Condition II : The state average

Condition III : Heavy or light rain and low temprátures have accoured in the last five days, when saturated soil.

3. Result and Discussion

3.1 Land Use Cover Changes (LUCC)

Protected area of Cipunten Agung watershed consists of protected forest in Pulo Sari Mountain and shoreline area. Many of DAS Cipunten Agung area used as agricultural land. BPS Pandeglang (2015) state that most of 24.75 % family cultivates crops and 24.54 % cultivates horticulture. In 1995, either local community or government has not use Cipunten Agung resource optimally. There were residential development on cultivated tree area and floodplains with a distance 20 m from riparian area, it becomes flooding area in 2010, 2012, 2013, 2015, and 2016. Therefore LUCC were analyze for identify the increases of run off volume which can cause flooding.

The classification process use 15 training area in each class, namely water bodies, forest, cultivated tree, settlement, and paddy field. Classification of 1995 result shows (Figure 3), the smallest area is water bodies 38.98 ha and the widest area is cultivated tree 2570 ha. Land use as cultivated tree and paddy field was intensive for daily needs supply. The classification result for 2005 (Figure 4) shows the widest area is cultivated tree 2,286.01 ha or 57.44 % of total area. Cultivated tree was decreasing while paddy field was increasing to 969.9 ha or 24.37 % of total area. The dense area of settlement from 1995 indicates the communities tend to build at downstream.

The classification result for 2015 (Figure 5) shows the widest area is cultivated tree area with size 2,420.5 ha or 61.41 % of total area. LUCC from paddy field into settlement area many occur from 2005 to 2015 with size 64.62 ha. LUCC graphic (Figure 6) show the decreases of cultivated area and increases of paddy field from 1995 to 2005 and 2015. LUCC into paddy field many occur at Jiput sub district (Middle stream). LUCC into settlement area many occur at 2005 to 2015 with size 195.21 ha. The LUCC in rural landscape particularly change from forest area to paddy fields and gradually change to settlement area. The changing of rural landscape will also impact to the water quantity and quality in agricultural field [5].

IOP Conf. Series: Earth and Environmental Science **91** (2017) 012002 doi:10.1088/1755-1315/91/1/012002

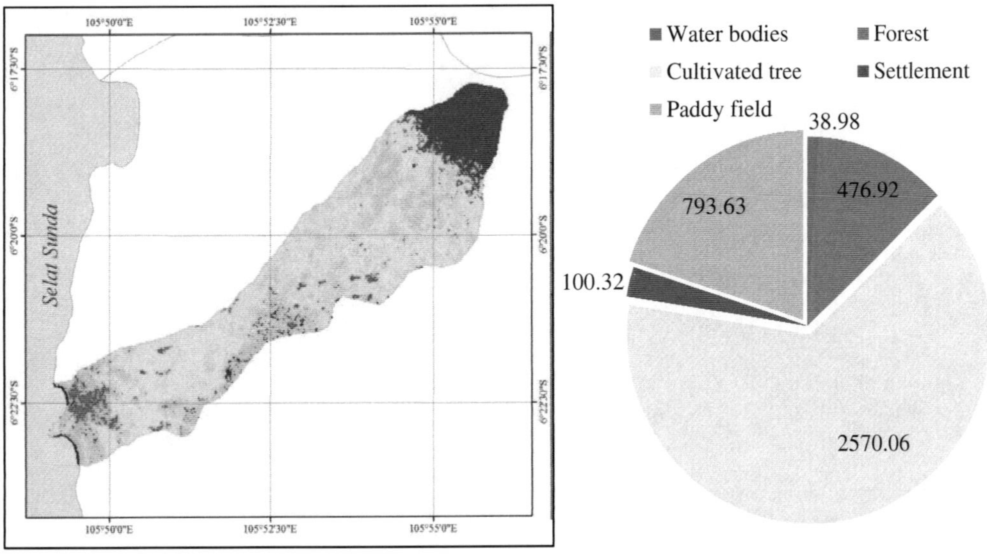

Figure 3. Land Use Cover classification of 1995 (unit area: ha)

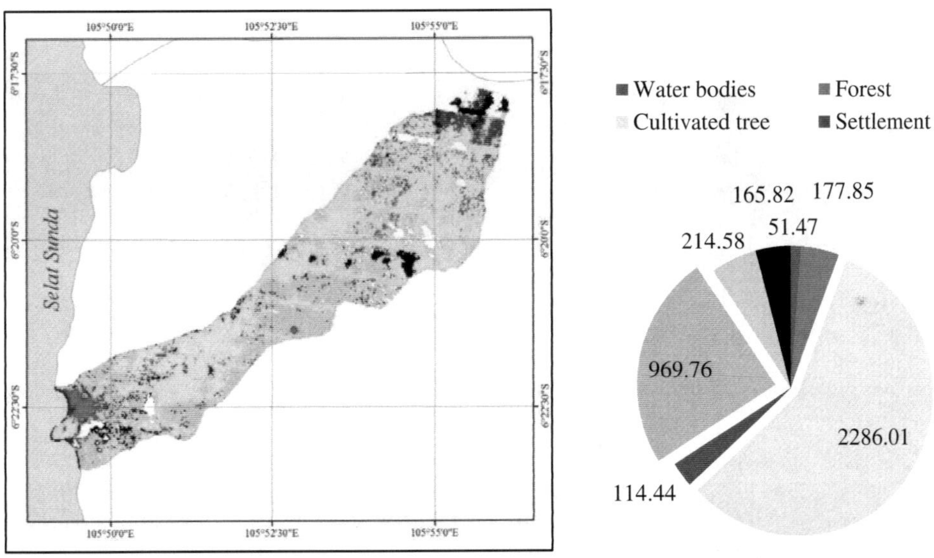

Figure 4. Land Use Cover classification of 2005 (unit area: ha)

Over all, the LUCC analysis from 2005 to 2015 shows there are several cultivated area inside the protected area with size around 49 ha. The size of protected area until 2011 is 885.80 ha (22.54%) of total area. It's not suitable with Indonesia decree number 26 of 2007 clause 17 paragraph 5 which state a watershed must have a minimum of 30% forest area. Therefore a robust action should be arranged in order to reduce the acceleration of the LUCC in this watershed. The LUCC from forest to open field or settlement can cause run off which potential for flooding at downstream area [2, 6]. Run off volume changes could be identified through the Curve Number (CN) value in hydrological analysis. In addition, to reduce the CV number can be achieved by landscape services management in agricultural landscape. The management of landscape services due to landscape simplification can only be addressed by a concerted effort to fundamentally redesign agricultural landscapes [7]

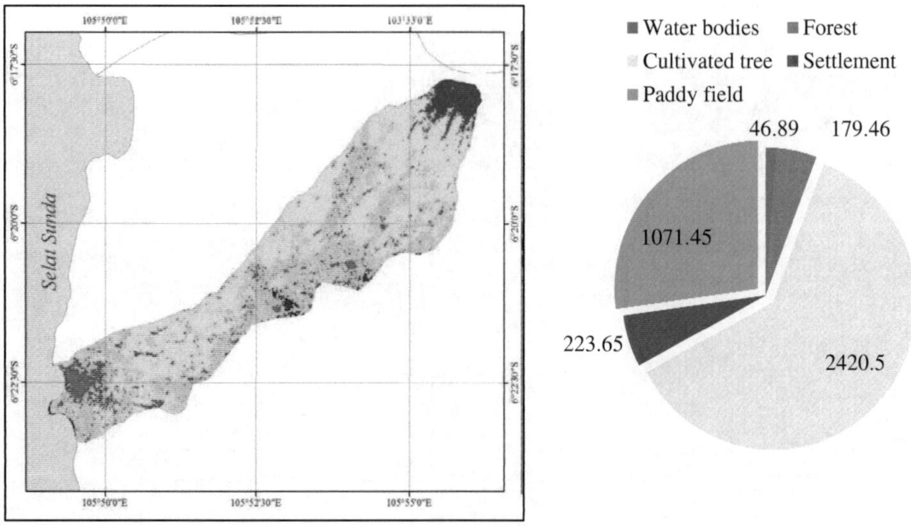

Figure 5. Land Use Cover classification of 2015 (unit area: ha)

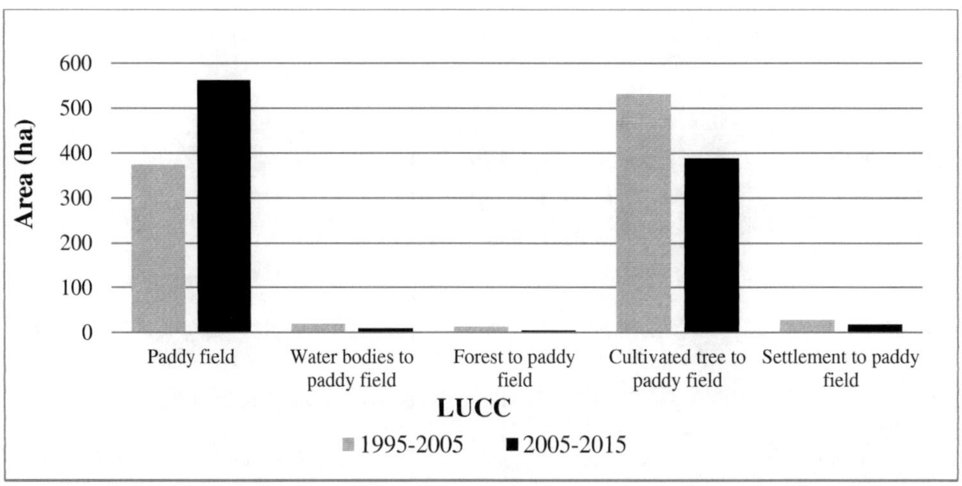

Figure 6. The decreases of cultivated tree and increases of paddy field in 1995-2005

3.2 Curve Number

Run off volume estimation start with the Curve Number calculation, it consists of Hydrology Soil Group (HSG), AMC, and land cover area. HSG indicate the infiltration potential after dampness in certain time [1]. Cipunten Agung watershed widest HSG is group B (latosol type) with area size 2,967 ha or 75.75% of total area, while group C (podsolik type) and D (alluvial type) has area 628 ha and 322 ha or 16.03% and 8.22% of total area. Podsolik and alluvial soil type has infiltration rate at 0-4 mm/ hours, it indicates small capability to infiltrate the water and easy to run off, so it can cause floodplains when heavy rain. In AMC II condition the highest CN for land cover type is 100.00 for water bodies and 84.00 for paddy field while the lowest CN is forest 55.00. Land cover size in each year and CN for each land cover type was calculate and generate watershed curve number (CN DAS) of each year, 70.95; 72.47; and 72.81 (Table 1). The lower value of CN indicates better infiltration capability, in the other side the high value of CN indicates there has been a decreases in the ability of soil to retain the water, so the rain water falls onto the surface more into runoff.

Table 1. Watershed Curve Number (CN DAS) of 1995, 2005, and 2015

Land Use Cover type (1995)	HSG	Area (ha)	AMC II	CN
Water bodies	B	15.90	100.00	1,590.4
	C	7.67	100.00	767.00
	D	0.32	100.00	32.00
Forest	B	467.94	55.00	25,736.70
	C	0.00	70.00	0
	D	0.00	77.00	0
Cultivated tree	B	1,826.46	69.00	126,025.70
	C	520.99	77.00	40,116.23
	D	237.35	83.00	19,700.05
Settlement	B	68.28	83.00	5,667.24
	C	8.87	89.00	789.43
	D	19.12	91.00	1739.92
Paddy field	B	585.32	73.00	42,728.36
	C	90.10	81.00	7298.10
	D	65.23	84.00	5,479.32
CN DAS				70.95

Land Use Cover type (2005)	HSG	Area (ha)	AMC II	CN
Water bodies	B	17.02	100.00	1,702.00
	C	8.35	100.00	834.73
	D	10.63	100.00	1,062.80
Forest	B	162.39	55.00	8,931.45
	C	0.00	70.00	0
	D	0.00	77.00	0
Cultivated tree	B	1,710.50	69.00	118,024.50
	C	433.48	77.00	33,377.96
	D	162.34	83.00	13,474.22
Settlement	B	95.37	83.00	7,915.37
	C	5.13	89.00	456.89
	D	8.98	91.00	816.86
Paddy field	B	685.66	73.00	50,053.18
	C	144.05	81.00	11,668.05
	D	110.83	84.00	9,309.72
CN DAS				72.47

Land Use Cover type (2015)	HSG	Area (ha)	AMC II	CN
Water bodies	B	37.70	100.00	3,770.14
	C	2.03	100.00	202.95
	D	0.14	100.00	14.00
Forest	B	178.03	55.00	9,791.71
	C	0.00	70.00	0

	D	0.00	77.00	0
Cultivated tree	B	1,769.87	69.00	122,121.30
	C	462.58	77.00	35,618.76
	D	191.23	83.00	15,871.94
Settlement	B	145.65	83.00	12,088.67
	C	27.22	89.00	2,422.46
	D	41.65	91.00	3,790.17
Paddy field	B	833.65	73.00	60,856.81
	C	135.83	81.00	11,002.28
	D	88.96	84.00	7,472.58
CN DAS				72.81

3.3 Run off volume

Run off volume estimation influence by rainfall, CN DAS, and the maximum potential of retention when runoff start (S) (Figure 7). Heavy rain about 84 mm/day was happen in January 18th, 1995 when cultivated tree and paddy field were dominated with area size 2,570.06 ha and 793.63 ha. Based on the CN DAS calculation the result is 70.90 and it's lower than 2005 and 2015. The maximum retention of 1995 is 104.00 it could be identify after CN calculation. It indicates that land cover has potential to retaining the run off stream. In 1995 highest run off volume potential is 23.95 mm. Even though there were wider area of cultivated area, but it has not been able to reduce the run off potential volume that affected by heavy rainfall.

In 2005, land cover was dominated by cultivated tree and paddy field with area size 2,286.01 ha and 969.76 ha or 57.44 % and 24.37 % of total area. The maximum daily rainfall is 76 mm/day, based on the CN DAS calculation the result is 72.47 and the maximum retention value is 96,49. It indicates the degradation of water retention capability from 1995. Highest run off volume calculation result in 2005 is 20.99 mm.

In 2015, land cover was dominated by cultivated tree and paddy field with area size 2,420.5 ha and 1,071.45 ha or 61.40 % and 27.18 % of total area. The maximum daily rainfall is 60.80 mm/day, based on the CN DAS calculation the result 72.81 and the maximum retention value is 94.85. It indicates the degradation of water retention capability from 2005. Highest run off volume calculation result in 2015 is 12.80 mm. It is predicted that the run-off will gradually occurred more frequently and denser.

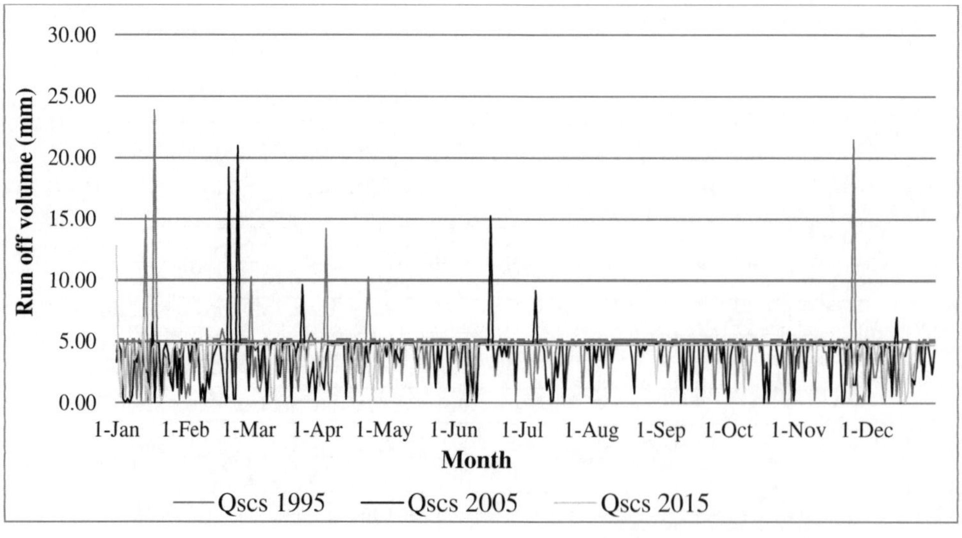

Figure 7. Run off volume calculation result of 1995, 2005, and 2015

3.4 LUCC vs Run off volume

Daily run off volume summation could be a water yield or total of water volume production from run off in one year period (Decree of the Director General of Land Rehabilitation and Social Forestry). Total water yield of 1995, 2005, and 2015 are 1,588.96; 1,442.59; and 1,575.80 mm. it could be increases in line with daily rainfall and causing flood, even though physically flooding causes are sedimentation and riparian constriction. Department of Water Resources and Resettlement of Banten Province state the initial steps to deal with flooding is reducing sedimentation in estuary area and water control construction. Land use cover and run off volume could be the consideration material in watershed management, especially the reduction of flooding and water quality improvement [8].

Based on run off volume analysis, it shows that run off volume more affected by daily rainfall than by the size of cultivated tree area or LUCC. The effects of LUCC to run off volume illustrated through the linear correlation with the Curve Number of each land cover type. The correlation graphic shows positive relation with R^2 0.99. It indicates CN value will be closed to 100 when the land cover type can not absorb the water well. The application of CN and Qscs method for run off estimation using remote sensing data (TRMM and Landsat) has a good potential to be developed as an early warning system against flooding, especially in predicting when the floods that can be determined from surface run off hydrograph.

4. Conclusion

LUCC affected run off volume was illustrated by the increases of watershed Curve Number. This indicates that the ability of Cipunten Agung watershed store water was decreasing with the intensive use of agricultural land and settlement. In 1995, cultivated tree was dominated land cover, but in that time still produce higher volume of run off than any period affected by the heavy rain. If Cipunten Agung watershed have a heavy rain with the decreases of cultivated tree area, then at downstream area will have a high potential of flooding.

5. Acknowledgements

This study was conducted under the project entitled Management of Agroforestry Landscape as Landscape Services Provider towards Low Carbon Landscape of PUPT IPB 2016. This study was supported by Kemenristek Dikti of Republic of Indonesia.

6. References

[1] Dramstad W, Olson JD, Forman RTT. 1996. *Landscape Ecology Principles in Landscape Architecture and Landuse Planning.* Washington (US): Island Press.
[2] Aroengbinang, BW, and RL Kaswanto. 2015. Driving Force Analysis of Landuse and Cover Changes in Cimandiri and Cibuni Watersheds. Procedia Environmental Sciences 24:184-188.
[3] Kodoatie RJ, Robert, Rustam S. 2006. *Pengelolaan Bencana Terpadu : Banjir, Longsor, Kekeringan dan Tsunami.* Jakarta (ID): Yusuf Watampone Press.
[4] Asdak C. 2007. *Hidrologi dan Pengelolaan Daerah Aliran Sungai.* Yogyakarta (ID): Gadjah Mada University Pr.
[5] Kaswanto RL, Arifin HS and Nakagoshi N. 2012. Water Quality Index as A Simple Indicator for Sustainability Management of Rural Landscape in West Java, Indonesia. International J of Enviromental Protection 2 19.
[6] Sofan P, Febrianti N, Prasati I.2014. *Estimasi Limpasan Permukaan dari Data Satelit untuk Mendukung Peringatan Dini Bahaya Banjir di Wilayah Jabodetabek. J Penginderaan Jauh.* 11(1): 43-62.
[7] Landis DA. 2017. Designing agricultural landscapes for biodiversity-based ecosystem services. Basic Appl Ecol 18:1-12.
[8] Karima A, and RL Kaswanto. 2017. Land Use Cover Changes and Water Quality of Cipunten Agung Watershed Banten. IOP Conference Series: Earth and Environmental Science 54:012025.

The Habitat Susceptibility of Bali Starling (*Leucopsar rothschildi*Stresemann 1912) Based on Forest Fire Vulnerability Mappin in West Bali National Park

F Pramatana[1], L B Prasetyo[2], and S B Rushayati[3]

[1,2,3] Forest Resource Conservation and Ecotourism Departement, Faculty of Forestry, IPB

Email: fadlan.pramatana@gmail.com

Abstract. Bali starling is an endemic and endangered species which tend to decrease of its population in the wild. West Bali National Park (WBNP) is the only habitat of bali starling, however it is threatened nowadays by forest fire. Understanding the sensitivity of habitat to forest & land fire is urgently needed. Geographic Information System (GIS) can be used for mapping the vulnerability of forest fire. This study aims to analyze the contributed factor of forest fire, to develop vulnerability level map of forest fire in WBNP, to estimate habitat vulnerability of bali starling. The variable for mapping forest fire in WBNP were road distance, village distance, land cover, NDVI, NDMI, surface temperature, and slope. Forest fire map in WBNP was created by scoring from each variable, and classified into four classes of forest fire vulnerability which are very low (9 821 ha), low (5 015.718 ha), middle (6 778.656 ha), and high (2 126.006 ha). Bali starling existence in the middle and high vulnerability forest fire class in WBNP, consequently the population and habitat of bali starling is a very vulnerable. Management of population and habitat of bali starling in WBNP must be implemented focus on forest fire impact.

1. Introduction

According to International Union for Conservation of Nature and Natural Resources (IUCN) red list, Bali starling (*Leucopsar rothschildi* Stresemann) [22] is categorized to the critically endangered species.The Convention on International Trade in Endangered Species of Wild Fauna and Flora (CITES) put the Bali starling in Appendix 1 which is categorized as a critically endangered species and it is not allowed for free traded. Bali starling is also protected by government of Indonesia trough the Minister of Agriculture decision No. 421/Kpts/Um/8/70on 26 August 1970. In addition, Bali starling is one of the 14 Indonesian priority species which the population in the wild has toincrease 3% a year. However, the Bali starling population tend to decrease every year which lead to be categorized as a rare and critically endangered species [22].

West Bali National Park (WBNP) is the only one habitat of Bali starling. Long time ago, Bali starling can be found widly in the most WBNP area, however, only 6 individues of Bali starling were recorded on 2001 [16]. Forest fire is suggested as the factor causing the constricting home range of Bali starling. It may contribute to unsecure environment for Bali starling reproduction believed One of them that causes bali starling home range decreased is forest fire and on 2001, the species only 6 in the wild, so stated for secure reproduction very difficult [6].

Forest fire have occured almost every year in WNBP. In the period of 2009-2014, forest fires destroy about 32 ha of WNBP [20].Forest fire has negative impact on local and national development, such as to the biodiversity, livelihood, and healthy [18]. The study of forest fire maping, home range and habitat of Bali starling in WBNP is very important for the in situ conservation of Bali starling.Geographic Information System (GIS) can be used for maping the vulnerability of forest fire. This study aimed to analyze the causes of forest fire, and to map out the vulnerability level of forest fire in WBNP, in relation to and to estimate the habitat threatened of Bali starling. It hopes to contribute to as basis for the habitat management of Bali starling in WNBP.

2. Methods

2.1. Materials

Research field was conducted in West Bali Natioal Park (WBNP) Bali Province on March 2016.Supporting materials used for collecting and processing data were camera, binocular, global positioning system, administration bali map, landcover map, bali road map, bali village map, landsat 8 path/row 117/66, software ArcGIS 10.3, and erdas imagine 2014, DNRGPS.

2.2. Map Interpretation and Field Observation

Interpretation of map data was carried out to obtain the map of as follows: surface temperature, village distance, road distance, Normalized Difference Vegetation Index, Normalized Difference Moisture Index, slope, and vegetation cover.Datas of fire area groundcheck, landcover groundcheck, and bali starling distribution were collected by field observation.

2.3. Data Processing and Analysis

2.3.1. Distance Map

Distance map used in the mapping of forest fire was road distance and village distance to WNBP. Asuming, the risk of forest fire will be low when forest area is far from road, village, and river [3].

2.3.2. Land Cover

Land cover map was created by processing the landsat 8 path/row 117/66 with supervised classification in software Erdas imagine 2014. The map accuracy was evaluated by groundcheck observation.

2.3.3. Surface Temperature

The surface temperature map was created base on the band 10 from landsat 8. The surface temperature was calculated by the formulas as follow:

$$L\lambda = (Radiance\ mult\ band\ 10\ x\ Digital\ number) + Radiance\ add\ band\ 10$$

$$T = \frac{K2}{ln\frac{K1}{(L\lambda + 1)}}, and\ C = T - 273{,}15$$

Remarks:

T : Efective temperature (K)
K_2 : Calibration constants 2
K_1 : Calibration constants 1
L_λ : Radiation spectral (W/(m2*ster*µm))
C : Temperature (^{o}C)

2.3.4. Normalized Difference Vegetation Index (NDVI)

Vegetation Index (NDVI) is an index describingthe vegetation level at the certain location [18]. NDVI map was obtained based on the formula as follow :

$$NDVI = (NIR - RED) / (NIR + RED)$$

Remarks:

NDVI : Normalized Difference Vegetation Index
NIR : Band 5 on landsat 8
RED : Band 4 on landsat 8

2.3.5. Normalized Difference Moisture Index (NDMI)

Moisture index was used for evaluation of element moisture defferences in the landscape [7]. NDMI value wasobtained from normalized band 4 and band 5 on landsat 7 [13]. From landsat 8, NDMI value was obtained from normalized band 5 and band 6 with the formula as follow:

$$NDMI = (NIR - SWIR\ 1) / (NIR + SWIR\ 1)$$

Remarks:

NDMI : Normalized Difference Moinsture Index
NIR : Band 5 on landsat 8
SWIR 1: Band 6 on landsat 8

2.3.6. Slope

Slope map wasobtained by processing STRTM data with spatial analyst slope in software ArcMap 10.3 and it was devided by class.It was indicated that WBNP area with low level of slope and high accessibility tend to be more risk to forest fire.

2.3.7. Vulnerabelityof Forest Fire

Variable and scoresystem that used for mapping of forest fire vulnerabelity was referred toArianti and Hadi M [3] which is presented in Table 1.

Tabel 1. Variable and Score for Vulnerabelity Level of Forest Fire in WBNP

Variable	Value (%)	Class	Score
Road distance	18	Distance<500 m	5
		500 ≤ distance< 1000 m	4
		1000 ≤ distance< 1500 m	3
		1500 ≤ distance< 2000 m	2
		Distance ≥ 2000 m	1
Village distance	18	Distance< 1000 m	5
		1000 ≤ distance< 2000 m	4
		2000 ≤ distance< 3000 m	3
		3000 ≤ distance< 4000 m	2
		Distance ≥ 4000 m	1
Land cover	13	Savanna (Very dry)	5
		Coastal forest and monsoon forest (Dry)	4
		Lowland forest (moist)	3
		Mangrove (Dampish)	2
		Water (Wet)	1
Variable	Value (%)	Class	Score
Temperature	18	Temperature< 20 °C	1
		20 ≤ temperature< 22 °C	2
		22 ≤ temperature< 24 °C	3

		$24 \leq$ temperature $< 26\,^{o}$C	4
		Temperature $\geq 26\,^{o}$C	5
NDVI	12	NDVI < 0.36	5
		0.36 ≤ NDVI < 0.43	4
		0.43 ≤ NDVI < 0.5	3
		0.5 ≤ NDVI < 0.57	2
		NDVI ≥ 0.57	1
NDMI	14	NDMI < 0.25	5
		0.25 ≤ NDMI < 0.30	4
		0.30 ≤ NDMI < 0.35	3
		0.35 ≤ NDMI < 0.40	2
		NDMI ≥ 0.40	1
Slope	7	Slope< 8%	5
		8% ≤ slope< 15%	4
		15% ≤ slope< 25%	3
		25% ≤ slope< 40%	2
		Slope ≥ 40%	1

Vulnerabelity of forest fire value wasobtained by this formula :

NKK = a*S(RD) + b*S(VD) + c*S(LC) + d*S(ST) + e*S(NDVI) + f*S(NDMI) + g*S(S)

Remarks:
NKK : Vulnerabelity of forest fire score
(a - g) : Variable value
S(RD) : Road distance
S(VD) : Village distance
S(LC) : Land cover
S(ST) : Surface temperature
S(NDVI) : Normalized Difference Vegetation Index
S(NDMI) : Normalized Difference Moisture Index
S(S) : slope

Final map was created base on the variable combination and it was analyzed by multiplicate the variable value and score value of each variable. The result was devided to vulnerabelity class by formula as follow:

$$Class\ interval = \frac{maksimum\ value - minimum\ value}{Total\ class}$$

The forest fire vurnerability was classified in to different level based the scoring as follow:

Vulnerability forest fire very low level : $100 \leq x \leq 200$
Vulnerability forest fire low level : $200 < x \leq 300$
Vulnerability forest fire middle level : $300 < x \leq 400$
Vulnerability forest fire high level : $400 < x \leq 500$

2.4. Bali Starling Distribution

The distribution data of Bali starling was obtained by Ad hoc survey method. Ad hoc method was carried out by searching information ofBali starling from local people in WBNP.Next, field survey of

Bali starling in the wild was done to verify the existence of the species [12]. The data of bali starling distribution was combined with previously obtained map. The final map expressed the habitat and population susceptibility of Bali starling in WBNP.

3. Results and Discussion

3.1. Variable Factor for the Evaluation of Forest Fire Vurnerability in WNBP

3.1.1. Road Distance

WBNP is located in the area which is accesible to public transportation. WBNP was passed by the main road from Gilimanuk port to Tabanan and Denpasar city. Consequently, it lead to the very easy access to get WBNP.The area and proportion road distance value to WBNP total areawas presented in Table 2.

In the recent five years period (2009-2014), the forest fire in WNBP mostly occured in the area with has interval road distance <500 m (88.24%). It is the fact that un-managed local community surounding the WNBP area might be as an important factor for the forest fire vulnerability in WNBP.

Table 2. The area and proportion road distance value in WBNP

No.	Road distance (m)	Area (Ha)	Proportion (%)	Forest Fire
1.	Distance < 500	5 005.113	26.29	30
2.	500 ≤ distance < 1000	4 352.032	22.86	4
3.	1000 ≤ distance < 1500	2 991.992	15.72	0
4.	1500 ≤ distance < 2000	2 250.028	11.82	0
5.	Distance > 2000	4 436.063	23.30	0

3.1.2. Village Distance

The area of WBNP covered "enclave" of some villages, such as Gilimanuk village and Sumberklampok village, which are exist before WNBP is defined as national park. Furthermore, WNBP area was also surounded by many villages, such as Pejarakan village, Melaya village, and Blimbingsari village, which all of them commonly have direct acces to WNBP. The area and proportion village distance value was presented in Table 3.

Table 3. The area and proportion village distance value in WBNP

No.	Village distance (m)	Area (Ha)	Proportion (%)	Forest Fire
1.	Distance < 1000	2 474.040	12.99	12
2.	1000 ≤ distance < 2000	3 709.119	19.48	1
3.	2000 ≤ distance < 3000	3 613.733	18.98	1
4.	3000 ≤ distance < 4000	3 022.745	15.88	6
5.	Distance > 4000	6 219.888	32.67	14

During period of 2009-2014, the forest fire in WNBP mostly occured in the area withinterval village distance >4000 mwhich it is about 41.18%.It was an interesting that forest fire occured in the longest distance from the villages. It is indicated that WNBP is categorized as an open access area which local people can enter MNBP area very easily.

3.1.3. Land Cover

According to landsat 8 OLI/TIRS analysis on 19th January 2016, vegetation cover on WBNP was classified into 6 classes; those are savanna, water, lowland forest, mangrove, monsoon forest, and data deficient; with the accuracy assessment value is 85.05%. The accuracy assesment was in accordance with the United States Geological Survey, which state that overall accuracy value must be greather than 85% [2].The area and proportion vegetation cover in WBNP was presented on Table 4.

Table 4. The area and proportion vegetation cover in WBNP

No.	Vegetation Cover	Area (Ha)	Proportion (%)	Forest Fire
1.	Data deficient	113.239	0.59	0
2.	Savanna	1 573.765	8.17	18
3.	Water	4 786.330	24.86	0
4.	Lowland forest	1 380.634	7.17	0
5.	Mangrove	392.067	2.04	0
6.	Monsoon forest	11 008.296	57.17	16

In the past five years (2009-2014), forest fire in WBNPmostly occured at savanna (52.94%) and monsoon forest (47.06%). This finding was reasonable phenomenon, because savanna commonly is characterized as dry area whihc is very easy to burned-out. The forest fire in monsoon forest also occured mainly in the dry season. By long dry season, vegetation is getting dry and it more accessible to get fire.

3.1.4. Surface Temperature

According to band 10 (thermal) analysis from landsat 8 OLI/TIRS on 19th January 2016, surface temperature value in WBNP was in the interval 16.15-31.07 °C. The area and proportion ofsurface temperature value in WBNP was presented on Table 5.

During years of 2009-2014, forest fire in WBNP occured in the area with interval surface temperature 20-22 ^{0}C. Nurdiana and Risdiyanto [11] stated that hotspot will appeare on the surface temperature interval 18-28 ^{0}C, especially 24-26^{0}C. Consequently, most of the WNBV area has high risk to become hotspot.

Table 5, The area and proportion surface temperature in WBNP

No.	Surface Temperature	Area (Ha)	Proportion (%)	Forest Fire
1.	Temperature < 20 ^{0}C	1 858.727	9.78	0
2.	20 ^{0}C ≤ temperature < 22 ^{0}C	9 793.651	51.51	6
3.	22 ^{0}C ≤ temperature < 24 ^{0}C	5 304.998	27.90	15
4.	24 ^{0}C ≤ temperature < 26 ^{0}C	1 365.168	7.18	9
5.	Temperature > 26 ^{0}C	689.654	3.63	4

3.1.5. NDVI (Normalized Difference Vegetation Index)

Interval of NDVI value is between -1 to 1. The vegetation has NDVIvalue between 0 to 1, and non vegetation has NDVI value between -1 to 0. The potential burned-out of vegetatio increase when the level of NDVI value is low. Based on the Table 5, NDVI value in WBNP area was in the interval -0.27 until 0.60.The area and proportion the distribution of NDVI value in WBNP was presented on Table 6.

Table 6. The area and proportion of NDVI value in WBNP

No.	NDVI Value	Area (Ha)	Proportion (%)	Forest Fire
1.	NDVI < 0.36	7 111.110	37.36	19
2.	0.36 ≤ NDVI < 0.43	3 120.733	16.39	12
3.	0.43 ≤ NDVI < 0.50	7 464.974	39.22	3
4.	0.50 ≤ NDVI < 0.57	1 241.472	6.52	0
5.	NDVI > 0.57	2.391	0.01	0
6.	Data Deficient	95.255	0.50	0

Latest five years on 2009-2014, the most groundcheck forest fire in WBNP on interval NDVI value < 0.36 with the proportion 55.88%. Forest fire groundcheck which the NDVI value < 0.5 have a high potention to burned-out. This potention will be increased if the located at west region in WBNP, cause the rainfall value smaller than east region in WBNP [20].

3.1.6. NDMI (Normalized Difference Moisture Index)
NDMI variable is one of the vegetation index that use to know moisture value in the reasearch location [9].NDMI value in WBNP region at interval -0.151 until 0.401.The area and proportion distibution NDMI value in WBNP will be described on Table 7.

Table 7. The area and proportion distibution NDMI value in WBNP

No.	NDMI Value	Area (Ha)	Proportion (%)	Forest Fire
1.	NDMI < 0.25	10 454.358	54.92	32
2.	0.25 ≤ NDMI < 0.30	5 723.189	30.07	2
3.	0.30 ≤ NDMI < 0.35	2 718.040	14.28	0
4.	0.35 ≤ NDMI < 0.40	45.216	0.24	0
5.	NDMI > 0.40	0.205	0.001	0
6.	Data Deficient	95.172	0.50	0

In the period of 2009-2014, the forest fire in WBNP was oocured in the area that has NDMI value <0.25 with the proportion 94.12%. The NDMI value of <0.25 was categorized as low level score.If the level of NDMI value is low, the potential forest fire is increased.

3.1.7. Slope
Slope was included as variable of forest fire vurnerability in WBNP. Slope value was analyzed base on the SRTM data. Asuming, if the slope value is low level or the area has high open access to local people, the potential forest fire might be increased.The area and proportion distribution slope value in WBNP was described on Table 8.

Table 8. The area and proportion distribution slope value in WBNP

No.	Slope	Area (Ha)	Proportion (%)	Forest Fire
1.	Slope < 8%	10 977.122	57.66	24
2.	8% ≤ slope < 15%	3 174.247	16.67	2
3.	15% ≤ slope < 25%	2 219.675	11.66	8
4.	25% ≤ slope < 40%	1 884.360	9.90	0
5.	Slope > 40%	782.490	4.11	0

In the period of 2009-2014, the most forest fire in WBNP occured in the area that has slope value <8% with the proportion 70.59%. Based on the land cover, the forest fire occured in the savanna and monsoon forest which have low level of slope. The level of slope is also realted to the accesibilty. The lower slop level, the easier of accessibility, and consequently the potential forest firest migt be increased.

3.2. Forest Fire Vulnerability
Based on the scoring of each variable, the fores fire vulnerability was classified into four classes, there are very low, low, middle, and high class (Table 9). According to hotspot clasification by NASA active fire data, the hotspot value in WBNP was <75%. It is indicated that the forest fire in WNBP was not because of hotspot point, but due to the open access factor and the vegetation characteristic which was very dry. Table 9 showed that the most of forest fire occured in the middle level of forest fire vulnerability (6 778.656 ha). The area and proportion forest fire level in WBNP will be described on Table 9.

Table 9. The area and proportion forest fire level in WBNP

No.	Forest Fire Vulnerabelity Level	Area (Ha)	Proportion (%)	Forest Fire
1.	Very low	9.821	0.07	0
2.	Low	5 015.718	35.76	0
3.	Middle	6 778.656	48.34	16
4.	High	2 126.006	15.16	18
5.	Data deficient	94.245	0.67	0

In the period of 2009-2014, the forest fire in WBNP occured in the area that has a midlle and high level of forest fire vulnerability classes, and it was 100% confirmed by groundcheck observation. The result showed that model of fores fire vulnerability in WBNP can be used as a predicted tool for the forest fire assesment in WBNP with high validation value (>80%). The susceptible area in WBNP is 8.905 ha which was classified to middle and high level of forest fire vurnerability. The vegetation characteristic and open access area of WBNP should be managed properly for endemic species habitat.

3.3. Bali Starling Distribution

Bali starling population in the wild is very low and it is included into the criticallty endangered specieslist [14].To support the sustainable wildlife, it needs the region which have all of the necessary things; such as food, water, healty air, mineral, cover, breeding site, and caring site [15]. The trees which the most used by Bali starling for daily activity, such as pearch and sleep, are pilang (*Acacia leucophloea*) and walikukun (*Schoutenia ovata*). Furthermore, talok (*Grewia koordersiana*), tekik (*Albizia lebbeckioides*), kemloko (*Phyllantus emblica*), and kesambi (*Schleira oleosa*) were used by Bali starling for getting food; such as caterpillar, ant, and termite [15]. Based on Rianto [16], Bali starling was found in Teluk Brumbun on 1998 and 2002-2005 where the area was dominated by pilang (*Acacia leucophloea*), walikukun (*Schoutenia ovata*), and talok (*Grewia koordersiana*) trees.

Based on the survey in the wild area, Bali straling appeared in some location, such as Lampu Merah resort, Teluk Brumbun resort, Cekik Gilimanuk resort, and Labuan Lalang. Bali starling distributein the wider area due to the addition of food shelter distribution and breeding activitiy.As the result, the population and distribution of Bali starling increased. The habitat distribution and population susceptibility of Bali starling map in WBNP was showed in Figure 1.

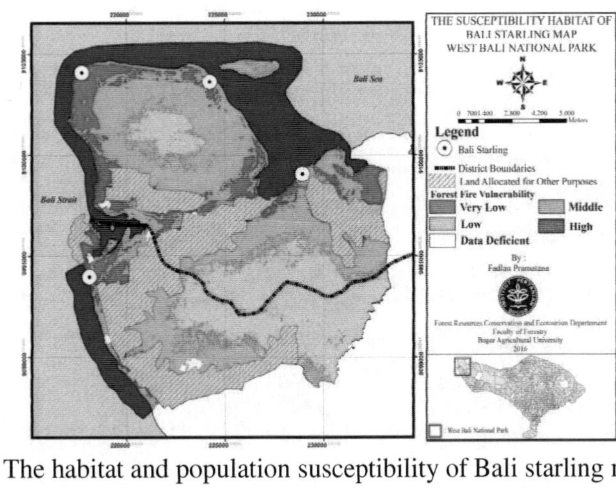

Figure 1. The habitat and population susceptibility of Bali starling map.

IOP Conf. Series: Earth and Environmental Science **91** (2017) 012003 doi:10.1088/1755-1315/91/1/012003

Figure 1 showed that Bali starling was found in four location. One of them appeared in the area that has middle level of forest fire vulnerability, and the rest of three appeared in the area that has high level of forest fire vulnerability. Accordingly, the management of forest fire is needed in order to conserve the wild life of Bali starling in WBNP. The forest fire will contribute to the extinction of species. Basically, the prevention of forest fire is better than handling of fire. Whenthe forest fire factor has been identified, the forest fire prevention should be to do in the first step [17]. If the forest fire have already done in the worse condition, it is very hard to handle eventhough it uses the high technology [1]. The information of fire source is a key factor for the high successfull of forest fire prevention [1]. Type of forest fire in WBNP is surface fire. According to De Bano *et al.* [4], surface fire is fire that burn the litter, undergrowth, plant on the forest floor, waste, and the other fuel on the forest floor.

One of prevention activity for the forest fire management is participation of local people. Local people is very important in the forest fire management, because local people may play as source of forest fire, as well as may act asagent of forest fire early prevention [1]. Management strategy for forest fire base on local people was used for alternative solution in Afrika and Asian country [8].

4. Conclusions and Recommendations

4.1. Conclusion

- The most influence factor for forest fire vulnerabelity in WBNPwere road distance, slope value, and land cover.
- The vulnerability of forest fire in WBNP was classified into four classes, i.e. very low level (9 821 ha), low level (5 015.718 ha), middle level (6 778.656 ha), and high level (2 126.006 ha).
- Bali straling appeared on four location, one of them was found in the middle level of vulnerability firest fire area, and the rest of three were found in the high level of forest fire vulnerability area.

4.2. Recomendation

- Vulnerability forest fire prevention must base on local people.
- Local people in WBNP must be given the counseling about forest fire, environment, and forest fire regulation.
- It needs special attention in the dry season due to increased potential forest fire in savana and moonson forest of WBNP.
- It needs supporting facilities and infrastructure for prevention and handling of forest fire in WBNP, especially in the area that has high level of forest fire vulnerability.

References

[1] Akbar A, Sumardi, Purwanto RH, Sabarudin MS 2011 Studi sumber penyebab terjadinya kebakaran dan respon masyarakat dalam rangka pengendalian kebakaran hutan gambut di areal Mawas Kalimantan Tengah. *Jurnal penelitian hutan tanaman* 8(5):287-300

[2] Amalina P 2015 *Pemetaan kerawanan kebakaran hutan di Taman Nasional Way Kambas* [undergraduate theses] (Bogor: Bogor Agricultural University)

[3] Arianti I 2006 *Pemodelan Tingkat dan Zona Kerawanan Kebakaran Hutan dan Lahan Menggunakan Sistem Informasi Geografis Di Sub Das Kapuas Tengah Propinsi Kalimantan Barat* [master theses] (Bogor: Bogor Agricultural University Sekolah Pascasarjana Institut Pertanian Bogor)

[4] DeBano, L.F., D.G. Neary and P.F. Ffolliott 1998 *Fire's effects on ecosystems* (New York: John Wiley & Sons Inc) p 333

[5] Hadi M 2006 *Pemodelan spasial kerawanan kebakaran di lahan gambut: studi kasus kabupaten bengkalis, provinsi riau* [master theses] (Bogor: Bogor Agricultural University)

[6] Heikkila T V, Gronovist R, and Jurveius M 1993 *Hanbook on forest fire control. Forestry training programme publication 21* (Finland: National board of education of the governmentof finland)

[7] Herbei M, Dragomir L, Oncia S 2012 Using satellite images LANDSAT TM for calculating normalized difference indexes for the landscape of Parang Mountains *GeoCAD.* 13: pp 158–167.

[8] Karki, S 2002 *Community involvement and management of forest fires in southeast asia* (Bogor: Project fire fight southeast asia)

[9] Mukti A 2015 *Pemetaan kerawanan kebakaran hutan di Taman Nasional Alas Purwo* [undergraduate theses] (Bogor: Bogor Agricultural University)

[10] Miardini A and Nugroho N P 2013 Pemetaan kawasan rawan kebakaran hutan dengan menggunakan teknologi penginderaan jauh dan sistem informasi geografis *Teknologi pengelolaan DAS* (Surakarta)

[11] Nurdiana A and Risdiyanto I 2015 Indicator determination of forest and land fires vulnerability using Landsat-5 TM data (case study: Jambi Province) *Procedia Environmental Science* 24: pp 141-151

[12] Nurwatha P F 2013 Modul pelatihan identifikasi, inventarisasi, dan monitoring avifauna (burung) *Pelatihan inventarisasi & monitoring floran & fauna* (Bandung)

[13] Price M and Tinant J 2000 *Prediction of thistle infested areas in Badlands National Park using a GIS model. Rapid City, SD, South Dakota School of Mines and Technology* [Internet]. [diunduh 2016 Juli 31]. Tersedia pada: http://sdspacegrant.sdsmt.edu/nasaepscor/PriceMThistleGISPaper.htm.

[14] Putra I K A, Watiniasih N L, Nuyana I N 2014 Perilaku harian burung jalak bali(*Leucopsar rothschildi*) periode *breeding* pada relung yang berbeda di Bali Bird Park, Gianyar, Bali. *Jurnal Biologi* 18(1): pp 1-4.

[15] Riany C F and Aunurohim 2013 Populasi burung jalak bali (*Leucopsar rothschildi*, Stresemann 1912) hasil pelepasliaran di Desa Ped dan hutan Tembeling Pulau Nusa Penida, Bali. *Jurnal sains dan semi POMITS* 2(2): pp E186-E190.

[16] Rianto T 2006 Review faktor pembatas ekologi dalam upaya pengembalian populasi liar jalak bali (*Leucopsar rothschildi*) taman nasional bali barat. Cekik. Bali.

[17] Saharjo B H 2003 Pengendalian Kebakaran Hutan dan Lahan Yang Lestari: Perlukah Dilakukan?. Laboratorium Kebakaran Hutan dan Lahan. Silviculture Departement. Faculty of Forestry. Bogor Agricultural University.

[18] Samsuri 2008 *Model spasial tingkat kerawanan kebakaran hutan dan lahan (studi kasus di wilayah propinsi Kalimantan Tengah)* [master theses] (Bogor: Bogor Agricultural University)

[19] Sudiana A and Diasmara E 2008 Analisis indeks vegetasi data satelit NOAA/AVHRR dan TERRA/AQUA-MODIS. *Seminar on Intelligent Technology and Its Applications.*

[20] TNBB 2013 *Evaluasi review rencana pengelolaan taman nasional bali barat* (Gilimanuk)

[21] _____ 2015 *Statistik tahun 2014 balai taman nasional bali barat* (Cekik)

[22] Yunanti B D 2012 *Teknik penangkaran dan analisis koefisien inbreeding pada jalak bali (Leucopsar rothschildi Stresemann, 1912) di Mega Bird Orchid Farm Bogor, Jawa Barat* [undergraduate theses] (Bogor: Bogor Agricultural University)

The Role Of Wetland Ecosystems To Reduce The Concentration Of Nitrate In Groundwater

A J Sutrisno[1,2],Z Han[2], S Satake[2],and K Fukumoto[2]

[1]Departement of Landscape Architecture, Bogor Agriculture University, Dramaga 16680, Indonesia
[2]Graduate School of Horticulture, Chiba University, Matsudo 271-0092, Japan

Email: alfredjsmanurung@gmail.com

Abstarct. Nitrate is a compound very soluble in water. Ichikawa city is one of the cities that have a pear farming. In this city, there was a wetland where this area was surrounded by pear farming in upland areas. There were 3 sites (S4, R2, and S14) in this area.S4 and S14 were close to pear farming, but R2 was in the middle of wetland. We used piezometers with different depth (1m, 2m, and 3m) in each site. The focus in this study is identification the role of wetland ecosystem as a green infrastructure to reduce nitrate concentration. Groundwater flow in the wetland comes from pear farming with 2 flow direction. First, groundwater comes from S4 with average nitrate concentration was 185.73 mg/L and average DO concentration 6.37 mg/L. Second, groundwater comes from S14 with average nitrate concentration was 190.29 mg/L and average DO concentration 7.00 mg/L. All this direction flow to the middle of wetland (R2) with average nitrate concentration was 3.46 mg/L and average DO concentration 1.52 mg/L.

1. Introduction

Groundwater is a one of water resources flowing in the soil. Groundwater is important natural resources that are facing serious pollution [4,3,5].One of pollution is nitrate concentration. Nitrate contamination of shallow groundwater is a recognized problem resulting from diffuse or point sources associated with intensive agriculture [1].But, Wetlands in agricultural landscapes can reduce nutrient loadings and sediments, i.e., phosphorus and nitrogen [10].

The World Health Organization sets a limit of 10 mg/L nitrate for drinking water[11]. Two ecosystem processes have primarily been described to be relevant for nitrogen removal in wetlands. Firstly, plant uptake associated with incorporation into biomass and secondly, denitrification by anaerobic microorganisms in the soil [8].

In this study, we are focus on the second ecosystem process. The purpose of this study is how the wetland ecosystem as a green infrastructure can reduce of nitrate concentration and how the relationship wetland ecosystem with nitrate concentration. So, we will be used linier regression as a method in this study.

2. Site Description and Methodology

2.1. *Site description*

This study was conducted at typical headwater wetland located in Ichikawa city with elevation about 16m above sea level. There were 3 sites (S4, R2, and S14) in this wetland with different depth (1m, 2m, and 3m). Every year, nitrogen load in pear orchard of the upland is estimated about 400 kg/ha.

IOP Conf. Series: Earth and Environmental Science **91** (2017) 012004 doi:10.1088/1755-1315/91/1/012004

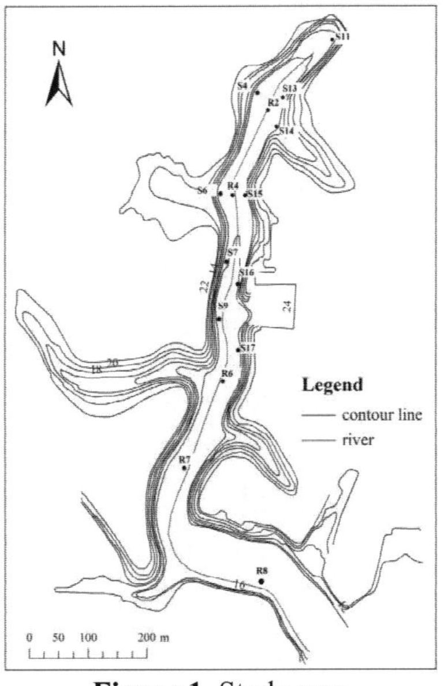

Figure 1. Study area

2.2. *Sampling and analysis*

Water samples collected from piezometers in September 2015, October 2015, and December 2015. We collected 50 mlwater samples in each site and brought back to laboratory for analyzing with Ion Chromatography (LC-10A, Shimadzu, Japan). Before analyzing, we stored water samples in refrigerator with temperature under 4°C. we used ion chromatography to determine the concentrations of Ca^{2+}, K^{+}, Mg^{2+}, Na^{+}, Cl^{-}, NO_3^{-} and SO_4^{2-} and to determine the concentration of HCO_3^{-}, we filtered 20 ml water sample with a drop of mixing solution (1% bromocresol green in 95% alcohol and 1% methyl red in 95% alcohol were mixed with the volume ratio is 1:1) and then titration with 0.01 mol/L dilution hydrochloric acid in laboratory.

2.3. *Data analysis*

We used linier regression for data analysis in this study. Linier regression can determined the relationship parameter with nitrate concentration. We used DO and ORP as a parameter and identified the relationship with nitrate concentration.

3. Result

3.1. *Geochemical*

The average concentration of HCO_3^{-}, Ca^{2+}, K^{+}, Na^{+}, Cl^{-},and SO_4^{2-} in each site was not too many different from September 2015 to December 2015. But, the average concentration of Mg^{2+} and NO_3^{-} was different. In site S4,the concentration of Mg^{2+}ranged from 30.04 to 45.44 mg/L and the concentration ofNO_3^{-} ranged from 150.51 to 248.14 mg/L. In site S14, the concentration of Mg^{2+}ranged from 36.70 to 39.14 mg/L and the concentration ofNO_3^{-} ranged from 181.08 to 203.56 mg/L. In site R2 the average concentration of Mg^{2+} and NO_3^{-} was very low. the concentration of Mg^{2+} was< 5.80 mg/L and the concentration of NO_3^{-} was < 4.69 mg/L (Table 1).

Table 1. Geochemical of the groundwater in the wetland

Site	Depth (m)	Period	Avarage concentration (mg/L)							
			Anion				Cation			
			HCO_3^-	Cl^-	NO_3^-	SO_4^{2-}	Na^+	K^+	Ca^{2+}	Mg^{2+}
S4	1, 2, 3	Sep-15	227.77	50.66	248.14	9.98	17.39	1.30	86.11	45.44
	1, 2, 3	Oct-15	239.97	41.52	150.51	32.51	36.01	1.27	53.15	30.04
	1, 2, 3	Dec-15	254.21	41.59	158.54	37.83	17.70	1.71	85.30	44.75
R2	1, 2, 3	Sep-15	244.04	30.58	1.90	24.31	10.39	1.05	45.62	5.80
	1, 2, 3	Oct-15	92.53	37.43	3.80	31.28	10.18	0.46	46.19	5.62
	1, 2, 3	Dec-15	97.62	37.05	4.69	34.32	9.87	0.38	46.52	5.51
S14	1, 2, 3	Sep-15	89.48	22.44	181.08	12.19	16.13	0.27	39.14	39.14
	1, 2, 3	Oct-15	92.53	26.60	203.56	12.65	15.00	n.d[a]	33.16	38.60
	1, 2, 3	Dec-15	101.68	27.13	186.22	11.85	14.49	n.d[a]	31.49	36.70

[a]concentration = 0 mg/L

3.2. *Environmental parameters*

In site S4, the highesttemperature was20.5^{0}C (September 2015), temperature of groundwater decreased 18.8^{0}C in October 2015 and 12.4^{0}C in December 2015, decreasing temperature of groundwater not only in site S4 but also in site R2 and S14. In all site, the pH value ranged from 6.85 to 7.22. In site S4 and site S14, The DO concentration were high. DO concentration ranged from 4.34 to 8.54 mg/L in site S4and in site S14 ranged from 5.46 to 8.07 mg/L. But, The DO concentration in site R2 was low with ranged from 1.98 to 4.22 mg/L. The average ORP in site S4 and site S14 were> 156 mV, but in site R2 the average ORP ranged from -24 to 16 mV. The EC of groundwater ranged from 26.6 to 77.9µs/m (Table 2).

Table 2. Environmental parameters of the groundwater in the wetland

Site	Depth (m)	Period	Average parameters				
			EC (µs/m)	ORP (mV)	DO (mg/L)	pH	Temperature ^{0}C
S4	1, 2, 3	Sep-15	77.9	228	5.85	7.22	20.5
	1, 2, 3	Oct-15	60.5	194	4.34	7.08	18.8
	1, 2, 3	Dec-15	56.9	190	8.54	6.90	12.4
R2	1, 2, 3	Sep-15	26.6	13	1.37	7.21	21.3
	1, 2, 3	Oct-15	27.1	16	1.98	6.90	18.5
	1, 2, 3	Dec-15	31.7	-24	1.22	6.93	12.0
S14	1, 2, 3	Sep-15	47.2	287	6.11	6.64	19.4
	1, 2, 3	Oct-15	47.0	156	5.46	6.87	17.2
	1, 2, 3	Dec-15	77.3	201	8.07	6.85	12.8

3.3. *Groundwater flow, distribution of nitrate and DO*

Groundwater flow direction in September 2015, October 2015, and December 2015 were same.Groundwater flow has 2 direction. First direction was from site S4 to R2 and second direction was from site S14 to R2. Distribution of nitrate (Figure 2.) and DO (Figure 3.) always follow the groundwater flow direction. We choose September 2015 data for knowing the distribution of nitrate and DO, because the pattern of distribution in another period was same.

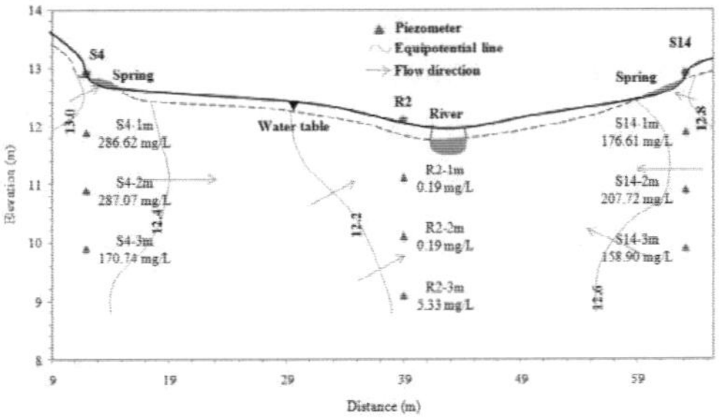

Figure 2. Groundwater flow and distribution of nitrate concentration

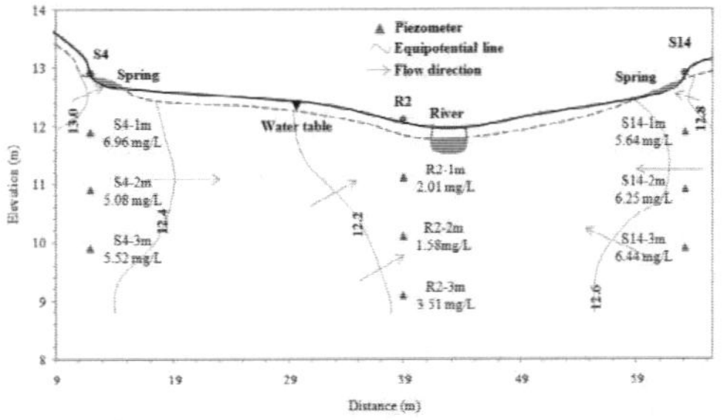

Figure 3. Groundwater flow and distribution of DO concentration

3.4. *Relationship analysis*

This analysis for identified the relationship nitrate concentration with DO concentration and ORP. There is a positive relationship nitrate concentration with DO concentration (Figure 4.) and ORP (Figure 5.). When DO concentration and ORP were high then nitrate concentration was high and when DO concentration and ORP were low then nitrate concentration was low.

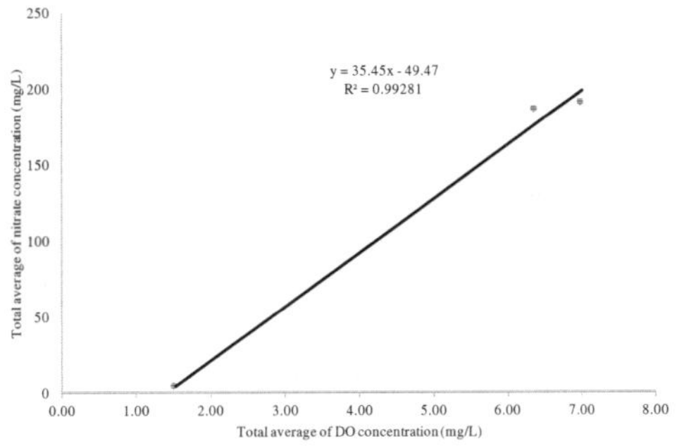

Figure 4. Relationship nitrate concentration with DO concentration

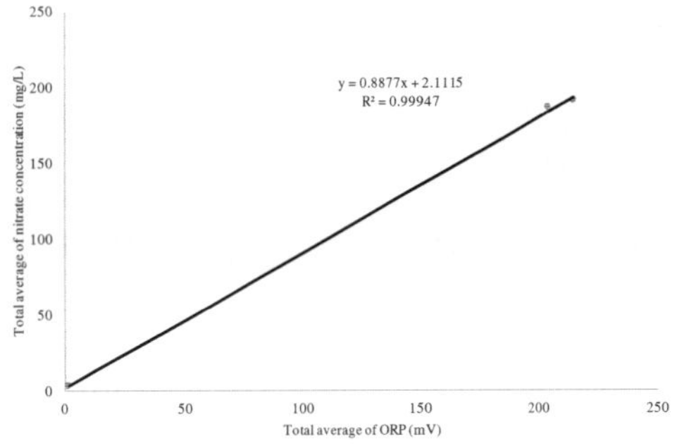

Figure 5. Relationship nitrate concentration with ORP

4. Discussion

Wefound that groundwater flowing from upland area (pear orchard) as a recharge area to wetland as a discharge area and groundwater flow has 2 direction, first direction from S4 to R2 and second direction from S14 to R2. In groundwater flow, we can understand not only movement of groundwater but also movement of concentration. We found that According to some previous study that groundwater flow systems have been found to be related to many aspects such as geological, chemical and biological processes [2,6,9].In denitrification process, microbial was very important to remove nitrate and microbial growth requires low oxygen content or no oxygen.Denitrification process would be restricted when DO concentration was lower than 2 mg/L [7].

Wefound in site R2 DO concentration was lower than 2 mg/L and when DO concentration was low, nitrate concentration was low too. We found that nitrate concentration was lower than 5.In first direction, total average of nitrate concentration (September 2015, October 2015, and December 2015)was 185.73 mg/L with total average of DO concentration 6.37 mg/L in site S4 become decreased in site R2 withtotal average of nitrate concentration was 3.46 mg/L and total average of DO concentration 1.52 mg/L. In second direction, total average of nitrate concentration was 190.29 mg/L with total average of DO concentration 7.00 mg/L and when groundwater flowing to site R2 total average of nitrate concentration and DO concentration become decreased. A positive relationship nitrate concentration with DO concentration (R^2 = 0.993, $P<0.05$) and ORP (R^2 = 0.999, $P<0.05$) proved that there is a function of microbial in this ecosystem, so that it can reduce the nitrate concentration in groundwater.

5. Conclusion

In this study, we found that one of ecosystem services in the wetland was reduction process of nitrate concentration.Wetland ecosystem as a green infrastructure can decreasing nitrate concentration from agriculture area and increasing the water quality. This function was affected by environmental parameters in wetland. We found thata positive relationship of NO_3^- concentration between ORP and DO concentration proved that there was a role microbial in wetland.

Acknowledgment
This study was supported by Wetland Park of Ichikawa city. Wetland Park of Ichikawa city have contributed in giving permission to use the wetland as a study area.

Reference

[1] Aravena, R., Evans, M.L., Cherry, J.A., 1993. *Ground Water***31** 180–186
[2] Batelaan, O., De Smedt, F., Triest, L., 2003. *J. Hydrol.* **275** 86–108.
[3] Bowmer, K.H., 2011. *J. Hydrol.***403** 176–185
[4] Goss, M., Richards, C., 2008. *J. Environ. Manag.***87** 623–632
[5] Hurley, T., Sadiq, R., Mazumder, A., 2012. *Water Res.***46** 3544–3552
[6] Marchetti, Z.Y., Carrillo-Rivera, J.J., 2014. *J. River Res. Appl.* **30** 166–179.
[7] Randall, C.W., Buth, D., 1984. *J. Water Pollut. Con. Fed.***56** 1039–1044.
[8] Sutton-Grier, A.E., Wright, J.P., Richardson, C. 2013*J. Plant Soil***365**41–57
[9] Tóth, J., 1999. *J. Hydrogeol.* **7** 1–14.
[10] Woltemade, C., 2000. *J. Soil Water Conserv.* **55** 303–309
[11] World Health Organisation (WHO), 2006. *Guidelines for Drinking Water Quality*vol. 2, Recommendations.-3rd ed. (Geneva:World Health Organization) p 417

Evaluation of Aesthetic Function and Thermal Modification of Vertical Greenery at Bogor City, Indonesia

B Sulistyantara[1] and R Sesara[2]

[1]Lecturer, Dept. of Landscape Architecture, Bogor Agric. Univ.
[2]Student, Dept. of Landscape Architecture, Bogor Agric. Univ.

E-mail: renosesara@gmail.com

Abstract. Bogor city currently develops vertical greenery due to counter the decreasing of green space quantity. Vertical greenery is a planting method using vertical structure similar to retaining walls. There are some benefits of vertical greenery, such as providing aesthetics value of the landscape, to protect from the heat, to reduce noise, and to reduce pollution. The purpose of this study were to identify thermal modification by vertical greenery in Bogor city, to assess the aesthetics value from vertical greenery, and to provide a recommendation in attempt to manage and improve the quality of vertical greenery in Bogor city. The study was conducted using Scenic Beauty Estimation method, and was done by providing questionnaires to the respondents in order to assess the aesthetics value of vertical greenery. Infrared thermometer was also used to measure the surface's temperature to evaluate thermal modification function of the vertical greenery. The result of study proved that vertical greenery in the Bogor city has considerably good aesthetic. It also showed that there is a decreasing in surface temperature of the vertical greenery structure.

1. Introduction

Green open spaces become a most important constituent element of the city's the function of aesthetic, ecological, social, even in economic terms. Decreasing availability of green open space in urban areas lead to serious problems such as urban heat island, flooding, and so on. The decreasing availability of land for building a green open space made the community and the government chose vertical spaces as an alternative to increasing the quantity of green open space in the city. Planting is done on vertical structures such as embankments or retaining walls commonly called vertical greenery. According Blanc [4], the vertical garden concept provides benefits, among other things: add to the natural beauty of the environment, creating a beautiful garden in a limited area, hold heat from the outside, reducing noise levels, reduce air pollution, capture particles of dirt, reduce the effects of run off, and increase the supply of oxygen. Another function of planting this way is to make a wall or slope is more attractive and can create a habitat for wildlife [2].

One of the city today that is developing a vertical greenery is the Bogor city. Vertical greenery is expected to improve the visual quality of the landscape and improve comfort for its users. The function of vertical greenery can not be felt significantly. This is caused by various problems like crop conditions are relatively bad as well as the loss of some plants on vertical greenery. Evaluation of the effectiveness of vertical greenery in the city to accommodate the aesthetic function needs to be done. The assessment is conducted to determine the usability aspects of the aesthetic function, thermal modification, and assist in the preparation of recommendations to the City Government, especially at

IOP Conf. Series: Earth and Environmental Science **91** (2017) 012005 doi:10.1088/1755-1315/91/1/012005

the Dinas Kebersihan dan Pertamanan Bogor city in managing and developing vertical greenery is already existing and planned.

2. Methodology

2.1. location and time
The location of the study (Figure 1) is in the province of West Java, precisely in the city of Bogor, which is in three locations namely Stasiun Bogor, Terminal Baranangsiang, dan *Underpass.*

(a) (b) (c)

Figure 1. (a) Bogor Station (b) Terminal Baranangsiang (c) *Underpass*

The time needed to conduct this research that for two months in June and July 2016. Furthermore, the process of data analysis and writing take one month from the month of August 2016.

2.2. Methodology

2.2.1. Preparation
The preparation phase is done with a selection site research and obtain licenses. The chosen location is Bogor Station, Underpass, and Terminal Lunch Baranang

2.2.2.Inventaritation
This stage is to conduct the activities carried out field survey and interview related agencies regarding the type of plant, structures, planting media and so on, to capture data in the field is divided into phases and stages questionnaires surface temperature calculation.

Stage questionnaire using a questionnaire distributed to 40 students of Landscape Architecture semesters 6 and 8. With a scale of 1 (worst) to 10 (most excellent) by displaying a slide that contains 10 photos. Phase measurement of surface temperature using a infrared thermometer GM320 Thermometer Temperature Laser Gun with range -50 ~ 330 ° C, 0.1 ° C resolution and an accuracy of ± 1.5 ° C for 0 ~ 330 ° C and 3 ° C to -50 ~ 0 ° C. Sampling will be conducted temperature for 7 days and with three repetitions on each day (morning, day, afternoon) to compare whether there is climate modification to the structure of the vertical greenery.

2.2.3.Analysis
The method of analysis used by the method of SBE (Scenic Beauty Estimation) to analyze the function of aesthetics with SBE formula of formula [5].

$$SBE_x = (ZL_x - ZL_s)x100$$

SBEx = Value SBE simulated images to x

ZLx = The average value of the simulation to the image Z x

ZLS = average value of Z picture simulations are used as standard

The control variable is the surface temperature does not have the plants and to the independent variables are each plant species found on the vertical greenery. Data obtained will be processed using t-test to see the effects of such vegetation, with an initial hypothesis affect plants in the modification of the surface temperature of vertical greenery.

3. Result and discussion

3.1. Analysis of Aesthetic Quality in Bogor city

3.1.1. Color freshnesh

SBE analysis on aspects of the freshness of color produces a diverse SBE value. Results of the assessment are grouped into an low aesthetic landscape (s.d -160.90 -79.74), moderate (-79.75 s.d 1.42), and high landscape (1.43 s.d 82.59) (Figure 2). Dominated by a picture with moderate value total of five images (G2, G3, G4, G5 and G6) with G4 as standard (most close to the value 0).

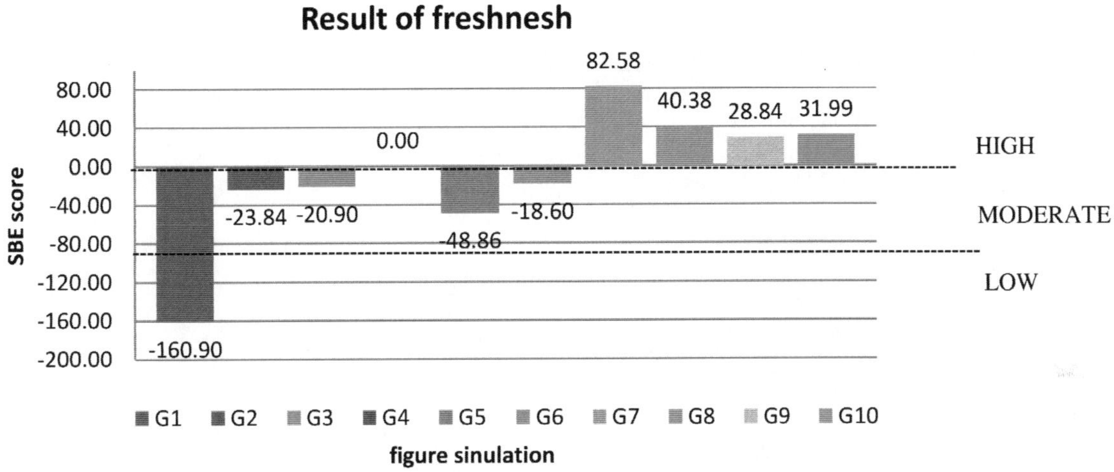

Figure 2. SBE color freshness value charts

According Metha A, the use of color combinations can make any colors used to be preferred color of people. But the bad combination of color can cause a decline in the value of aesthetics. Based Whiting and de Jong, the principles of design consists of unity, balance, simplicity and variety, emphasis, and the sequence.

On the vertical greenery in underpass G7 got the highest score (82.58) while the lowest value is G1 (-160.90). The magnitude of the difference is caused by a number of plant species in the Vertical greenery on the image G1 seen only two types of plants with the color of the plant's air-tone dull causing the impression of freshness of color is reduced, while the image G7 There are eight species of plants with three color variations: green red, purple and yellowish green color that contrasts create vertical greenery becomes more alive.

G6 (-18.60) located in Bogor Station which is one focus of this research was included in the group of high value. Basically Vertical greenery has a contrasting color (emphasis) between green and light green color yellow, but the lack of unity in it caused too massive of its *Asplenium nidus* causes the focus divided because basically yellowish-green tint looks pale and impressed unhealthy and there are some parts of the *Asplenium nidus*, which seem less healthy. So that reduced the value of unit.

Vertical greenery on Terminal Baranangsiang G9 (28.84) included in the group of high value, This is caused by G9 has a color and plant species varied set with a pattern of organic and arranged randomly makes the only vertical greenery visible sequence (rhythm / repetition) and looks so varied because large scale creates the impression of a smooth texture and color produced is much brighter.

The reason why respondent not gave maximal value on G5 (-48.86) is because by less bright purplish red color in plants and not the creation of the design principles. G3 (20.90) which is similar to G5 (-48.86) has a greater value because respondents tend to think of a fresh green color than red owned G3 and looks neat because of the size of the plants that are not too big and smooth looks impressed balance and simplicity. Figure G2 (-23.84) despite having more types of plants to the color of dark green and yellowish. But a merger between colors less felt their unity and was not impressed balance.

Figure G4 (0.0) which is used as a standard because it has ZLx value (average value of Z) by 0.09 most nearly zero. A mix of green and red colors create a balance but simple. However, when viewed from the freshness of color, the color in the center is one of the negative values in the G4. That's why the G10 (31.99) which is similar to the G4 enter the group of high value. Because the G10 is much more unity and more focussed on the bright green color than the G4. Figure G8 (40.38) contrasts with G9. Although it has the same position both tall and large-scale G8 has only one fresh green vegetation and greatly accentuate simplicity .but can get a value far greater than the G9.

3.1.2. Density
Analysis SBE on aspects density diverse generating value. Results of the assessment are grouped into an low aesthetic landscape (s.d -166.13 -82.13), moderate (-82.12 s.d 1.87), and high landscape (1.88 s.d 85.51) (Figure 3). Dominated by high-value images of five images (G2, G6, G7, G8 and G9) with G10 as standard (most close to the value 0).

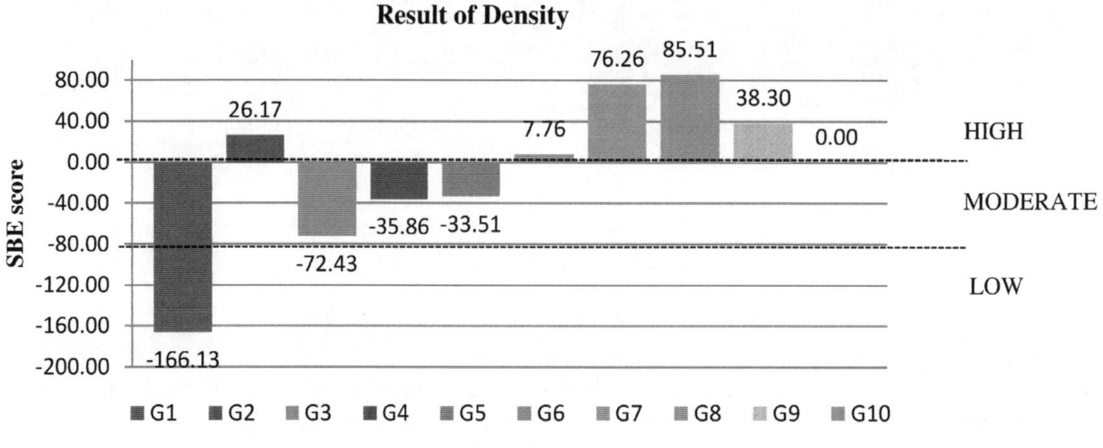

Figure 3. SBE density value charts

Density is very influential on aesthetics, according to the 2011 criteria Aprilis plants as aesthetic functions in the selection of plant species that form the canopy and branching very interesting, skalatis size, there are variations in color (stems, leaves, flowers and fruit) and interesting textures. SBE value for G1 (-166.13) located on the left Underpass is the only value in the group of low value. G1 only had two types of plants, *Bromelia* sp. in the middle, and *Chlorophytum* sp. that is around. Comparison scale plants that are too small a spacing coupled with branching *Chlorophytum* sp. less makes causes a lot of holes that are created everywhere. As for the G7 (76.26) is the second highest value in the group tinngi value, this is because in addition to the diversity of plant species vertical greenery dominated by the Philodendron scandens plants that are vines so it looks very solid and tight.

Vertical greenery located in Bogor Station G6 (7.76) included in the group of high value because it has a plant with height canopy spread, but less precise planting layout causing poor plant growth because it was blocked from the sun and many plants that die making vertical greenery into the lack of value. G9 (38.30) located in Terminal Baranangsiang included in the group of high value because basically at a very large scale plants in it has ramifications that spread and looks very solid with a smooth texture, but the number of plants that are drought and die making it not beautiful.

As for the G4 (-35.86), G3 (-72.43) and G5 (-33.51) has a pot as cropping media. The distance created between pot is suspected to be the main cause less preferred G4, G3, and G5. G10 (0.00), which became the standard is basically similar to G4 for using the same pot and plant species are the same except for the G10 circular shape as basically causes no side formation and thus become more tightly and neatly.

G2 (26.17) included in the group of high value because the density of plant species plus branching plants that spread making it look very tightly, but because of too much diference size making it uneven because there are plants who blocked the smaller size plant and cause visiblelike a hole, G8 (85.51) is the highest value of SBE, despite only having one type of plant, but the plant is very solid for almost the entire surface is covered. Three research sites in the aspect of color only vertical greenery Bogor station is not included in the group of high value but the group is, while the density of the aspects of the entire vertical greenery in the group of high. This proves that the aesthetic value of the three vertical greenery is already very good.

3.2. Surface temperature

3.2.1.Bogor Station

The t-test results for the surface temperature on the vertical greenery on Bogor station which has seven types of plants. Data were divided into three groups: the data in the morning, day, and afternoon and see whether the effect is real or not as in table 1.

Table 1. the average value of the surface temperature in the morning, day, afternoon, and the t-test results on the vertical greenery Bogor City Station

Plant name	morning		result	day		result	afternoon		result
	$\bar{t}$	$\bar{t0}$		$\bar{t}$	$\bar{t0}$		$\bar{t}$	$\bar{t0}$	
Philodendron burle	23.64	25.29	*	33.10	42.70	*	28.60	32.44	*
Dracaena raflexa	23.76	25.29	*	32.27	42.70	*	28.19	32.44	*
Coleus blumei	23.77	25.29	*	32.20	42.70	*	27.63	32.44	*
Asplenium nidus	23.79	25.29	*	32.40	42.70	*	27.29	32.44	*
Callisia fragrans	23.74	25.29	*	32.40	42.70	*	28.14	32.44	*
Nephrolepis sp.	23.91	25.29	*	32.99	42.70	*	28.07	32.44	*
Dracaena marginata	23.84	25.29	*	32.26	42.70	*	28.94	32.44	*

Information: * shows affect significanly

Table 1 shows that the seven plants on vertical greenery Bogor City Station affect significantly on surface temperature vertical greenery where the t-test showed that (t Stat)> t-table (t Critical one-tail) or states that H0 is rejected. The most significant temperature differences occur during the day with a range of different temperatures of 9.60°C until 10.50°C, on the morning 0.46°C to 0.73°C, and at 3.50°C until 5.16°C in the afternoon. It is influenced by each difference in the intensity of the sun for the morning sun is not as strong as the sun during the day, and also the presence of dew on the plants in the morning that perhaps went wet the surface behind him, in the day due to begin the reduction in light and watering in vertical greenery that done at 12.00 as in Figure 4.

The most influential plant is the *Philodendron Burle* in the morning, the lowest temperature during the day is a *Coleus blumei* and in the day is *Aspelium nidus*. Thirdly this plant has the characteristic features are almost similar compared to other plants in vertical greenery is a pattern that spreads except for *Coleus blumei*, the leaves are large and wide. Powered by Grey and Deneke statement [6] stating Effectiveness plant in capturing solar radiation depends on the density of the leaves, leaf shape and branching pattern.

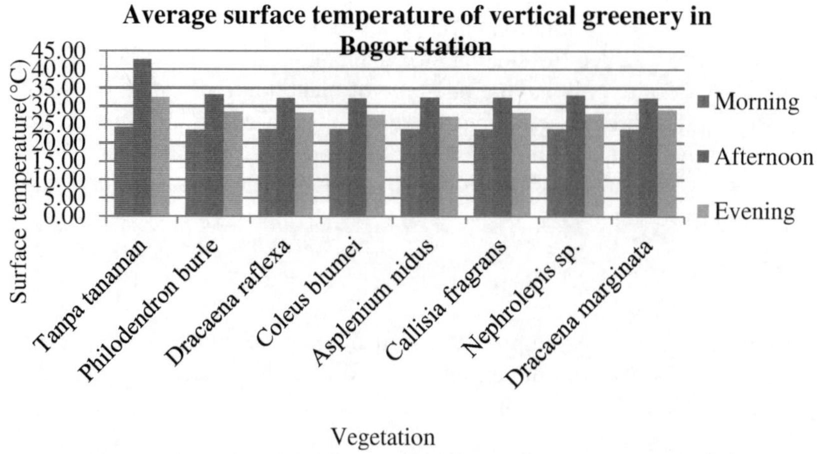

Figure 4. Average surface temperature charts in Bogor station

3.2.2. Underpass

The t-test results for the surface temperature on the vertical greenery on Underpass which has 8 plants. Data will be divided into three groups: the data in the morning, day, and afternoon and see whether the effect is real or not as shown in Table 2.

Table 2. the average value of the surface temperature in the morning, day, afternoon, and the t-test results on the vertical greenery Bogor City Station

Plant name	morning		result	day		result	afternoon		result
	$\bar{t}$	$\bar{t0}$		$\bar{t}$	$\bar{t0}$		$\bar{t}$	$\bar{t0}$	
Schefflera arboricola	24.59	25.29	*	31.39	37.43	*	26.59	28.79	*
Philodendron scandens	24.69	25.29	*	31.83	37.43	*	26.90	28.79	*
Bromelia sp.	24.37	25.29	*	30.46	37.43	*	26.04	28.79	*
Scindapsus sp.	24.21	25.29	*	30.49	37.43	*	25.81	28.79	*
Bromelia sp.	24.39	25.29	*	31.23	37.43	*	26.41	28.79	*
Dracaena reflexa	24.59	25.29	*	32.97	37.43	*	26.53	28.79	*
Carex morrowii	24.26	25.29	*	31.06	37.43	*	26.30	28.79	*
Chlorophytum sp.	24.06	25.29	*	30.57	37.43	*	25.50	28.79	*

Information: * shows affect significanly

Table 2 shows that the eight plants on vertical greenery Underpass significant effect on surface temperature vertical greenery where the t-test showed that (t Stat)> t-table (t Critical one-tail) or states that H0 is rejected. The temperature difference most significant occurred during the day with a range of different temperatures of 4.46°C to 6.97°C, for the morning at 0.60°C until 1.23°C, and by 1.89°C until 5.16°C in the afternoon as in the show to Figure 5.

Plants that are most influential in the morning was *Chlorophytum* sp., The lowest temperature during the day is a plant *Bromelia* sp. pink grass And in the afternoon is *Chlorophytum* sp. This is slightly different than the vertical greenery on Bogor Station as the most significant drop in temperature occurs in plants *Chlorophytum* sp. and *Bromelia* sp. pink grass which visually and physical characteristics although less suitable for Bromelia sp. pink grass leaves are thick and there are other plants that better meet the criteria of Grey and Deneke [6] As, *Philodendron scandens*. One is probably because the distance between vegetation and surface in his blekang very close and almost stuck to make the surface behind it perfectly protected and also because its position is surrounded by *Philodendron scandens*.

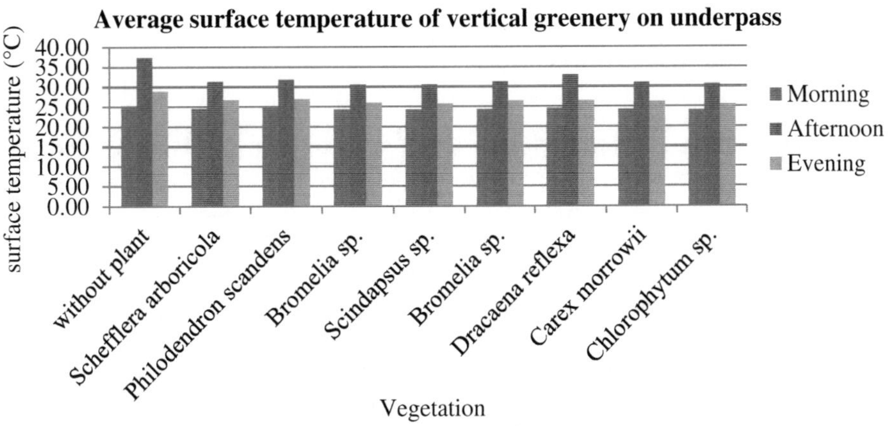

Figure 5. Average surface temperature charts in Underpass

According Wilmers [7] Greening the city can bring benefits to micro-climate through several physical processes, such as plants and shade trees can reduce solar heat gain in buildings, Shading reduces terrestrial radiation because over its surface temperature is lower, and the latent heat of the cooling in the atmosphere increases because the added moisture in the air through evapotranspiration of the plant. This statement is supported by Badrulzaman et al. [3] stated that the vertical greenery system that uses plants to a building can cool the building and the surrounding area through shading, which reduces the heat emitted by evapotranspiration. Shading produced by the plant became a major factor in the reduction in surface temperature. Based on the results above it can be concluded that the main factor affecting the decrease in temperature and shading is the density of leaves, leaf shape and branching pattern. Indirectly types of vegetation also influence the ability of vertical greenery in reducing the surface temperature.

4. Conclusion

There is a decrease in temperature on the surface of vertical greenery Bogor Station 0.46°C to 0.73°C in the morning, 9.60°C until 10.50°C during the day, and at 3.50°C until 5.16°C in the day, and the lowering the temperature on the surface of vertical surface greenery at Underpass 0.60°C to 1.23°C in the morning, 4.46°C to 6.97°C during the day, and by 1.89°C to 5.16°C in the day. It is caused by shading in plants that protect the surface behind it from direct sunlight. Shading can be affected by several factors in plants such as leaf density, leaf shape and branching pattern. The evaluation results indicate that the aesthetic value of the average vertical greenery in the city of Bogor has aesthetic value both in terms of the freshness of color and density aspects. Vertical greenery still need a lot of development and is focused on the types of plants with high density and good color freshness to increase the aesthetic value and thermal modifications.

References

[1] Aprilis P 2011 Penilaian Fungsi Pengaman dan Estetika Jalur Hijau Jalan Jenderal Sudirman Kota Pekanbaru Provinsi Riau [skripsi] (Bogor (ID): Institut Pertanian Bogor)
[2] Arifin H S, Munandar A, Arifin N H S, Pramukanto Q and Damayanti V D 2008 Sampoerna Hijau Kotaku Hijau Edisi Kedua, Cetakan Pertama. Maryanto C danKumalashinta [penyunting]. PT. Sampurna Hijau.
[3] Badrulzahman J, Ismail S, Mohd H R 2011 Evaluating the Impact of Vertical greenery System on Cooling Effect on High Rise Buildings and Surroundings: A Review. Senvar. 12:5b-02.
[4] Blanc P 2008 The Vertical Garden from Nature to the City (New York: Michel Lafon Publishin)
[5] Daniel TC and Boster RS 1976 Measuring Landscape Aesthetics: the Scenic Beauty Estimation Method USDA Forest Service Research Paper RM- 167.
[6] Grey G W and F J, Deneke 1978 Urban Forestry (New York: John Willey and Sons inc)
[7] Wilmers F 1991 Effects of vegetation on urban climate and buildings Energy and Buildings (5):507-514

Productive Urban Landscape In Developing Home Garden In Yogyakarta City

S N R Irwan[1] and A Sarwadi[2]

[1]Department of Agronomy, Faculty of Agriculture, Universitas Gadjah Mada
Jl. Flora Bulaksumur Yogyakarta 55281 Indonesia
[2]Departement of Architecture and Planning, Faculty of Engineering, Universitas Gadjah Mada, Jl. Grafika Yogyakarta 55281 Indonesia

Email: rofiqoirwan@ugm.ac.id

Abstract. Home garden is one type of agroecosystem that supports ecosystem services even in the urban settlement. The studies involved literature references and field survey along with a framework of the productive urban landscape that support ecosystem services in home garden. Productive urban landscape provided environmentally, socially and economically benefits that contained in ecosystem services. Problems on limited space in the urban settlement have to be managed by modified home garden system in order to work for ecosystem service in developing productive landscape. This study aimed to assess home garden (Pekarangan) system in a cluster of high density settlement in Yogyakarta City. Structured interview and vegetation identification of home garden have been conducted on 80 samples in Rejowinangun Kotagede District, Yogyakarta City. People showed enthusiasm in ecosystem services provided by home garden "Pekarangan Produktif" through developing productive urban landscape. Some benefits on ecosystem services of home garden were revealed on this study consisted of food production for sale (4.7%), home industry (7.69%), aesthetics (22.65%), food (14.10%), biodiversity (10.68%), ecosystem (12.82%), education (2.56), social interaction (11.54%), recreation (4.70%), and others (8.55%). Nevertheless, vegetation and other elements of home gardens have been managed irregularly and in particularly, the planned home gardens were only 17.07%. Actually, home gardens provided a large set of ecosystem services including being cultural services those are the category most valued. The urban people almost hided the understanding of the cultural benefit of ecosystem services of home garden, even though Yogyakarta has known the cultural city. Thus, urban home garden, as way as "Pekarangan Produktif" in the limited space that managed and planned sustainably, provide many benefits of ecosystem services in a productive urban landscape.

1. Introduction

Urban green space covers a range of landscape types of varying complexity and morphology that influence of comfort living. Urban people have high effort to consider any kind of limited space around their life as well as their home. Productive urban landscapes have accommodated the intensity of quality ecosystem that involve benefits of ecosystem services. Ecosystem services were correlated by the landscape productive even any ecological problems in city. Home gardens were existed on limited space of urban settlement that has constructed people think about linking to ecosystem. Even though the ideas of utilized space have been not supported yet by recognizing of productivity and effectiveness of utilization of home gardens.

Productive urban landscape is open urban space planted and managed in such a way as to be environmentally, socially and economically productive, for example, providing food from urban agriculture, pollution absorption, the cooling effect of trees or increased biodiversity from wildlife corridors [9]. Productive Urban Landscapes performed the elements that support ecosystem services. Ecological intensification worked by prioritizing environmental urban layers, which were either connected to open space use, or to implementing sustainable technology and activity patterns. Usually, these environmental layers were then superimposed with other locally appropriate layers, such as economic, social, cultural, historical, etc. Urban home garden system provides tightly key ecosystem services that may be conceptualized as supplying, regulating, cultural, and supporting services. In the wide meaning, home gardens serve as habitat for a large range of flora and fauna and thus help in biodiversity conservation. Provisioning services are those resources supplied by home gardens to human communities, which include food products, natural medicines, aromatic, relaxing, and educating.

Yogyakarta Province provides opportunity on living comfort. Otherwise, Yogyakarta is a favorite tourism destination in Indonesia. Some reasons have supported Yogyakarta as well as having ecologically life quality. It means people should have home sweet home as well home garden with all supporting life. Activities, productiveness, services, all human well-being needs should be provided by surrounding home. In specifically, home garden "Pekarangan Produktif" were revealed as response of the human well-being needs, by integrated conceptual productive urban landscape and ecosystem services. Recently, Yogyakarta local settlements have changes to urbanized settlement that indicate the rate houses increase to high price, followed by smaller and smaller place for home garden. Some reasons have to be considered by thinking productive home garden (Pekarangan Productive) on the limited space of urbanized settlement because of the needs of human well-being in which supported by ecosystem services. Because of several reasons to reveal the understanding of productive urban landscape and ecosystem services applied by home garden, this study aimed to asses potential development of productive home garden system in a cluster of high density settlement in Yogyakarta city.

2. Methods

The research was conducted in high density settlement in Rejowinangun Kotagede district, Yogyakarta city. Based on research objective, the methodological approach for this work included observations, interviews and literature study. Observations and interviews were implemented at 80 samples of household from two clusters of neighborhood (Rukun Tetangga) at Rejowinangun Kotagede district Yogyakarta (Figure 1). Observations covered vegetation, space situation (size, form) in home garden/pekarangan. Interviews were conducted by questioners that contained questions regarding to variables of research. Data were collected by census of all samples from the two clusters. The survey methods were conducted on May – June 2015, continued on May – June 2016. Data were analyzed by statistical descriptive and facts correlation matrix that showed home garden system at the study area. The matrix showed interrelation home garden element, mainly vegetation, to development of productive home garden and ecosystem services. Literature study was involved to understand productive urban landscapes aspects that support ecosystem services.

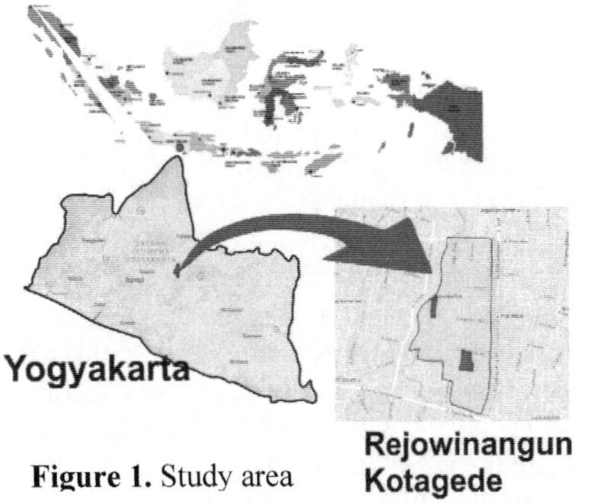

Figure 1. Study area

We tried to refer the Millennium Ecosystem Assessment framework [5], showed by Figure 2, to be related to CPULs (Continuous Productive Urban Landscapes) concept [9] that be applied in home garden system in study area. The interaction among various services provided by the home garden system and environmental benefits were evaluated to understanding home garden level system. The research variables of ecosystem services were referred to Calvet-Mir, *et.al.*, (2012) research [2]. The most valued ecosystem service, as indicator of research, was provision of food, followed by hobby, maintenance of landraces, heritage value of home gardens and associated traditional ecological knowledge and enjoyment of home gardens aesthetic features.

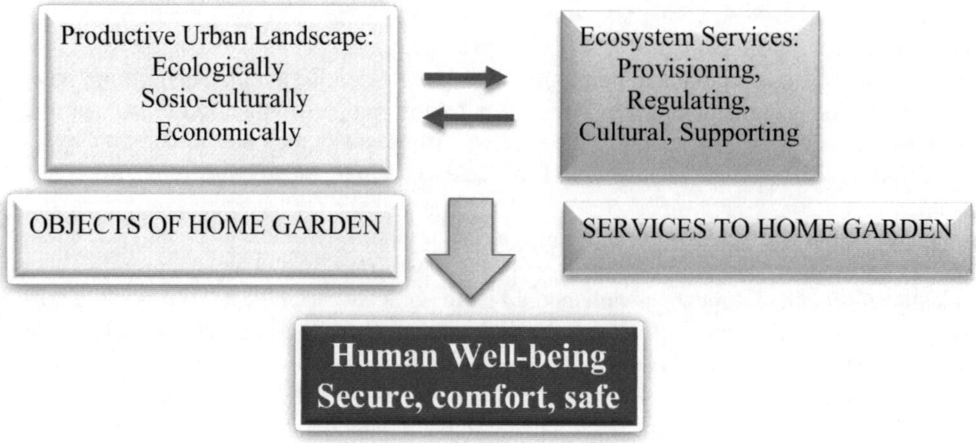

Figure 2. Linkage of analizing methods aspects between Productive Urban Landscapes and Ecosystem Services to human well-being in home garden system [5] [9]

3. Results and Discussion

Yogyakarta city has been famous as tourism destination and cultural place in Indonesia and also known as place for high education. People come from any places of Indonesia come to Yogyakarta for recreation and it has been like a magnet for people surrounding the city. Therefore, Yogyakarta with 32.5 km2 has come urbanized area with high density population of 421,704 inhabitants. Study area was an urban high density settlement with limited space around house as home garden.

Previous research by Irwan, *et. al.*, 2015, home gardens were grouped by garden size, as way; small (<50 m^2), medium (50-100 m^2) and large (>100m^2) [3]. People though to plant fruit trees, vegetables, and aesthetical plants. Based on garden sizes, it was relatively no difference in reasons of vegetation choice, but there was difference on quantity of trees. People have planted less trees in small garden than medium or large size. Vegetation were planted pots or polybag (Figure 3), because of limited space in small garden, even a few small garden were also planted trees or small trees. In medium and large garden people were easier to arrange trees or fruit trees and shrubs, such as vegetable and aesthetic plants.

Urban trees are considered for productive landscape elements because a major factor in providing cooling, via shade and evapotranspiration. The extent to which gardens contribute to cooling is unclear, because varieties of home garden situation, such as size, vegetation, and maintenance. The trees function was also productive. Productive urban landscape is urban green space planted and managed in such a way as to be environmentally, and economically productive, for example, providing food from urban agriculture, pollution absorption, the cooling effect of trees or increased biodiversity from wildlife corridors. Ecological intensification worked by prioritizing environmental urban layers, which were either connected to open space use, or to implementing sustainable technology and activity patterns. Usually, these environmental layers were then superimposed with other locally appropriate layers, such as economic, social, cultural, historical, etc [9]. It means productive home garden needs conceptual development of environment, socio-culture and economics. Actually, Continuous Productive Urban Landscapes (CPULs) started to develop in London around

2005. Forty years later they were everywhere, having reached a maturity that enables us to study their success in relation to the very initial design intention.

Figure 3. Limited space of home garden at study area

Figure 4 showed answers of questioner about functions of home garden. They said aesthetics is important (22.65%) but actually the planned home gardens were only 17.07%. Several reasons can explain that people may not care to design, people don't have knowledge and skills, people think no need design in limited space. It was interesting, people think about food for eating (14.10%) and social economic (11.54% and 4.7% food for selling) facilitated by home garden. Local culture play important roles to influence people think about home garden. Thus Pekarangan Produktif has been a cultural activities of Yogyakarta City, also as Indonesian culture. Unfortunately, limited space in cities face a big question mark for home garden. How can Pekarangan Produktif can be applied. Otherwise, other functions on home garden showed at Figure 4.

Urban settlement of area study covered from $28m^2$ to $1650\ m^2$ of home area size. Home garden filled the area from 0% (no space) to 80.15% ($134\ m^2$), see Figure 3. There are many houses without space around house. Some reasons try to develop the no space home garden. Models of vertical garden covered vertical space (wall) and around house as narrow street/alley (gang). Problems on the alley and limited spaces were not optimum of sun light catching for vegetation photosynthesis [4]. Recreation and education were some kind of activities in home garden, whether with family members or neighbors. Provisioning food, showed in Figure 4, were delivered for home industry (like handmade snack), for eating and for sale in neighborhood. People have enthusiasm on Pekarangan Productive in selling and buying vegetable, fruits and herbal. In this case, social interaction, educational, healthy activities were conducted in settlement. Experiences have been held in study area that on Sunday morning people gathered with family on field to do exercises, jogging or other sports. After the activities, they purchases some fresh vegetable, fruits and herbal. Other way, nutrients contained in fresh materials have richness in vitamin, fiber and enzyme. People also met each other in home garden area talk anything and to care children any way. It was seen daily activities in local characteristics of Kampong in study area.People used limited and alley around house in urban high density in study area. Other people surrounding the study areas have often visited Kampong Pilahan, Rejowinangun

Kotagede for buying fresh vegetable. The Kampong was familiar with "Kampung Wisata Agro" (Agrotourism Kampong).

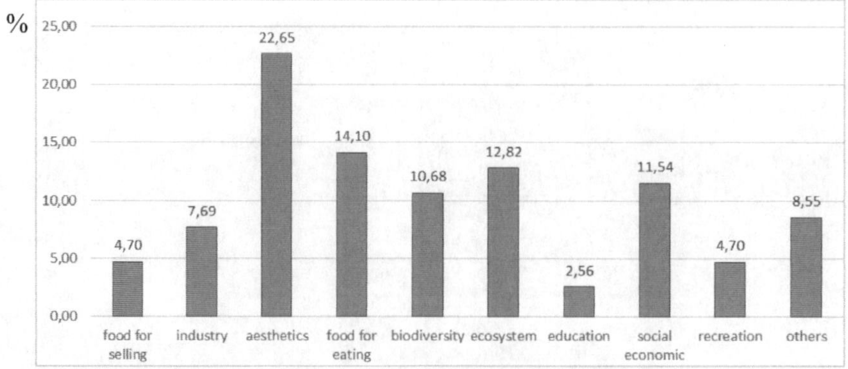

Figure 4. Respondent response about home garden

Figure 2 shows methods for analyzing, ecosystem services aspects (Figure 5) are relevant in connecting to productive landscapes. Those differ that conceptual productive landscape turn in objects or elements and ecosystem services focus on services. Both of those concept are supporting each other in human well-being as well. Linkages between Ecosystem Services and Human Well-being [5] is connecting four broad strands of ecosystem service functions (provisioning, regulating, cultural and supporting) are incorporated, although much of the literature is mainly focussing on the first three of these (provisioning, regulating and cultural). The services are connecting to security, basic material for good life, health, good social relations. Regulating involves most strong intensity of linkages to security, basic material for good life and health. Then, provisioning only has high potential for mediation by socio-economic to security and basic material for good life. In case, Manggo trees, in home garden, showed services of provisioning fruits. It strongly influenced health and potential for selling (socioeconomic factors). In developing productive urban landscape, Manggo trees in pekarangan produktif also in charge in environmentally (ecosystem), economically and socio-culturally.

Future challenges for further research include how to enhance the resilience of home garden systems against socioeconomic and global climate changes by integrating traditional home garden systems, modern technology, and the global economy [7]. Ross *et.al*. [8] resumed researches on domestic garden, including home garden, regarding functions for regulating of climate, water and soil in ecosystem services [8]. Due to the close proximity to the home and hence accessibility for many, possibly the greatest benefit of home garden is on human health and well-being, Beyond trees, a wider range of vegetation has potential to improve cooling, e.g. through planting next to a house. There is a suggestion that the strategies to conserve energy in temperate climates cannot be generalized, due to the complex interactive effects of shade and wind reduction. Little information specifically identifies the role of the home garden in carbon dynamics, but the effectiveness of garden plants and soils for carbon storage will strongly depend on design. Vegetation, trees especially, intercept intense precipitation, hold water temporarily within their canopy thus reducing peak flow and easing demand on urban drains.

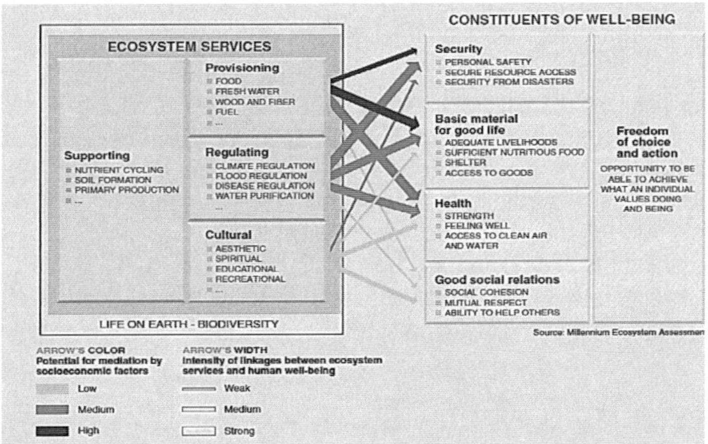

Figure 5. Linkages between Ecosystem Services and Human Well-being [6]

Varieties of vegetation were dominated by papaya, chili, longan (lengkeng), mango, banana, guava and lime. The vegetation were edible plants and have services of provisioning food. The fruit tress serve ecologically to human well-being. The fruits are favorite for the owner and other people in neighborhood, it means economically can reduce family expenses. As kampong area, people often change fruits or other home garden production (barter) that will socially interact each other and as a local culture. Even Yogyakarta has known as a cultural city, there are cultural plants that should be exist in any kind of place, also in urban settlement. Unfortunately, cultural tree of Yogyakarta, Kepel tree (*Stelechocarpus burahol*), was planted by 2 respondents/houses from 80 respondents/houses. The reason was not recognized, people might not care to plant or did not understand the cultural trees. Other cultural plants of Yogyakarta malay maple Jambu darsono (*Syzygium aqueum*) was not exist in study area.

Developing productive urban landscape (PUL) on home garden has potential in study area. People need improvement knowledge and skills. Local government support is important to correlate the programs of community empowerment to the development home garden. As ecosystem services, home gardens are an essential element of life providing opportunity for engagement with nature, self-actualization, creativity or wellbeing; to others, they are at best a parking lot, or worse, represent an additional chore to an already busy lifestyle. Conversely, there may be little point in every house having a garden if their potential is not realized. Educational processes may be required to encourage engagement, especially as society becomes increasingly displaced from the natural world and food cultivation, but evidence is required of the clear benefits before such activity could be undertaken.

According to people needs to develop productive home garden model concept (Pekarangan Produktif), linkage between productive urban landscape (PUL) and ecosystem services (ES) could be thought in arranging and design of home garden. Components of PUL and ES should be combined as PULES to proceed the model concept of productive home garden (Pekarangan Produktif) to continue the research, see Table 1.

Table 1. Linkage between Productive Urban Landscape and Ecosystem Services in Productive Home Garden (Pekarangan Produktif)

		Ecosystem Services			
		Provisioning	Regulating	Cultural	Supporting
Productive Urban Landscape	Economically	High linkage	Low linkage	High linkage	Low linkage
	Socially	High linkage	Low linkage	High linkage	Low linkage
	Ecologically	High linkage	High linkage	Low linkage	High linkage

Low linkage High linkage

IOP Conf. Series: Earth and Environmental Science **91** (2017) 012006 doi:10.1088/1755-1315/91/1/012006

4. Conclusion

Productive urban landscape (PUL) in developing home garden in Yogyakarta has components that it has provided ecosystem services (ES) in urban settlement environment. Home gardens have shown ecologically, socially and economically potential to be developed as Pekarangan Produktif, even limited space. The limited and or narrow space around home can use vertical (green wall) and horizontal (ground) garden. This study also found that some benefits on ecosystem services of home garden were food production for sale, home industry, aesthetics, biodiversity, ecosystem, education, social interaction, and recreation. The detail and quantitative study on productive urban landscape in relation to ecosystem services should be carried out in order to identity the potential to develop the home garden and alley (gang) area in urban settlements as a more productive urban landscape.

References

[1] Badan Pusat Statistik D.I. Yogyakarta 2016 http://yogyakarta.bps.go.id/linkTabelStatis/view/id/70. Diunduh 24 Oktober 2016.

[2] Calvet-Mir, Laura, Erik Gómez-Baggethun, Victoria Reyes-García 2012 Beyond food production: Ecosystem services provided by home gardens. A case study in Vall Fosca, Catalan Pyrenees, Northeastern Spain. Ecological Economics **74** 53–160.

[3] Irwan, S.N. Rofiqo and Ahmad Sarwadi 2015 Lanskap Pekarangan Produktif di Permukiman Perkotaan Dalam Mewujudkan Lingkungan Binaan Berkelanjutan, Prosiding Seminar Nasional Sains dan Teknologi 2015 Fakultas Teknik Universitas Muhammadiyah Jakarta, 17 November 2015. Website: jurnal.ftumj.ac.id/index.php/semnastek

[4] Irwan, S.N. Rofiqo and Ahmad Sarwadi 2016 Pemanfaatn Ruang Terbatas Sekitar Rumah di Permukiman Perkotaan Melalui Pengembangan Lanskap Produktif, Prosiding Seminar Nasional Sains dan Teknologi 2016 Fakultas Teknik Universitas Muhammadiyah Jakarta, 8 November 2015.

[5] Millennium Ecosystem Assessment 2005 *Ecosystems and Human Well-being: Synthesis*. Island Press, Washington, DC.

[6] Millennium Ecosystem Assessment 2007 *A Toolkit for Understanding and Acion.* Island Press, Washington, DC.

[7] Mohri Hideyuki, ShrutiLahoti, OsamuSaito, Anparasan Mahalingam, Nimal Gunatilleke, Irham, Van Thang Hoang, Gamini Hitinayake, Kazuhiko Takeuchi, Srikantha Herath 2013 Assessment of ecosystem services in home garden systems in Indonesia, Sri Lanka, and Vietnam. Ecosystem Services **5** e124–e136.

[8] Ross W.F. Cameron, Tijana Blanuˇsa, Jane E. Taylor, Andrew Salisbury, Andrew J. Halstead, Beatrice Henricot, Ken Thompson 2012 The domestic garden – Its contribution to urban green infrastructure. Urban Forestry & Urban Greening **11** 29– 137.

[9] Viljoen Andre dan Katrin Bohn 2005 An introductory glossary: Continuous Productive Urban Landscape. Designing Urban Agriculture for Sustainable Cities. Oxford. Architectural Press [an imprint of Elsevier]. 2005

Correlation of Carbon Stock and Biodiversity Index at the Small Scale Agroforestry Landscape in Ciliwung Watershed

M B S Choliq[1] and R L Kaswanto[2]

[1]Research Scholar, Department of Landscape Architecture, Bogor Agricultural University, West Java, Indonesia.
[2]Lecturer, Department of Landscape Architecture, Bogor Agricultural University, West Java, Indonesia.

Email address: bagussapi@gmail.com

Abstract. *Pekarangan* is part of a complex of small-scale agroforestry landscape. Pekarangan have 3 functions i.e. ecological, economic, and social. ecological function, for providing landscape services such as carbon stock and biodiversity; economic function, can supplies foods and nutrition; and social function, for building low carbon communities and increasing the environmental awareness. Therefore, this research aims to correlate carbon stocks and biodiversity index of *Pekarangan* in Ciliwung Watershed. This study has measured 48 samples which were divided in three stream, namely upstream, midstream, and downstream. The samples were divided into four groups, G1 (*pekarangan* size less than 120 m^2 and doesn't have other agricultural land (no other agricultural land - OAL), G2 (<120 m^2 with OAL < 1000 m^2), G3 (120-400 m^2 with no OAL) and G4 (120-400 m^2 with OAL < 1000 m^2). The results show that correlation between carbon stock and biodiversity index value is $R^2 = 0.05$. The results showed no correlation between carbon stocks and biodiversity index could be due to the amount of *Pekarangan* owners who prefer potted plants than plant a tree, so that the carbon sequestered in the Pekarangan only slightly.

1. Introduction

Pekarangan or home garden is a part of a complex small scale agroforestry landscape, usually located around the house and its range is only about 0.1 to 0.3 hectares [1]. *Pekarangan* have 3 functions i.e. ecological, economic, and social [1]. Ecological function is for providing landscape services such as carbon stock and biodiversity; economic function can supplies foods and nutrition; and social function is for building low carbon communities and increasing the environmental awareness. The biodiversity index of area has a great contribution to the preservation and the diversity of ecosystems, thus *pekarangan* as agroforestry systems provide a significant contribution to carbon storage area while also enhancing rural livelihoods [2] [3]. This research aims to determine correlation between carbon stock and biodiversity index at *pekarangan* in Ciliwung Watershed. In addition, to figures out whether biodiversity index inside *pekarangan* can influeces of carbon stock value and vise versa. Both carbon stock and biodiversity index are type of landscape services which also be considered as the asset and ecosystem sharing value to improve rural community welfare [1].

2. Methods

2.1. Location, Tools, and Materials

This study is located in Ciliwung Watersheds (Figure 1). The tools used in this study consist of hardware and software, such as digital camera, stationery, measuring tools, laptop, and spreadsheet software. Mataerials used in this study is formula of carbon stock and biodiversity index (Table 1) and other materials obtained directly in the field.

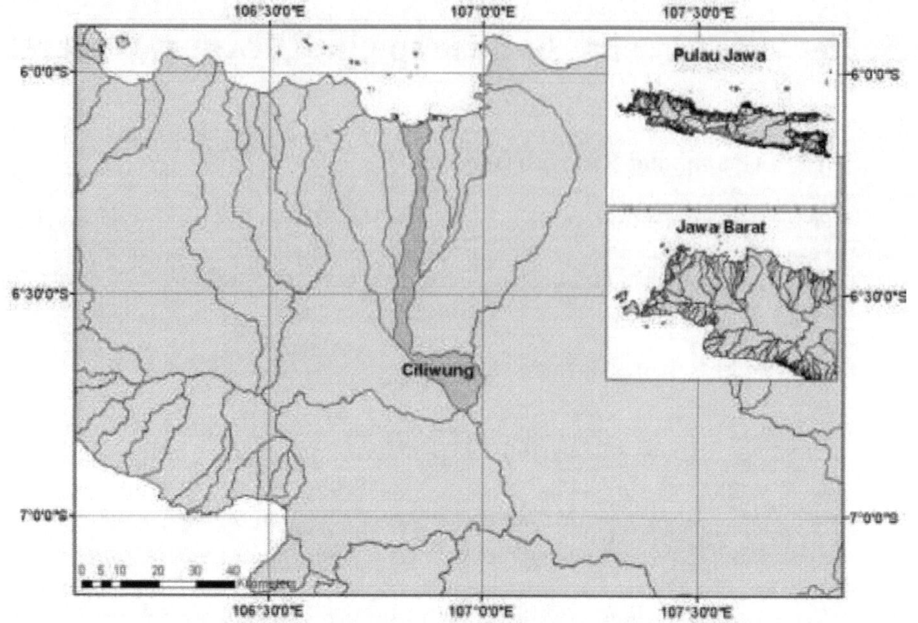

Figure 1. Ciliwung Watersheds Location

Table 1. Carbon stock and Biodiversity Indeks Formula

Carbon stock formula	
Allometric	(Tabel 2)
Biomass [4]	*Wtc = Y x 0,5*
Biodiversity formula	
Shannon-Wiener [5]	$H' = -(\sum pi \ln pi)$
Correlation	
Correlation	$y = ax + b$

Table 2. Allometric Equation for Biomass

Plant	Allometric
Tectona grandis [6]	$Y = 0{,}153\ D^{2{,}39}$
Musa sp [5]	$Y = 0{,}0303\ D^{2{,}1345}$
Multispecies D <5 [7]	$\ln(AGB) = -3{,}50 + 1{,}65 \ln(D) + 0{,}842 \ln(H)$
Another tree [3]	$Y = 42{,}69 - 12{,}8\ D + 1{,}242D^2$

2.2. Methods

The method used in this study using survey and measurement on the field to get carbon stored and heterogeneity. There are 48 samples *pekarangan* and each 16 samples are in upstream, midstream, and downstream. Samples were taken by purposive random sampling (Figure 2). *Pekarangan* group were divided into four groups, namely as G1 (*pekarangan* size less than 120 m^2 and doesn't have other agricultural land (no *other agricultural land* - OAL), G2 (<120 m^2 with OAL < 1000 m^2), G3 (120-400 m^2 with no OAL) and G4 (120-400 m^2 with OAL < 1000 m^2).

IOP Conf. Series: Earth and Environmental Science **91** (2017) 012007 doi:10.1088/1755-1315/91/1/012007

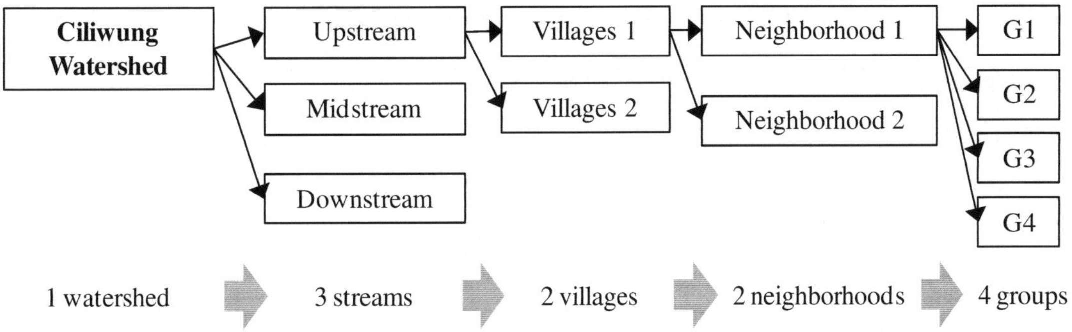

Figure 2. Flowchart on this study that using purposive random sampling

3. Result and Discussion

3.1. Correlation of Carbon Stock and biodiversity index

Based on the formula in Table 1 and Table 2, carbon stock at *pekarangan* in Ciliwung watershed are around 0.13 Mg C/ha up to 27.15 Mg C/ha, the highest average in midstream reach to 7.71 Mg C/Ha, then up stream is 1.51 Mg C/ha and downstream 1.83 Mg C/ha (Table 3). The biodiversity index (Shannon Wiener Index) can be categorized from low to medium and ranges from 0.77 to 2.51, the highest average in downstream with 2.00, then upstream is 1.69 and midstream 1.77 (Table 4). Pekarangan in the downstream could be has higher biodiversity because the household tend to plant the ornamental plants for the aesthetic purposes [3].

Table 3. Carbon Stock (Mg C/ha)

UPSTREAM			**MIDSTREAM**			**DOWNSTREAM**		
V1: Megamendung			V1: Kedunghalang			V1: Pasar Minggu		
KP. Pondok Gede	G1	2.76	RW 03	G1	1.14	RW 07	G1	0.68
	G2	0.76		G2	0.31		G2	2.09
	G3	0.58		G3	3.73		G3	1.42
	G4	0.18		G4	1.36		G4	1.29
KP. Pasir Muncang	G1	6.58	RW 05	G1	2.13	RW 11	G1	1.24
	G2	0.13		G2	0.39		G2	6.44
	G3	4.51		G3	6.76		G3	1.71
	G4	4.53		G4	1.15		G4	1.27
V2: Cisarua			V2: Nanggewer			V2: Tebet		
KP. Tugu Utara	G1	0.32	RW 01	G1	13.07	RW 05	G1	1.97
	G2	0.28		G2	22.06		G2	3.31
	G3	1.03		G3	10.43		G3	0.72
	G4	1.27		G4	27.15		G4	0.65
KP. Tugu Selatan	G1	0.19	RW 07	G1	24.61	RW 03	G1	0.85
	G2	0.40		G2	4.34		G2	1.19
	G3	0.26		G3	1.88		G3	0.47
	G4	0.39		G4	2.81		G4	4.04
Average		1.51	Average		7.71	Average		1.80
Max		27.20						
Min		0.13						
Average All		3.68						

Table 4. Biodiversity index (Shannon Wiener Index – H')

UPSTREAM		
V1: Megamendung		
KP. Pondok Gede	G1	1.73
	G2	1.58
	G3	2.49
	G4	2.01
KP. Pasir Muncang	G1	2.31
	G2	1.95
	G3	2.48
	G4	2.11
V2: Cisarua		
KP. Tugu Utara	G1	1.29
	G2	1.17
	G3	2.13
	G4	1.24
KP. Tugu Selatan	G1	1.09
	G2	1.05
	G3	1.27
	G4	1.13
Average		1.69
Max		2.51
Min		0.77
Average All		1.82

MIDSTREAM		
V1: Kedunghalang		
RW 03	G1	1.79
	G2	1.31
	G3	1.37
	G4	1.91
RW 05	G1	2.06
	G2	1.21
	G3	1.97
	G4	0.77
V2: Nanggewer		
RW 01	G1	1.99
	G2	1.28
	G3	2.43
	G4	2.47
RW 07	G1	2.28
	G2	1.87
	G3	1.87
	G4	1.75
Average		1.77

DOWNSTREAM		
V1: Pasar Minggu		
RW 07	G1	2.12
	G2	2.15
	G3	2.27
	G4	1.73
RW 11	G1	2.17
	G2	1.75
	G3	2.44
	G4	1.95
V2: Tebet		
RW 05	G1	1.64
	G2	1.53
	G3	2.51
	G4	2.24
RW 03	G1	1.85
	G2	1.39
	G3	2.42
	G4	1.78
Average		2.00

Based on the data Table 3 and Table 4 the correlation were obtained is $R^2 = 0.058$ (Figure 3). Based on this value the correlation of carbon stock and biodiversity index was positively related even though the value of which is derived relative small, this is because biodiversity index is very diverse while the carbon stock obtained relative small. Things like this happen as more plants were grown in pots as houseplants than trees which incidentally have more carbon stock. Tropical homegardens or pekarangan with high agrobiodiversity have high potential for carbon (C) sequestration, especially under changing environments [8], and particularly in West Java [9] [3].

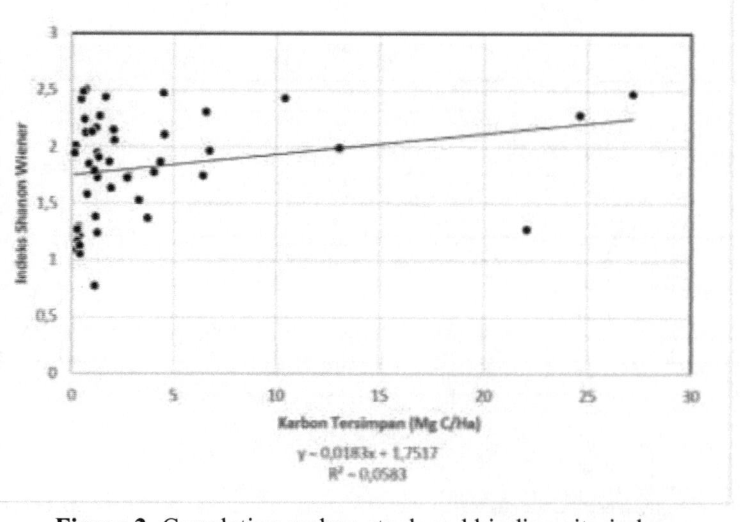

Figure 3. Correlation carbon stock and biodiversity index

4. Conclusion

Based on this study, it can be concluded that the correlation of carbon stock and biodiversity index in Ciliwung Watershed is still relatively low. This is due to increased levels of development in Ciliwung Watershed which impacted to the open space area and the household prefer to potted their plant on *pekarangan*. The land use change in Ciliwung Watershed is believed tend to reduce the carbon stock and at the same time degrade the biodiversity condition [10] [11]. This condition should be overcome by integrated landscape approach [12] for increasing the community awareness [13]. The community should improve their understanding about the important of home garden or *pekarangan*. *Pekarangan* can provide ecological, economical and social function, thus it can be determine as the household capital [14]. Furthermore, pekarangan can be utilized not only for production but also provide landscape services asset [15].

5. Acknowledgements

This study was conducted under the project entitled Management of Agroforestry Landscape as Landscape Service Provider towards Low Carbon Landscape (LCL) of PUPT IPB 2016. This Study was supported by Kemenristek Dikti of Indonesia.

6. References

[1] Kaswanto RL, Nakagoshi N. Landscape Ecology-Based Approach for Assessing Pekarangan Condition to Preserve Protected Area in West Java. In: Nakagoshi N, A. Mabuhay J, editors. Designing Low Carbon Societies in Landscapes. Ecological Research Monographs: Springer Japan; 2014. p. 289-311.

[2] Christanty L, Abdoellah OS, Marten GG, Iskandar J. Traditional agroforestry in West Java: the pekarangan (homegardens) and kebun talun (annual-perennial rotation) cropping systems. In: Marten GG, editor. Traditional agriculture in South-east Asia: Westview, Boulder; 1986. p. 132-58.

[3] Kaswanto, Nakagoshi N. Revitalizing Pekarangan Home Gardens, a Small Agroforestry Landscape for Low Carbon Society. HIKOBIA 2012;16:161-71.

[4] Brown S, Food, Nations AOotU. Estimating Biomass and Biomass Change of Tropical Forests: A Primer: Food and Agriculture Organization of the United Nations; 1997.

[5] Nagendra H. Opposite trends in response for the Shannon and Simpson indices of landscape diversity. Applied Geography. 2002;22(2):175-86.

[6] Hairiah K, Rahayu S. Pengukuran karbon tersimpan di berbagai macam penggunaan lahan. Bogor World Agroforestry Centre-ICRAF, SEA Regional Office, University of Brawijaya, Unibraw, Indonesia. 2007.

[7] Ali A, Xu M-S, Zhao Y-T, Zhang Q-Q, Zhou L-L, Yang X-D, et al. Allometric biomass equations for shrub and small tree species in subtropical China2015.

[8] Kumar B. Carbon sequestration potential of tropical homegardens. In: Kumar B, Nair P, editors. Tropical Homegardens. Dordrecht: Springer; 2006. p. 185-204.

[9] Tatag Aisyah F, Regan Leonardus K. Carbon stock and plants biodiversity of pekarangan in Cisadane watershed West Java. IOP Conference Series: Earth and Environmental Science. 2017;54(1):012024.

[10] Arifasihati Y, Kaswanto. Analysis of Land Use and Cover Changes in Ciliwung and Cisadane Watershed in three Decades. Procedia Environmental Sciences. 2016;33:465-9.

[11] Kaswanto, Nakagoshi N, Arifin HS. Impact of Land Use Changes on Spatial pattern of landscape during two decades (1989–2009) in West Java region. HIKOBIA. 2010;15:363-76.

[12] Reed J, van Vianen J, Barlow J, Sunderland T. Have integrated landscape approaches reconciled societal and environmental issues in the tropics? Land Use Policy. 2017;63:481-92.

[13] Kabir ME, Webb EL. Can homegardens conserve biodiversity in Bangladesh? Biotropica. 2008;40(1):95-103.

[14] Westerink J, Opdam P, van Rooij S, Steingröver E. Landscape services as boundary concept in landscape governance: Building social capital in collaboration and adapting the landscape. Land Use Policy. 2017;60:408-18.

[15] Termorshuizen JW, Opdam P. Landscape services as a bridge between landscape ecology and sustainable development. Landscape Ecology. 2009;24(8):1037-52.

IOP Conf. Series: Earth and Environmental Science **91** (2017) 012008 doi:10.1088/1755-1315/91/1/012008

Identification of Potential Wild Herbal as parts of Landscape Elements

Bambang Sulistyantara[1] and Nio Mentari[2]

[1]Lecturer, Dept. of Landscape Architecture, Bogor Agric. Univ.
[2]Student, Dept. of Landscape Architecture, Bogor Agric. Univ.

E-mail: niomentari@gmail.com

Abstract. *Many landscape plants can grow on their own without cultivated by humans. They are type of plants that can be found anywhere, so they can be categorized as wild plants. The economic value of wild plants are easy to obtain and their maintenance costs are low. Because wild plants not widely known even a just a few of people that aware of their existence, it is necessary to do a study to learn the potential of the wild plants to be used as an element of landscape. This research aims to identify the species that have potential to be used in landscape design, to describe the benefits of the their implementation as a landscape element, and to recommend the wild plants that have functional value and visual. This research used a scoring method based on the functional and visual criteria, and questionnaires were conducted to 50 students of Landscape Architecture IPB who have completed Landscape Plants courses. Based on the research, there are 150 species of wild plants that found in the study site, and 60 of them are recommended as landscape elements. Then all of the species were arranged as a recommendations book so they can be used as alternative landscape plants.*

Keywords: functional, alternative landscape plants, wild plants, medicinal plants, visual

1. Introduction

Plants are landscape element that used as functional and aesthetic purposes. Potential plant that can be used as landscape plants, one of them is wild plants. Unfortunately, wild plants has not been used by people, moreover it considered as weed. Besides that, several wild plants has aesthetic value and it easier to maintain than other decorative plants. Therefore, the provision of wild plants as landscape element would be helpful to solve park maintenance problems.

Not all plants has cultivated or planted deliberately, many plants has grown by itself. That kind of plants can be found in many potential places, so that can be categorized as wild plants. Wild plants are indigenous plants or local plants, original plants as introduction plants that developed and known by people in certain areas (Rachman *et. al.* 2002).

The existence of the plants need to be conserved because of it has economic value and other benefits, such as to be consumed or as pharmaceutical ingredients. Economic benefits of wild plant are easy to find and lower cost to maintain it. Because of wild plants were not widely known yet, moreover only few people are aware of its existence, so necessary to do an study about potential wild plants to be used as landscape elements.

This research objectives are to identify wild plants species that potential to be used in landscape design, to describe other benefits for wild plants owner in its application as landscape elements, and to recommend wild plant species and its design form that can be used in landscape element.

IOP Conf. Series: Earth and Environmental Science **91** (2017) 012008 doi:10.1088/1755-1315/91/1/012008

2. Methodology

2.1. Location and time
This research was located in Pos Kesehatan Dusun (Poskedus) Tangkilan, Sidoarum Village, Subdistrict Godean, District Sleman, Special Region of Yogyakarta. This research was conducted in June 2015 until February 2016.

Figure 1. Research location

2.2. Research methods
This research consist of five stages, that is: field survey, species identification, presentation and questionnaire, assessment, and recommendation arrangement in gallery book. About 150 wild plant species randomly selected from field survey, then identified by using literature study. Functional quality assessment was conducted by comparing the morfology of each species with four criterias of plant function assessment. That four criterias are architectural function, engineering function, climate amelioration, and aesthetics. Plant morfology that used were: type of plant, plant height, canopy shape, and leaves texture. Visual quality assessment was conducted by presentation and questionnaires to 50 Respondents. Those respondents are Landscape Architecture students (Department of Landscape Architecture, IPB), who has passed Plant in The Landscape course. Respoondents were requested to assess visual quality of wild plants based on 150 photos that presented by researcher. The value obtained in questionnaire divided to four criterias, that is form, texture, color, and scale. Visual quality value was obtained with calculating average value of each criteria per species, then it was described in plant description. In recommendation arrangement stage, each of species were given description about functional and visual quality values based on assessment that was obtained, and arranged in an gallery book of landscape plants by using application *File Maker Pro 11*.

2.2.1. Functional quality criterias. Plant function in the landscape consists of architectural function, engineering function, climate amelioration, and aesthetics (Carpenter 1975). The criterias of functional quality assessment presented in Table 1 and Table 2.

Table 1 Criteria of plant architectural function

No.	Function	Planting Criteria
Articulation or Forming		
1	Floor	• Plants that be used are grasses and ground cover plants

2	Wall	• Plants that be used are climber plants, shrubs, bushes, and trees • Shrubs/bushes can be used as border plants (>1.25 m) • Climber can be places on wire, woods structure, bamboos, or pergola.
3	Divide bigger space to smaller one	• Plants that be used are, shrubs, bushes, and trees • Plants can be planted in line that can be make a space
4	Circulation	• Plants that be used are, shrubs, bushes, and trees
5	Corridor	• Plants that be used are, shrubs, bushes, and trees
6	Connecting space	• Plants that be used are, shrubs, bushes, and trees
7	Physical Border	• Plants that be used are, shrubs, bushes, and trees
Screen		
1	Screening the bad view	• Plants that can be used are shrubs and bushes (>1.25 m) • Can be planted in high density line, and allowed to be grow higher.
Give the private place		
1	Making private place	• Partial arranged, that is separated but still can be seen (>1.25 m) • Full arranged, that is fully separated and can not be seen (>1.25 m)

(Source: Taihuttu 2013)

Table 2 Criteria of other functions

No.	Functions	Planting Criteria
Engineering functions		
1	Erosion control	• To be as controller erosion by water and air. • To be planted so that be cover ground surface • Using ground cover plants, climber plants, shrubs, bushes, and trees.
2	Noise control	• Using shrubs and bushes that planted in several high density line (>1.25 m)
3	Sunlight reflector and filter	• Plants that be used are, shrubs, bushes, and trees
4	Manage traffic in the space room	• Plants that be used are, shrubs, bushes, and trees • Planted according to desired pattern and arrangement.
5	Setting the accoustic	• Plants that be used are shrubs and bushes • Planted in several high density line
Climate Amelioration		
1	Controlling the sunlight radiation	• Using ground cover plants, climber plants, shrubs, bushes, and trees.
2	Wind Control	• Plants as wind barrier, wind breaker, and wind floow control • Plants that be used are trees
3	Air filter	• Plants that be used are shrubs, bushes, and trees • Will be effective if the plants are coniferous plants, plants that has wide leaf, or rough surfaced leaf plants.
Aesthetics purposes		
1	Two dimensions elements	• Plant aesthetics can be seen from two dimensions, such as its shadow. • Plants that be used are climber plants, shrubs, bushes, and trees
2	Three dimensions elements	• Can be seen from plant shape, size, colour, and texture. • Plants that be used are climber plants, shrubs, bushes, and trees
3	Plant as backdrop, frame, topiary, or espalier	• Plants that be used are ground cover plants, climber plants, shrubs, bushes, and trees

(Source: Taihuttu 2013)

2.2.2. Functional Quality Assessment. Functional quality assessment was conducted by comparing the morfology of each species with function criterias that have been appropriate. Plant morfology that used were: type of plant, plant height, canopy shape, and leaves texture.

The result of functional quality assessment was as follows: Score 1 (Bad) if ≤40% of criteria were fulfilled, thats means from 22 criterias, only 1-7 criterias were fulfilled. Score 2 (Medium) obtained if 41-60% criteria were fulfilled, that means from 22 criterias, 8-12 criterias were fulfilled. Score 3 (Good) obtained if 61-80% of criteria were fulfilled, that means from 22 criterias, 13-16 criterias were fulfilled. Score 4 (Very good) obtained if ≥81% of criterias were fulfilled, that means from 22 criterias, 17-22 criterias were fulfilled (Wungkar 2005).

2.2.3. Visual Quality Criteria. The assessment was conducted by spreading the questionnaire to 50 respondents. Visual quality of the wild plant was rated based on four criterias, that is colour, form, texture, and scale. To obtain the total score from each criteria, all score that has been obtained must be summed, then divide to 50. To obtain the total visual quality score, we must be sum the total score of each criteria, then divide to 4 (Figure 2).

$$\overline{x}_{response\ criteria} = \frac{x_1 + x_2 + x_3 + x_4 + x_5 + \cdots + x_{50}}{50} = \frac{\sum x_n}{n}$$

$$\overline{x}_{visual} = \frac{\overline{x}_{colour\ response} + \overline{x}_{form\ response} + \overline{x}_{texture\ response} + \overline{x}_{scale\ response}}{4}$$

Figure 2. Visual quality asessment formula

3. Result and discussion

3.1. General description

Sidoarum village located 7 KM from west of Sleman and 8 KM from the city of Yogyakarta. Geographical coordinates between 5° 46' 18" LS - 5° 49' 16" LS and 110° 17' 24" E - 110° 19' 35" E, with UTM coordinates 423 076 425 820 MT and MT-MU 9,138,318 - 9,142,432 MU (Hendriono 2013). Sidoarum village has 8 hamlets, namely Cokro Konteng, Bantulan, Beji, Cokro Bedog, Kramat, Potrowangan, Tangkilan, and Sebaran. Sidoarum village has lower topography with altitude 110 mdpl, temperature 26° C and average of rainfall about 1180 mm/year (Hendriono 2013). The river which is through the village Sidoarum are Kontheng and Bedog River and the upper reaches are in the North of Mount Merapi heading to the South and empties in the Progo River, so that the average of groundwater depth in this village less than 7 meters. Tangkilan located in the Sidoarum village, Godean, Sleman. Total area of Godean is 306.37 Ha. Tangkilan hamlet rich of medicinal plant and the development center known as Pos Kesehatan Dusun (Poskesdus). Poskesdus cultivate more than 501 medicinal plants including wild plants, and they won first prize as a medicinal plant cultivators in provincial level (Hendriono 2013).

3.2. Species identification

Obtained 150 wild plant species identified from the results of field survey in Pos Kesehatan Dusun (Poskesdus) Tangkilan.

Table 3 Species identification

FAMILI	LOCAL NAME
Acanthaceae	dlujru
Acanthaceae	sambiloto
Acanthaceae	daun madu
Acanthaceae	landep
Acanthaceae	dandang gendis
Acanthaceae	gandarusa

FAMILI	LOCAL NAME
Acanthaceae	gempur batu
Acanthaceae	kejibling
Amaranthaceae	jarong lelaki
Amaranthaceae	bayam dempo
Amaranthaceae	kremah
Amaranthaceae	bayam duri

FAMILI	LOCAL NAME
Amaranthaceae	boroco
Amaranthaceae	jengger ayam
Amaranthaceae	kaki kuda
Amaranthaceae	bunga kenop
Amaryllidaceae	bakung
Apiaceae	semanggi gunung

FAMILI	LOCAL NAME
Apocynaceae	pulosari
Apocynaceae	tapak doro putih
Apocynaceae	pule pandak
Apocynaceae	ginje
Araceae	sente
Araceae	keladi warna
Araceae	talas
Araceae	sambeng
Araliaceae	daun mangkokan
Araliaceae	cikra-cikri
Araliaceae	poncosudo
Asteraceae	daun seribu
Asteraceae	bandotan
Asteraceae	purwaceng
Asteraceae	ajeran
Asteraceae	daun sembung
Asteraceae	kenikir
Asteraceae	urang-aring
Asteraceae	tapak liman
Asteraceae	tempuh wiyang
Asteraceae	jabung
Asteraceae	bribil
Asteraceae	umyung
Asteraceae	beluntas
Asteraceae	wedelia
Asteraceae	stevia
Asteraceae	kembang kertas
Basellaceae	binahong
Basellaceae	gendola
Bignoniaceae	bungli
Bixaceae	kesumba keling
Boraginaceae	ekor anjing
Brassicaceae	sawi putih
Bromeliaceae	nanas-nanasan
Cactaceae	duri tengtong
Campanulaceae	kitolod
Cannaceae	bunga tasbih
Capparaceae	mamang besar
Caprifoliaceae	kerak nasi
Combretaceae	melati belanda
Commelinaceae	gewor
Commelinaceae	nanas kerang
Convolvulaceae	kangkung air
Convolvulaceae	kangkung alas
Convolvulaceae	songgo langit
Crassulaceae	sosor bebek
Cyperaceae	rumput teki
Cyperaceae	jukut pendul
Dioscoreaceae	gadung
Elaeocarpaceae	talok
Euphorbiaceae	anting-anting
Euphorbiaceae	ekor kucing
Euphorbiaceae	akalifa bintik
Euphorbiaceae	buni
Euphorbiaceae	patikan kebo
Euphorbiaceae	zig-zag
Euphorbiaceae	cerme
Euphorbiaceae	meniran
Euphorbiaceae	daun katu
Fabaceae	saga rambat
Fabaceae	daun kupu-kupu
Fabaceae	secang
Fabaceae	kedinding
Fabaceae	kembang telang
Fabaceae	jenu
Fabaceae	daun duduk
Fabaceae	putri malu
Fabaceae	ketepeng kebo
Fabaceae	turi
Iridaceae	kuma-kuma
Lamiaceae	gringsingan
Lamiaceae	lenglengan
Lamiaceae	kemangi
Lamiaceae	iler
Lamiaceae	tem
Lauraceae	tali putri
Lauraceae	adem ati
Leeaceae	girang
Liliaceae	andong
Limnocharitaceae	genjer
Malvaceae	waru lengis
Malvaceae	sidoguri
Malvaceae	pulutan
Marantaceae	bamban
Marantaceae	garut
Melastomataceae	senggani
Menispermaceae	camcau rambat
Moraceae	awar-awar
Moraceae	murbei
Myrsinaceae	daun lempeni
Nyctaginaceae	bunga pukul 4
Onagraceae	krangkong
Oxalidaceae	krambilan
Oxalidaceae	calincing
Passifloraceae	permot
Phytolaccaceae	getih-getihan
Piperaceae	suruhan
Piperaceae	cabe jawa
Plantaginaceae	sendokan
Plumbaginaceae	daun encok
Poaceae	jali
Poaceae	sere dapur
Poaceae	jukut jampang
Poaceae	alang-alang
Poaceae	rumput bambu
Polygonaceae	aseman
Polypodiaceae	sisik naga
Polypodiaceae	simbar layangan
Pontederiaceae	eceng gondok
Pontederiaceae	eceng padi
Portulacaceae	krokot
Portulacaceae	som jawa
Rosaceae	gucen
Rubiaceae	remek watu
Rubiaceae	mengkudu
Rubiaceae	nusa indah putih
Rubiaceae	rumput mutiara
Rubiaceae	daun kentut
Rutaceae	kemuning
Rutaceae	inggu
Saururaceae	amisan
Schrophulariaceae	jaka tuwa
Selaginellaceae	cakar ayam
Solanaceae	ciplukan
Thymelaeaceae	makuta dewa
Urticaceae	lateng
Verbenaceae	kembang bugang
Verbenaceae	bunga pagoda
Verbenaceae	tembelekan
Verbenaceae	pecut kuda
Violaceae	antanan kembang
Zingiberaceae	kecombrang

IOP Conf. Series: Earth and Environmental Science **91** (2017) 012008 doi:10.1088/1755-1315/91/1/012008

3.3. Description of quality plant

3.3.1. Functional quality rating. Plants that have a functional quality with grade 1 (poor) are 17 species, grade 2 (medium) are 5 species, grade 3 (good) are 65 species, and grade 4 (very good) are 63 species (Figure 3). From 150 species, plants which is getting a grade 1 (bad) are types of groundcover, because it only serves as an articulation and aesthetics. While the plants which is getting a grade 4 (very good) are types of shrubs and trees because they fulfill many functions.

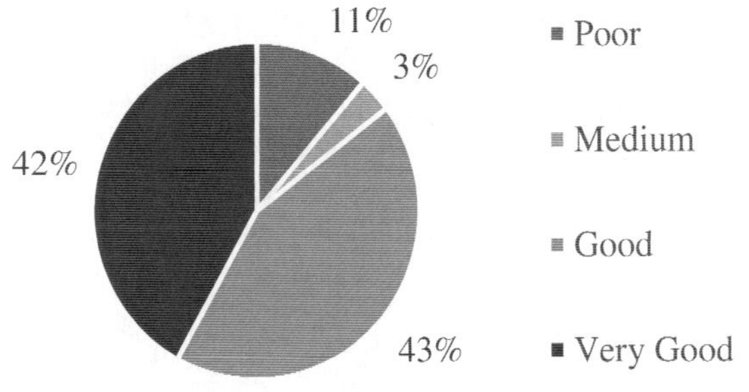

Figure 3. Functional quality

3.3.2. Visual quality assessment. Visual quality assessment consists of 4 criteria: color, shape, texture, and scale. Grades obtained based on the results of questionnaires from respondents.

The average value obtained of visual quality assessment criteria of color is 6.3. Wild plant species with a score above the average are 61 species (Figure 4). From the results of the assessment it is known that there are 41% species of wild plants whose color is considered attractive, whereas 59% considered less attractive.

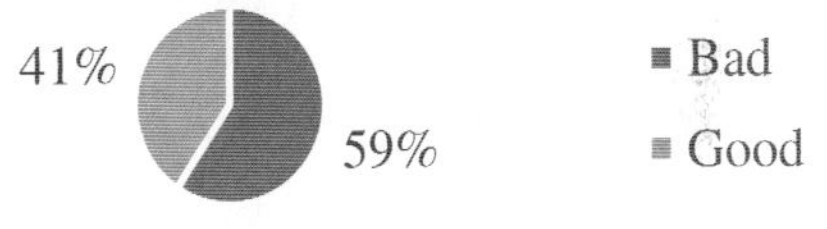

Figure 4. Visual quality of color

The average value obtained of visual quality assessment criteria of form is 6.17. Wild plant species with a score above the average are 78 species (Figure 5). From the results of the assessment it is known that there are 52% species of wild plants whose shape is considered attractive, whereas 48% considered less attractive.

Figure 5. Visual quality of form

The average value obtained of Visual quality assessment criteria of texture is 6.13. Wild plant species with a score above the average are 76 species (Figure 6). From the results of the assessment it is known that there are 51% species of wild plant whose texture is considered attractive, whereas 49% considered less attractive.

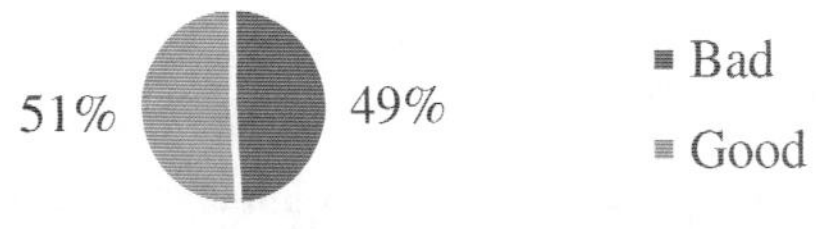

Figure 6. Visual quality of texture

The average value obtained of Visual quality assessment criteria of scale is 6.33. Wild plant species with a score above the average are 66 species (Figure 7). From the results of the assessment it is known that there are 44% species of wild plants whose size is considered attractive, whereas 56% considered less attractive.

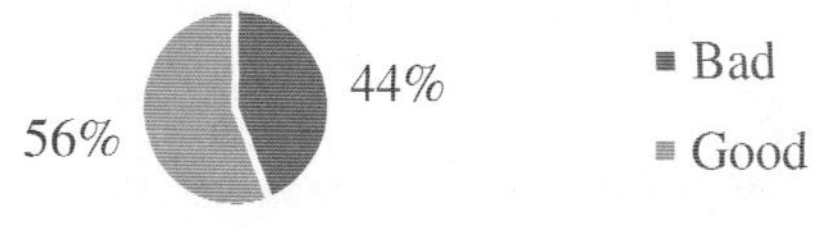

Figure 7. Visual quality of scale

3.4. Recommendation

The next step is make recommendation based on the results of functional and visual quality assessment by using *FileMaker Pro 11* application. The content that provided include: latin name, family, synonym, local name, description of plants (high maximum growth, morphology of the whole body of plants and how to multiply), habitat (open/with shade and grow in lowland/highland), chemical content, description of visual quality and functional owned and recommendation as an element of the landscape, crop images, as well as additional information (can not be consumed because they are poisonous or can consumed as vegetables and medicines). The following are 60 plant species that recommended as landscape element. Those plant species has functional and visual quality score above the average.

Table 4 Recommentation list

Local Name	Func	Vis	Local Name	Func	Vis	Local Name	Func	Vis
ajeran	4 (Vg)	6,58	daun katu	4 (Vg)	6,575	kembang kertas	4 (Vg)	6,925
akalifa bintik	4 (Vg)	7,515	daun kentut	4 (Vg)	6,695	kembang telang	4 (Vg)	7,12
alang-alang	4 (Vg)	6,405	daun kupu-kupu	4 (Vg)	6,965	kemuning	4 (Vg)	6,78
amisan	3 (G)	7,22	daun madu	4 (Vg)	7,065	kenikir	3 (G)	6,68
andong	4 (Vg)	7,925	daun mangkokan	4 (Vg)	6,41	kerak nasi	4 (Vg)	7,73
antanan kembang	3 (G)	7,63	dlujru	4 (Vg)	6,47	mamang besar	4 (Vg)	6,835
aseman	4 (Vg)	7,01	eceng padi	3 (G)	6,445	melati belanda	4 (Vg)	7,725
bakung	3 (G)	6,95	ekor anjing	3 (G)	6,41	nanas-nanasan	3 (G)	6,845
bayam dempo	3 (G)	6,67	ekor kucing	4 (Vg)	6,905	pule pandak	3 (G)	7,018
boroco	3 (G)	6,75	gempur batu	3 (G)	7,005	putri malu	3 (G)	6,58
bribil	3 (G)	7,445	getih-getihan	4 (Vg)	6,245	secang	3 (G)	6,655
bunga kenop	3 (G)	7,97	gewor	3 (G)	6,355	sente	3 (G)	6,51
bunga pagoda	4 (Vg)	8,39	iler	4 (Vg)	7,63	songgo langit	3 (G)	6,97
bunga pukul 4	3 (G)	7,25	jabung	3 (G)	6,26	suruhan	3 (G)	6,93
bunga tasbih	4 (Vg)	7,905	jarong lelaki	3 (G)	6,29	talok	4 (Vg)	6,69
buni	4 (Vg)	6,755	jengger ayam	3 (G)	6,975	tapak doro putih	3 (G)	6,995
cakar ayam	3 (G)	6,73	kangkung alas	3 (G)	7,225	tembelekan	4 (Vg)	6,925
calincing	3 (G)	6,95	kecombrang	4 (Vg)	7,88	turi	3 (G)	7,235
daun duduk	3 (G)	6,865	keladi warna	3 (G)	6,73	wedelia	4 (Vg)	7,61
daun encok	4 (Vg)	6,925	kembang bugang	3 (G)	6,54	zig-zag	3 (G)	6,705

4. Conclusions and suggestions

4.1. Conclusion

There are 150 wild plants identified in Pos Kesehatan Dusun (Poskesdus) Tangkilan, Sidoarum Village, Godean, Sleman, Yogyakarta. Consisting of ground cover types of plants, bushes, shrubs, and trees. Each species is described by latin name, family, synonym, local name, morphology, habitat, chemical content, visual and functional quality value obtained, on the landscape, crop images, and additional information fit the category. Additional information consists of non-herbal food crops that can or can not be consumed, as well as herbal food crops that can be consumed or can not be consumed. Wild plants that have functional value and visual quality above average are 60 species, so it can be recommended as an element of the landscape. The recommendations compiled in book form gallery landscape plants using *FileMaker Pro 11* application.

4.2. Suggestion

The results of this study can be input and consideration for the designers and managers in designing minimal management of park.

IOP Conf. Series: Earth and Environmental Science **91** (2017) 012008 doi:10.1088/1755-1315/91/1/012008

References

[1] Carpenter PL, Walker TD, Lanphear FO. 1975. Plant in The Landscape. San Fransisco (US): WH Freeman and Co.

[2] Hendriono HB. 2013. Yogyakarta di Tangkilan [internet]. [referred to 2016 Maret 8]. Available from: http://www.disanaproject.com.

[3] Rachman TA, Ramdlani S, Soekirno A. 2013. Penataan kembali kawasan pasar bunga dan pasar hewan (splendid) Kota Malang. Jurnal Mahasiswa Jurusan Teknik Arsitektur. 1(1):1-10.

[4] Taihuttu HN. 2013. Fungsi tanaman dalam desain lanskap Taman Makam Pahlawan Perang Dunia II – Australia di Kota Ambon. Jurnal Budidaya Pertanian. 9(2):92-94.

[5] Wungkar MM. 2005. Evaluasi aspek fungsi dan kualitas estetika arsitektural pohon lanskap jalan Kota Bogor [Tesis]. Bogor (ID): Institut Pertanian Bogor.

IOP Conf. Series: Earth and Environmental Science **91** (2017) 012009 doi:10.1088/1755-1315/91/1/012009

The Influence of Vegetation Function towards the Langsep Street Thermal Comfort

R Alfian[1], I Setyabudi[2] and R S Uran[3]

[1]Lecturer of Architectural Planning, Tribhuwana Tunggadewi University, Malang, Indonesia
[2]Lecturer of Architectural Design, Tribhuwana Tunggadewi University, Malang, Indonesia
[3]Student of Landscape Architecture, Tribhuwana Tunggadewi University, Malang, Indonesia

E-mail*:* rizkialfian87@gmail.com

Abstract. Streetscape is an important element for character building of the environment, spatial, and visual in order to provide an urban identity, especially in Malang City protocol streets. Langsep Street is one of the protocol streets in Malang City. Langsep Street famous with central education and offices area. This study aims (1) to identify vegetation of streetscape; (2) to analyze the thermal comfort of the streetscape, and (3) to evaluate the comfort level of Langsep Street. The method used was the THI approach. THI value that obtained was analyzed using the standard of Laurie (1990). Based on observations, the THI value of Langsep Street was 27.60. This was influenced by the trees canopy density and spacing of the trees on the streetscape. It can be concluded that streetscape required (1) shaded plants that have root systems that do not damage the construction of roads, (2) the branching plants are not easily broken and easy to maintain, and (3) the combination of trees, shrubs and ground cover.

1. Introduction

Streetscape is the character of the land or walk formed on a neighborhood street, both of which are formed from the natural landscape elements such as the shape and topography of land formed from man-made landscape elements adapted to the conditions of land [2]. Simonds [7] Streetscape was instrumental in establishing the character of the environment, spatial, and visual in order to provide an urban identity.

According to Utterman in Santyo, *et al.* [6], elements that affect the comfort in a pedestrian are: circulation, accessibility, natural style and climate, security, cleanliness and beauty.

Plants in the landscape view of the road serves as a controller, a physical barrier, climate control, erosion control, wildlife habitat, and aesthetics. Distribution and diversity of tree species in a landscape are important for improving the function of vegetation for environmental stability, both biotic and abiotic [1]. According to Laurie (1986) in Rahmiati [5], the standard moisture for human comfort in activities ranging from 40% - 70% with a temperature between 15ºC-27ºC and Hadi *et al* [3], states that the index of comfort in comfortable conditions ideal for man Indonesia in the range of THI (Temperature human Index) with a value of 20-26.

IOP Conf. Series: Earth and Environmental Science **91** (2017) 012009 doi:10.1088/1755-1315/91/1/012009

Today most of the streetscape on some streets in Malang City less attention to the physical condition and social landscape of the street, pavement element that dominates, the lack of availability of facilities for the streetscape, as well as the lack of the number, type, and maintenance of the vegetation on the landscape. Maintenance of the type and amount of vegetation on the streetscape is a problem that most influence on the comfort of street users and local residents. The street users are not comfortable for moving because of the street conditions are quite hot and high levels of pollution.
This article will discuss the level of comfort and its evaluation on streetscape on Langsep Street which is one of protocol street in Malang City. The comfort function is one thing that should be available on the landscape so that users can work well. Hence it is important to evaluate the comfort function of a streetscape so it can be analyzed the factors and the comfort level that affect the comfort of the street.

2. Method

The method used in this research was quantitative method with the following technical stages:

2.1 Preparation

The preparation stage was the stage of determining the location and preparation of tools and materials. From this stage determined the location of the research is Langsep Street that is the area of education and office center. The researcher process the permission to Public Works Department and the Department of Hygiene and Park, Malang City.

2.2 Survey

At this stage conducted observation and data collection of air temperature and humidity directly from the field. Data collection was performed by measuring the temperature and humidity at the observation point in distance of 50 meters (Figure 1). Point measurement of temperature and humidity occurs on road safety threshold and median road where there are plants and paving. Data collection was performed three times in one day at 06:00, 12:00, 17:00, for three days. Intake air temperature and humidity data is carried out only when the weather is sunny with using digital termohygrometer.

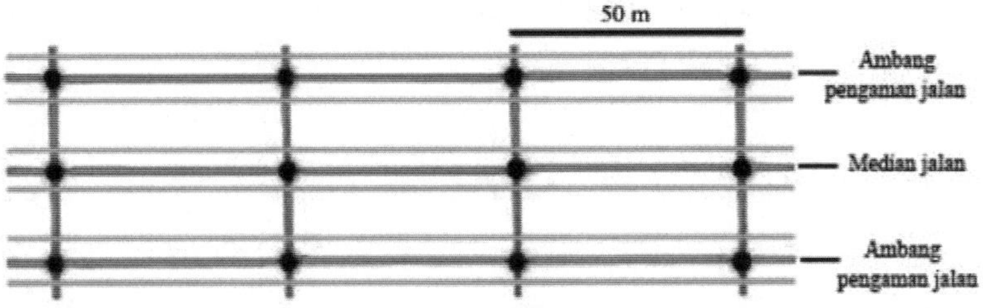

Figure 1 Illustration of observation point temperature and humidity on the streetscape

The size of the street, street structure and land use data obtained from study literature, reports, and standard regulations. The data collection of vegetation and plant spacing obtained from measurement technique, documentation, and collection of existing conditions. The social data of visitors' perception of Langsep Street used the questionnaire from 30 respondents.

2.3 Analysis

The method used in the analysis was the approach of THI comfort index. In this method, the temperature and humidity data was obtained from the measurement results then calculated by the THI formula value;

$$THI = 0,8T + (RH \times T)/500$$

THI: *temperature humidity index*
T: temperature (°C)
RH: humidity (%)

THI value obtained was analyzed using the standard of Laurie [4] which states that the ideal environment to have the air temperature 27-28 °C and humidity of 40-75%.

Answers about the purpose and motivation of respondents in each of the paths were analyzed descriptively. A questionnaire containing questions about the data themselves in general respondents (gender, age, education, place of residence), goals, motivations, and perceptions of respondents about the micro-climate comfort. Analysis of the results using chi-square questionnaire to determine the relationship between the answers perception of comfort microclimate street with the factor of gender, place of residence, education, and age of the respondents. Answers about the purpose and motivation of respondents in each park were analyzed descriptively. The results of the questionnaire and the calculation results are analyzed so that it can be used as material comfort evaluation and formulated ways to improve comfort microclimate in each street.

3. Result and Analysis

Langsep Street is one of busy street in Malang City, famous for its central areas of education such as schools and colleges. There are many offices scattered along the Langsep Street. Maintenance of plants on the Langsep street generally suitable with criteria plantings there are sufficiently shade plants on the street. The amount of shade plants on Langsep street enough and planted with appropriate spacing between trees that form a canopy shades helped influence the microclimate. Based on observations of Langsep streetscape have different types of shade trees include mahogany (*Swietenia mahagoni*), and Trembesi (*Samanea saman*).

Langsep street has a length of 1355 meters with an average spacing of trees 8 meters, 6.40 meters of street width, 2.50 meters of pavement width and 3.20 meter of median width. Surrounding Langsep streetscape are very few trees with rarely spacing, so high air circulation. In Alfian [1] stated Vegetation is very useful to manipulate the environment in urban aesthetic, controlling erosion and groundwater, reduce noise, waste water control, traffic control and glare, reduce light reflection, and reduce smell.

3.1 Micro climate Langsep Street

From mapping point observation of the street with a distance of 50 meters gained 27 points of data retrieval air temperature and humidity, based on the observations of temperature and humidity for three days of observation on Highway Langep obtained the following data (Figure 2, 3 and 4).

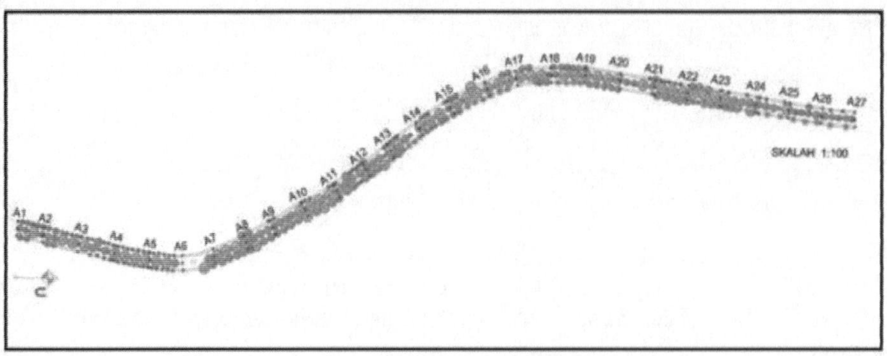

Figure 2. Observation point of temperature and humidity on Langsep Street

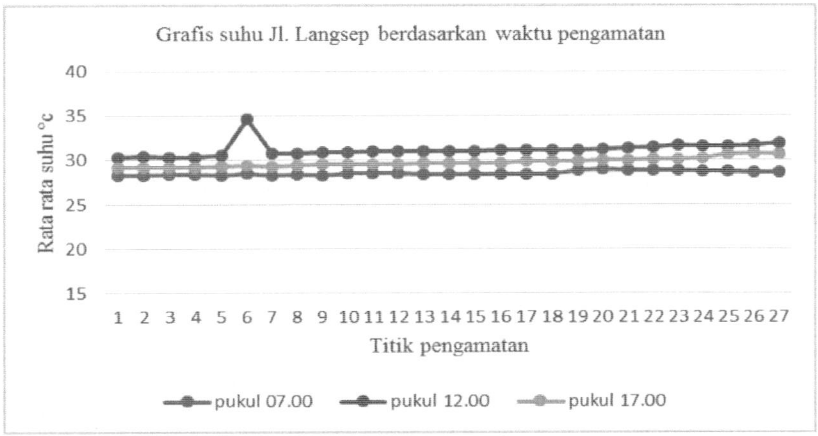

Figure 3. Graph of temperature observations on Langsep Street

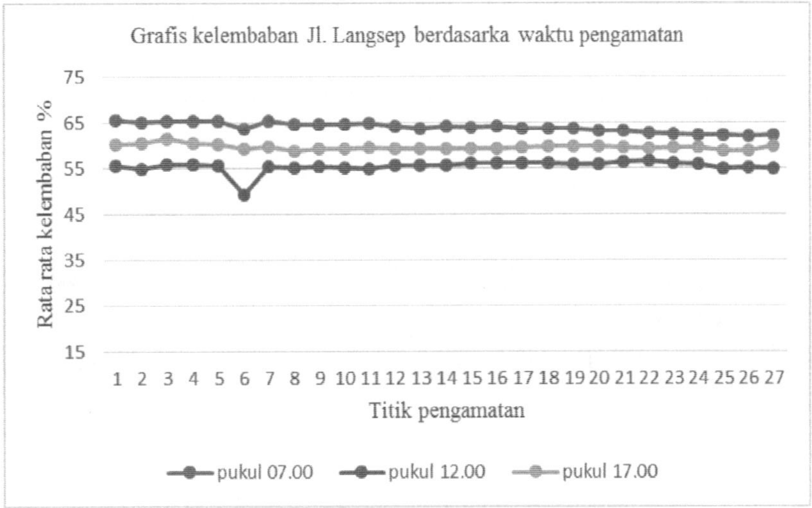

Figure 4. Graph of Humidity observation on Langsep Street

Based on the results of temperature measurements at Langsep street, the highest air temperature is at a point 6 at 12 pm. This point is located close to the intersection area or turn directions, and there is no trees so the air temperature becomes higher because the sunlight received directly in this area. In addition, the region is dominated form of asphalt pavement and road. Humidity value at a point 6 at 12 pm, have a low value because the area is dominated pavement so that direct sunlight reaches the road and cause higher evaporation rates as well as the wind factor that removes moisture evaporation results.

Air temperatures on Langsep street ranged between 28.28 - 34.59 °C with humidity ranged between 49.36 - 67.89%. The landscaping around Langsep street such as schools and shops. Schools and shops are neatly arranged with a shade tree titled big enough. The majority of the vegetation at Langsep street a Trembesi tree (*Samanea saman*) which has a fairly large tree canopy. Heading between trees that touch each pose shade that provides thermal comfort.

3.2. Analysis Humadity Temperature Index (THI) Langsep Street

To find the comfort of air temperature and humidity quantitatively used Temperature Humidity Index (THI). This method uses a factor of air temperature and humidity. The comfort level is a series of conditions on several factors. The results of some of these factors that affect the value of THI. Based on field observations THI value for Langsep street (27.60) with average temperatures ranging from

28.28 - 34.59 °C with humidity ranging between 49.36 - 67.89%. If analyzed according to the standard Laurie[4]. The value of THI at Langsep street out of the standard comfort. THI value by Laurie[4] categorized comfortable is between 21-27, and ideal climate for humans is the air temperature to the value of 27-28 °C and humidity 40-70%.

3.3. Analysis Questionnaire Results Highway langsep.

From the results of questionnaires on Langsep Street obtained street users is 56.67% men and 43.33% women. Street users education is at 13:33% secondary school, 30% high school, 6.67% Academic, 43.33% under graduate school and 6.67 the others.

Hypotheses for comfort relationship with the respondents' gender factor in Langsep Street are as follows:

H_0: Comfort road is not related to gender factor
H_1: There is a relationship between the comfort of roads with gender factor
Chi-table = 3.82
Chi-count = 0.54
Obtained chi-count <Chi-table accept H_0

The calculation is known that the comfort of the road on Langsep Street not related to the gender of street users (Table 1).

Table 1. Comfort respondent data by gender

No	Answer Choice	Gender		
		Man	Woman	Total
1	Comfort	17	10	27
2	Not Comfort	1	2	3
	Total	18	12	30

Table 1 shows that 90% of respondents (27 respondents) feel comfortable with a microclimate Langsep street. The proportion of the perception of comfort in terms of gender is quite comparable. Percentage opinion of a sense of comfort by male respondents was (94%) greater than female respondents (83%).

3.4. Leisure relationship with Respondent Education Factor

Hypotheses for comfort relationship with the respondents' education factor in Highway langsep are as follows:

H_0: Comfort road is not related to educational factors
H_1: There is a relationship between the comfort of roads by a factor of education
Chi-table = 3.82
Chi-count = 2,64
Obtained chi-count <Chi-table accept H_0

The calculation is known that the comfort of the street on Langsep street not related to street user education (Table 2).

Table 2. Data respondent convenience by education.

No	Education level	Comfort	Not Comfort
1	Secondary school	3	1
2	High School	8	1
3	Academic	2	-
4	Undergraduate school	12	1
5	The other	2	-
	Total	27	3

Table 2 shows that the percentage of respondents think that a sense of comfort by junior high school education by (75%) less than high school (89%), Academic (100%), universities (92%), and the others (100%).

4. Conclusion
Termal comfort on Langsep Street influenced by several factors, namely air temperature, humidity, wind and solar radiation. The results showed that the air temperature of the Langsep street ranged between 28.28 - 34.59 °C with humidity ranged between 49.36 - 67.89%. Average of air temperature on Langsep street 30.2 °C and an average humidity of 60%. Langsep street has a value of THI (27.60) that out of the comfort standards set by Laurie [4] at 21-27. In the street area that has a dense spacing of the plants tend to lower air temperature and humidity tends to be high. Vice versa on street area with rarely spacing of plants shows air temperature tends to be high with low humidity.

5. Suggestion
Air temperature and humidity can be modified by selecting the appropriate type of vegetation shade. To scale street in the city, can be used by the plant canopy of shade plants that have root systems that do not damage the construction of roads, branching plants are not easily broken and easy to maintain. Necessary to add more vegetation such as Tanjung (*Mimusops Elengi*), Kencana Ketapang (*Terminalia mantaly*), glodokan tiang (*Polyalthia longifolia*), trembesi (*Samanea saman*), Kiara Payung (*Fellicium Decipiens*) and Angsana (*Pterocarpus indicus*). By selecting the appropriate shade of vegetation and treatment is expected to help increase the level of comfort in every way. In addition to using vegetation shade, combining different types of landscape plants that have different growth patterns such as the use of groundcover and shrubs, manipulate the environment on the way to be more comfortable.

References
[1] Alfian R 2015 Pengaruh Bentuk Hutan Kota terhadap Kenyamanan Termal Di Sekitar Hutan Kota. Tesis. Institut Pertanian Bogor
[2] Direktorat Jenderal Bina Marga 1996 Tata cara Perencanaan Teknik lanskap Jalan. Departemen Pekerjaan Umum. Jakarta.
[3] Hadi and Rohman 2012 Evaluasi Indeks Kenyamanan Taman Kota (Lapangan Puputan Bandung I Gusti Ngurah Made Agung Denpasar,Bali).*E-Journal Agroekoteknologi Tropika Vol. 1 No. 1, Juli 2012*. Fakultas Pertanian. Universitas Udayana
[4] Laurie M 1990 *Pengantar Kepada Arsitektur Pertamanan*. Bandung: Intermata
[5] Rahmiati and Mimi 2009 *Studi Aspek Kenyamanan Ruang Pedestrian Dalam Rangka Peningkatan Efektivitas Penggunaannya pada Kawasan Jalan M.H Thamrin- Jend.Sudirman, Jakarta.*Tesis. Institut Pertanian Bogor

[6] Santyo, Dwi, Ardiansyah P, Fahmi, Parmono, Harwin and Dwi 2011 *Kajian Aspek Kenyamanan pada Jalur Pedestrian di Penggal Jalan Prof. Soedharto* (Ngesrep Patung Diponegoro-Gerbang Undip), Laporan Seminar, Universitas Diponegoro, Semarang
[7] Simonds J O 1983 *Landscape Architecture.* New York: Mc graw-Hill Book Company Inc.

IOP Conf. Series: Earth and Environmental Science **91** (2017) 012010 doi:10.1088/1755-1315/91/1/012010

Diversity and Utilization of Bamboo Plants in The Area of Hotel in Kedewatan Village, Ubud, Bali

N W F Utami[1] and N L M Pradnyawathi[2]

[1]Landscape Architecture Study Program, Faculty of Agriculture, Udayana University
Jl. P.B. Sudirman Denpasar, Bali, INDONESIA 80232
[2]Agroecotechnology Study Program, Faculty of Agriculture, Udayana University
Jl. P.B. Sudirman Denpasar, Bali, INDONESIA 80232

E-mail: wayan_febriana@unud.ac.id

Abstract. Bamboo or *tiying* (Balinese language) is a widely used non-timber plant in Indonesia especially in Bali. The presence of bamboo appertains to its ethno-botanical function of bamboo especially for rituals. However, there are other utilization of bamboo which is naturally grown or intentionally planted. Kedewatan as a famous place in northern Ubud, Bali have many lavish hotels with its natural environment and appealing place. The aims of this study is to invent bamboo species diversity and bamboo utilization on private areas of hotel in Kedewatan. Methods used in this study was field survey with observation and interview technic. Observation was implemented by purposive sampling methods by selecting hotel which adjacent to Ayung and Wos rivers. Interview was conducted with some key persons in charge on managing hotel garden. In addition, bamboo species identification was established through literature study. The results show that there are eleven bamboo species found on the survey area with most commonly employed species in the area were *tiying tali* (*Gigantochloa apus* (J.A. & J.H. Schultes) Kurz.) and *tiying gading* (*Phyllostachys sulphurea* (Carr.) A. e.t. C. Riv.) which were belong to exotic species. The areas which bamboo cultivated were welcome area as a hedgerow and near hotel lobby, between, outside and inside villa buildings, and naturally grown in the riverbanks with a good landscaping arrangement. Bamboo plantations were utilized to adorn and support the quality of the hotel building as well as to conserve soil and water along Ayung and Wos river canyons. The other utilization of bamboo was to facilitate ritual activity in Kedewatan village. They are allowed to ask for limited amount of bamboo culms with condition not to damage the physical appearance and function that desired by the hotel manager or hotel owner.

1. Introduction

Bali as one of famous tourist destinations locally and internationally has numerous tourist accommodation such as hotels, cottages and bungalows. The hotel growth rate in Bali was reaching up to 65% during five years since 2009 [1]. The existence of the hotel in tourist industry becomes a necessity for tourists who come and stay in Bali for specified period of time. There are several factors for tourist to determine the rate of accommodation facilities such as hotels. One of them is the quality of service and facilities. Therefore, the ability to maintain the quality of the environment, in this case is arrangement of landscape in the hotel, become major component to attract tourist to stay. The Environmental factors, particularly visual quality, has separate image for tourists especially for new visitors to determine lodging accommodation [2].

Ubud district is an area that has many lavish hotel accommodations which is equal to thirteen five-star hotels and 336 non-star hotel. The number was highest among the other districts in Gianyar. And Kedewatan village as part of Ubud district is also acknowledge to hotel landscaping to improve the quality of services for tourists. The hotel landscaping are composed by hardscape and softscape elements. Bamboo as type of softscape element has important involvement to the hotel beautification in general.

Bamboo or *tiying* in Balinese language is one of non-timber plantation that widely used by Indonesia people and Balinese community in particular. The presence of bamboo in Bali is essential for Balinese Hindu's ritual. There are at least 49 types offering product that utilize bamboo [3] or nearly 55 types of bamboo were utilized in the rituals of Balinese Hindu ceremony [4]. However, there are also other functions of bamboo plantation in addition to the functions mentioned above. It is interesting because of the abundant of bamboo found in Bali either deliberately grown on private land or naturally grown.

The focus of this study was to inventory bamboo plantations that are deliberately planted on private land of hotel and also to identify the function of bamboo plantation as softscape element in hotel, particularly for physical, ecological, and aesthetics functions. By composing the landscape, especially the utilization of bamboo be able to support the quality of the existing hotel. Besides, the geographical location of the village of Kedewatan ecologically appropriate for the growth of bamboo plants.

2. Methods

Kedewatan village is situated in northern of Ubud district, Gianyar regency, Bali province with geographical locations at latitude 8◦27'15" LS - 8◦27'50" S and longitude 115◦14'20" E - 115◦15'30" E (Figure 1).

Kedewatan is stretched from north to south and flanked by two big rivers namely Ayung river and Wos river. The village covered about 4,35 ha of land which is almost of area was agricultural land (56% of the total area). There is no forest found in Kedewatan, but the growing plants were forming small forest adjacent to both river [5].

There were three phase implemented in this study. First was pre-inventory and inventory phase. This phase begin by conducting interviews with local government officer of area (head of the village or *perbekel* and head of sub-village or *kelian banjar*) to determine the location of observations. Once the location was determined then it continued by selecting the sites. Later it followed by conducting visual observations in hotel and interviews with hotel managers to obtain information about the types of bamboo planted, recording morphological and ecological characteristics of observed bamboo as well as its function in the hotel landscape. Bamboo botanical names were checked through literature study.

The next step continued by recording ecological data for bamboo evaluation and assessment by recording bamboo functions in term of amenities (temperature, humidity, wind, glare), architectural (forming space, curtain, privacy), engineering quality of the environment (erosion, hydrology, air, aroma, wildlife) and aesthetic (complementary, unite, put pressure, welcoming, soften, framing). Data presented in tabulation and then analyzed descriptively. The last phase was synthesis which undertake to develop conclusions based on research objectives that have been made at the beginning of the study. Figure 2 describes briefly in respect to research framework base on phase 1, 2 and 3.

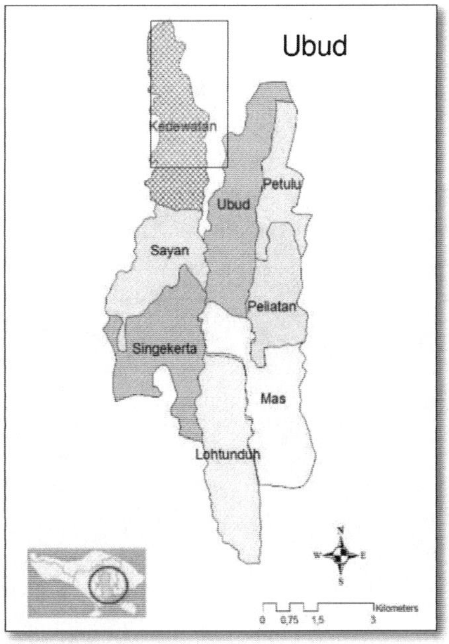

Figure 1. Location of the Kedewatan Village on the Ubud Dist

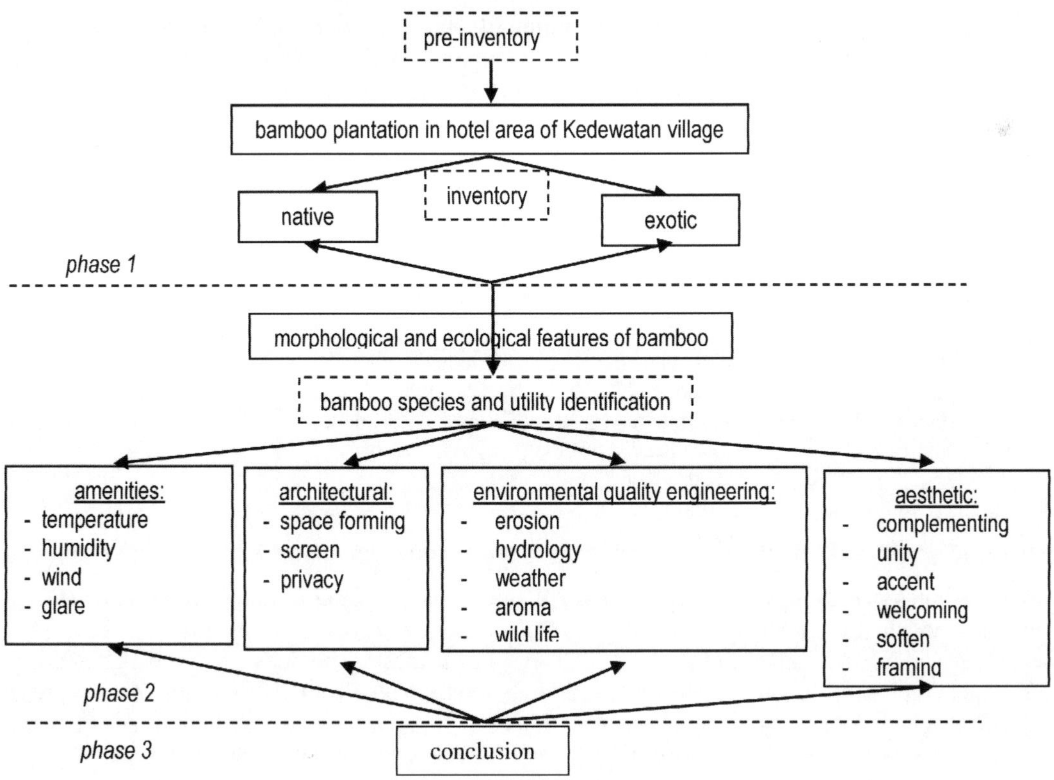

phase 1= inventory; phase 2= analysis; phase 3= synthesis

Figure 2. Research Framework

3. Results and Discussions

3.1. Study area and local people of Kedewatan

The population in Kedewatan village was reaching as much as 6.812 which 611 people constituted to productive age group (ranging from 35 to 39 years old). The main livelihood of the people were traders (1.197 people) and farmers (1.004 people). Because almost half of the people were farmer, it shaped the landscape of village become a rural agricultural landscape. Beside agriculture, forming small forest adjacent to the river had potential to support tourism industry of the village (Figure 3).

Figure 3. Natural and Agricultural Landscape View of Kedewatan village

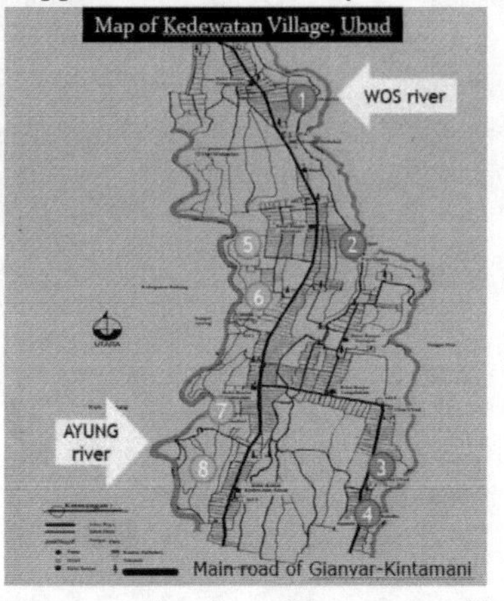

1=Komaneka,
2=The Payogan,
3=Uma Ubud,
4= Pitamaha,
5=The Royal Pitamaha,
6=Kupu-kupu Barong,
7=Amandari,
8= Mandapa

Figure 4. Bamboo Observation Site Selection

Observations showed that there were number of supporting infrastructure such as lodgings and other tourist facilities found in Kedewatan village. It categorized into four groups based on the amount of payments to the village government (Table 1).

Table 1. Hotel Classification Based on Amount of Payment

Hotel Category			
A	B	C	D
Hotel Amandari*	Komaneka*	Kori Ubud	Allison
Uma Ubud Resort*	Pitamaha*	Kirana Spa	Amir Rabik
Sobek Bali Utama	Puri Bunga	Ananda Cottages	
The Royal Pita Maha*	The Payogan Resort*	Botanika Spa	
Mandapa*	Ulun Ubud Hotel	Rest Pulau Kelapa	
	Bali Rich		
	Kupu-Kupu Barong*		
	Bar		

Table 1 illustrates that there were twenty tourism businesses in the form of hotels, home stays, as well as lodgings and bars that grouped into four categories, namely category A to category D. The category A contributed more than other categories and so on. From these groupings, then we selected eight hotels which represents hotel which has bamboo plantation also and the location also represent geographical location of the village which was adjacent to Wos river and Ayung river (Figure 4). Finally eight hotels were selected, four hotel located on eastern of the village or adjacent to Wos river (Komaneka, The Payogan, Uma Ubud, and Pitamaha) and the others were located on western of the

IOP Conf. Series: Earth and Environmental Science **91** (2017) 012010 doi:10.1088/1755-1315/91/1/012010

village or adjacent to Ayung river (The Royal Pitamaha, Kupu-kupu Barong, Amandari, and Mandapa).

3.2. Inventory and function of bamboo in hotel area

Overall, eleven bamboo species were identified and used by hotels in specific arrangement for managing the hotel landscape. The species found respectively: *bambu tali* (*Gigantochloa apus* (JA & JH Schultes) Kurz.), *bambu Jakarta* (*Bambusa glaucescens* (Willd.) Merr), *bambu kuning* (*Phyllostachys sulphurea* (Carr.) A. et C. Riv.)) , *bambu ampel gadang & ampel gading* (*Bambusa vulgaris* Schrad. ex JC Wendl.), *bambu Cina var. kuning/sudamala* (*Phyllostachys aurea* Carr. Ex.A.et C.Riv.), *bambu Jepang var. kuning* (*Pseudosasa japonica* (Siebold & Zucc. ex Steud.) Makino ex Nakai), *bambu kelisik/bambu Cina var. hijau* (*Bambusa multiplex* Lour. Raeusch. ex Schult), *bambu Cina var. varigata* (*Arundinaria auricoma*), *bambu hitam/selem* (*Gigantochloa atroviolacea* Widjaja), and *bambu ater/santong* (*Gigantochloa atter* (Hassk.) Kurz) (Table 2).

Table 2. Bamboo Inventory on Eight Observed Hotel in Kedewatan

No	Species and location	
1	Common name: Bambu tali (ID)/*tiying tali* (Balinese) Scientific name: *Gigantochloa apus (J.A. & J.H. Schultes)* Kurz. Location: 1, 3, 4, 5, 6, 7, 8	
2	Common name: Bambu Jakarta (ID)/*tiying Jakarta* (Balinese) Scientific name: *Bambusa glaucescens* (Willd.) Merr Location: 1, 2, 3, 4, 7, 8	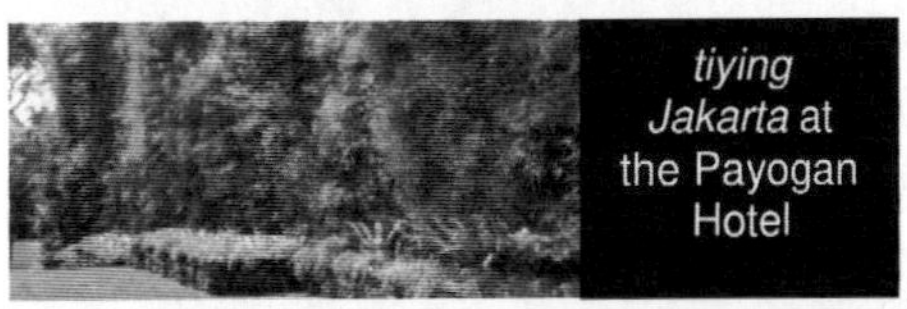
3	Common name: Bambu kuning (ID)/*tiying gading* (Balinese) Scientific name: *Phyllostachys sulphurea* (Carr.) Location: 1, 3, 4, 5, 6, 7, 8	
4	Common name: Bambu ampel hijau (ID)/*tiying ampel gadang* (Balinese) Scientific name: *Bambusa vulgaris* Schrad. ex J.C. Wendl. var. vulgaris Location: 1, 2, 4, 5, 6, 8	
5	Common name: Bambu ampel kuning (ID)/*tiying ampel gading* (Balinese) Scientific name: *Bambusa vulgaris* Schrad. Ex J.C. Wendl. var. striata Location: 1, 5	

IOP Conf. Series: Earth and Environmental Science **91** (2017) 012010 doi:10.1088/1755-1315/91/1/012010

No	Species and location	
6	Common name: Bambu Jepang var. kuning (ID)/*tiying Jepang* (Balinese) Scientific name: *Pseudosasa japonica* (Siebold & Zucc. ex Steud.) Makino ex Nakai Location: 4, 5, 6, 7	
7	Common name: Bambu cina kultivar kuning (ID)/*tiying sudamala* (Balinese) Scientific name: *Phyllostachys aurea* Carr. Ex.A.et C.Riv. Location: 6	
8	Common name: Bambu Kelisik/bambu cina var hijau (ID)/*tiying Cina* (Balinese) Scientific name: *Bambusa multiplex* (Lour.) Reauschel ex J.A. & J.H. Schultes Location: 5	
9	Common name: Bambu varigata (ID)/*tiying parigata* (Balinese) Scientific name: *Arundinaria auricoma* Location: 6	
10	Common name: Bambu hitam (ID)/*tiying selem* (Balinese) Scientific name: *Gigantochloa atroviolacea* Widjaja Location: 7	
11	Common name: Bambu ater (ID)/*tiying santong* (Balinese) Scientific name: *Gigantochloa atter* (Hassk.) Kurz Location: 2	

1=Komaneka; 2=The Payogan; 3=Uma Ubud; 4=Pitamaha; 5=The Royal Pitamaha; 6=Kupu-kupu Barong; 7=Amandari; 8=Mandapa

The results confirm that the presence of *tiying tali* species and *tiying gading* species were the highest among other species found in the observed hotel, except at the Payogan hotel. *Tiying tali* were common species planted in the hotel with adjacent to Ayung river and the Wos river. The function of the bamboos mainly to maintain the stability of the ground in a steep area. This finding corroborate results of other studies [6] that bamboo plants suitable to utilized in area with steep contours. Additionally, other studies shown that *tiying gading* species (*Phyllostachys sulphurea* (Carr.)) is usually used as an ornamental plant [7]. This findings suggests that *tiying gading* become important species for the aesthetic aspect of the landscape in hotel area such as complementing (*Bambusa vulgaris* Schrad. ex J.C. Wendl. var. vulgaris; *Phyllostachys sulphurea* (Carr.) A.et C.Riv), accent (*Bambusa vulgaris* Schrad var.striata; *Phyllostachys sulphurea* (Carr.) A.et C.Riv; *Pseudosasa japonica* (Siebold & Zucc. ex Steud.) Makino ex Nakai; *Arundinaria auricoma; Gigantochloa atroviolacea* Widjaja, welcoming (*Bambusa vulgaris* Schrad var.striata; *Phyllostachys sulphurea* (Carr.) A.et C.Riv; *Pseudosasa japonica* (Siebold & Zucc. ex Steud.) Makino ex Nakai; *Bambusa multiplex* (Lour.) Reauschel ex J.A. & J.H. Schultes; *Phyllostachys aurea* Carr. Ex.A.et C.Riv), and soften (*Bambusa multiplex* (Lour.) Reauschel ex J.A. & J.H. Schultes; *Phyllostachys aurea* Carr. Ex.A.et C.Riv; *Phyllostachys sulphurea* (Carr.) A.et C.Riv; *Arundinaria auricoma; Bambusa glaucescens* (Willd.) Merr; *Pseudosasa japonica* (Siebold & Zucc. ex Steud.) Makino ex Nakai; *Bambusa vulgaris* Schrad var.vulgaris; *Bambusa vulgaris* Schrad var. striata).

At observed hotel, diversity of bamboo then calculated using Shannon's-Wiener diversity index (H'). From the calculations, the index equal to 2.106. It represent that the value were moderate (1 <H '<3) or the site has moderate distributed bamboo species across the observed location. However, nine out of eleven bamboo species were belong to exotic species. Considering to the results, it can be understood that the main objective to planted bamboo was to arrange the hotel landscaping. The arrangement of the species, quantity and design patterns applied has been set by desires of the hotel owner or hotel manager. Some species were planted for aesthetic purpose while others were planted for river banks conservation. Consequently, the utilization of existing bamboo plants in protecting the and maintaining steep area of Ayung and Wos river banks was clarified by other study [8] that efforts to maintain the quality of the environment, in this case was maintaining the river banks, in the perception of international tourists as users provide satisfying reliability toward hotel service quality to factor of attributes of hotel surroundings and environmental.

3.3 Relation between bamboo and Balinese culture

It was inevitable that the presence of bamboo become basic requirement for the Balinese Hindu's community. There were three types of plants whose existence was essential to the sustainability of Balinese Hindu's rituals, namely coconut, banana and bamboo [9]. The other considering was bamboo become important and obligatory present in rituals such as in room decoration in religious ceremonies, in temporary construction of offerings buildings, in funerals, and other utilization [4]. In other word, people in Bali were highly dependent on this non timber commodities. Other study confirm [10] that there were two dominant species of bamboo used in Balinese Hindu rituals, one of which was *tiying tali*. This plant also has potential as traditional medicine because *tiying tali* has been known and used to treat a variety of diseases based on the book of traditional Balinese medicine of *Lontar Usada* [11].

From interviews, it can be inferred that bamboo plantations in the hotel were possible to use to facilitate the religious activities of community in the village of Kedewatan. For those purpose, people were allowed to ask for particular bamboo culms in limited quantities with condition not to damage physical appearance and function of bamboo that intended by the owner or hotel manager. With the result that the needs of particular bamboo culms for ritual activities can be satisfied from existing bamboo plantations at the hotel. However, the diversity of bamboo planted in hotel was not entirely species that obligatory used in the rituals. Further, some approach to vary the bamboo species, especially species that used for rituals, should be taken in to order increase diversity of ethno-botanical bamboo plantation on private lands including hotel area as an efforts to conserve genetic resources of bamboo in Bali and preserved the culture at once.

4. Conclusion

As concluding note to this study that there are eleven species of bamboo has been found in hotel. Nine of them are classified to exotic species i.e. *Gigantochloa apus* (JA & JH Schultes) Kurz., *Bambusa glaucescens* (Willd.) Merr, Phyllostachys sulphurea (Carr.) A. et C. Riv.), *Bambusa vulgaris* Schrad. ex JC Wendl, *Phyllostachys aurea* Carr. ex. A.et C.Riv., *Pseudosasa japonica* (Siebold & Zucc. ex Steud.) Makino ex Nakai, *Bambusa multiplex* Lour. Raeusch. ex Schult, *Arundinaria auricoma*. In contrary, *Gigantochloa atroviolacea* Widjaja is the native species to West Java, Indonesia and *Gigantochloa atter* (Hassk.) Kurz) is unknown species but widely cultivated in Java and several locations in Indonesia. The primary purpose for planting bamboo in hotel area is to achieve aesthetic function in addition to ecological functions in order to maintain the stability of the Ayung and Wos river banks. Moreover, bamboos which planted in the hotel area are possibly able to facilitate requirement of Balinese Hindu's religious rituals of community in the village of Kedewatan.

Acknowledgements
These investigations were supported by Non-tax Revenue Grant (PNBP) of Udayana University (grant no. 1245/UN14.1.23/PL/2016).

We thank to Mr. Dewa (Kedewatan village officer) for assistance during field investigation.

References
[1] BPS 2014 *Bali in Figures 2104* Central Bureau of Statistics of Republic of Indonesia http://bali.bps.go.id/tabel_detail.php?ed=611003&od=11&id=11
[2] Le Blanc G and Nguhen N 1996 *An Examination of the Factors That Signal Hotel Image to Travellers* Journal of Vacation Marketing Vol 3 No 1 pp 32-42.
[3] Sumantera I W 1995 *Bamboo in Balinese Rituals In* Rao I V R and Sastry C B 1995 Proceeding of the 5th International Bamboo Workshop and 4th of International Bamboo Congress pp 97-100
[4] Arinasa I B K and I N Peneng 2006 *Konservasi Bambu Tanaman Upacara Agama di Kabupaten Bangli-Bali* Prosiding Seminar Nasional Teknologi Bambu Terkini Jurusan Teknik Sipil FT UGM Yogyakarta (in Indonesian)
[5] Pemerintah Desa Kedewatan 2013 *Monography of Kedewatan in 2013* Government of Kedewatan Village (in Indonesian)
[6] Hasyim I S, Bayyid R S, Darel R P L P and Abiel F A 2015 *Rancangan Elemen, Sistem Sirkulasi, dan Tata Hijau Lanskap Pada Lahan Kontur di Hotel Padma Bandung* Jurnal Reka Karsa Vol. 3 No. 1 Februari 2015 Hal 1-12 (in Indonesian)
[7] Institute of Botany (State Key Laboratory of Systematic and Evolutionary Botany) Digital Plant Project Team 2004 Phyllostachys sulphurea (Carr) A.et C.Riv Academy of Sciences *Flora of China* Editorial Science Press http://frps.eflora.cn/
[8] Jawaheer T D 2004 *Exploring International Tourist's Perception of Hotel Operations by Using a Modified SERVQUAL Approach – A Case Study of Mauritius* Managing Service Quality Vol. 14 Iss: 5 pp. 350 - 364
[9] Eisman F B J 1992 *Bali Sekala and Niskala* Vol II Periplus Editions (HK) Ltd Singapore
[10] Arinasa I B K 2010 *Bamboo Diversity and Utilization in Balinese Rituals at Angsri Village-Bali, Indonesia* The Journal of American Bamboo Society 23 (1): 29-37
[11] Sujarwo W I B.K Arinasa I N Peneng 2010 *Potensi Bambu Tali (Gigantochloa apus J.A. & J.H. Schult. Kurz) Sebagai Obat di Bali.* Buletin Littro Vol 21 No 2 2010 halaman 129-137 (in Indonesian)

IOP Conf. Series: Earth and Environmental Science **91** (2017) 012011 doi:10.1088/1755-1315/91/1/012011

Comfort Evaluation of Some City Parks in Menteng Subdistrict Using Grid Method

P T Putra[1] and N Nasrullah[2]

[1]Department of Landscape Architecture, National Institute of Science and Technology
[2]Department of Landscape Architecture, Bogor Agricultural University

E-mail: pram.arl@istn.ac.id

Abstract. This article aims to evaluate the thermal comfort in three parks at Menteng Sub District, Central Jakarta. The method used in this research is grid method that mapped the parks into squares. Grid method indicated the spread of temperature and humidity value with the canopy coverage. This study purposed to identify the distribution of temperature and humidity in Menteng Park, Suropati Park, and Situ Lembang Park. From the results, the highest temperature was in Menteng Park indicated by the average value of temperature is 35.6°C compared with Situ Lembang Park (34.7°C) and Suropati Park (34.3°C). The highest humidity was in Suropati Park indicated by the average value of humidity was 52.6% compared with Menteng Park (50.5%) and Situ Lembang Park (48.2%).

1. Introduction

Major cities in Indonesia are currently experiencing a significant development progress, especially in Jakarta. The environmental conditions of Jakarta mostly are being suppressed by the massive urban development. One of the impacts of urban development is environmental problem. The existence of green open space in the city is a must to be able to provide comfort for the residents. The health of city dwellers can also be obtained by the presence of urban green open space. Based on Law No. 26 of 2007 explains that the minimum proportion of green open space in a city is 30%, consist of 20% owned by the government and 10% owned by the private. Parks ideally contained in any hierarchy of the city, starting at the level of neighborhood, village, district, up to the city.

Jakarta as the capital of Indonesia has a high population density but still do not have adequate green open spaces that meet to the standards. Based on data from the Parks and Funerals Department, Jakarta only has about 9% green open space. Spatial Plan of the City of Jakarta (RTRW Jakarta) targeted to increase the percentage of green open space in Jakarta. This indicates the importance of the existence of a city park as a form of green open space to compensate for the physical development of urban areas.

Thermal comfort is a term used to express state of the physical environment influences the atmosphere and climate on humans. Humans as users of the landscape need comfort as one of their primary need. Comfort can be presented with vegetation as microclimate controller [1]. Menteng Subdistrict in Central Jakarta is the center of government and business that has many city parks. Some of the city parks in Menteng Subdistrict are Menteng Park, Suropati Park, and Situ Lembang Park. It is necessary to evaluate those city parks to determine the level of comfort based from the influence of canopy coverage, the temperature, and the humidity itself. This study purposed to evaluate the distribution of temperature and humidity in Menteng Park, Suropati Park, and Situ Lembang Park in Menteng Subdistrict, Central Jakarta.

IOP Conf. Series: Earth and Environmental Science **91** (2017) 012011 doi:10.1088/1755-1315/91/1/012011

2. Methods

The research was conducted from March until April 2010 at Menteng Park, Suropati Park, and Situ Lembang Park at Menteng Subdistrict, Central Jakarta. The following methods were used quantitative method and analyzed descriptively [2]:

2.1 Preparation stage

The selected location for this study is Menteng Park, Suropati Park, and Situ Lembang Park at Menteng Subdistrict, Central Jakarta. Then, the permission for this study issued by the Parks and Funerals Department. The imageries data of Menteng Park, Suropati Park, and Situ Lembang Park was taken from Google Earth 2010 for the base map. Afterwards, the grid applied into the imageries data to determine the distribution of temperature and humidity inside the park connected with the canopy coverage. The grid method used the points as the representative's sample. The size of the grid was 20 x 20 m which has point in the center. Figure 1 illustrated the grid method used in this study. The grid plotted into the imageries data used AutoCAD 2007.

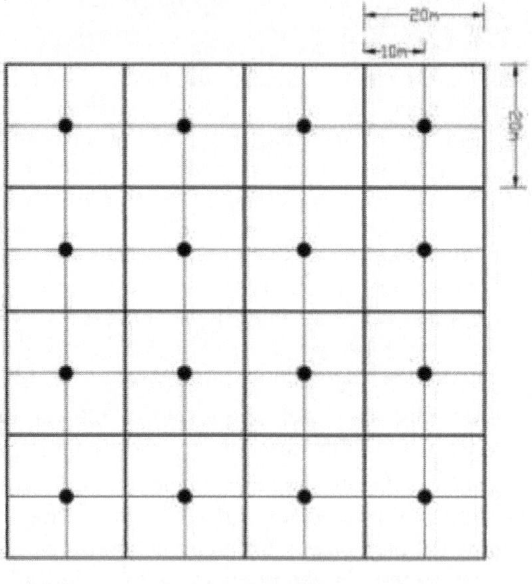

Figure 1. Grid method for the temperature and humidity data collection.

2.2 Survey stage

The quantitative data consists of temperature, humidity, and social data. The temperature and humidity data was taken during the midday when the weather was clear at 1 to 2 p.m used digital thermo-hygrometer. Each park was measured with the gauge in order to match it with the base map. It took 3 repetitions performed at each point. The map created by modeling the temperature and the humidity data to make grid function into *Surfer* version 8.0.

2.3 Analysis stage

Linear regression was conducted based on the data of average air temperature and humidity. Linear regression was used to assess the correlation between air temperature and humidity factor as the quantitative data with the canopy coverage of trees of each grid in the park. The linear regression equation is $y = ax + b$. Coefficient of determination (R^2) indicates the degree of the effect of canopy coverage to change the temperature or humidity in the park. The temperature and humidity values using analyzed with Temperature Humidity Index (THI). THI = 0.8T + (T x RH)/500, where T = temperature (°C), RH = relative humidity (%). THI value obtained from each park then was being analyzed using the standard of Laurie [3] which states that the environment has the ideal value of temperature ranged from

27 to 28 °C and humidity ranged from 40 to 75%. The temperature data and the humidity data from each park became the basis for the isotherm map and iso-hygro map to illustrate of the distribution of temperature and humidity value.

3. Results and Discussions

Menteng Subdistrict consists of settlement area, office buildings, and parks. Menteng Subdistrict was located in the heart of Jakarta and was designed by the Dutch when ruled Indonesia. The surrounding where the park located also affected the microclimate. Menteng Park located near Cokroaminoto Street and Prof. Moch. Yamin Street and it is quite crowded. Suropati Park and Situ Lembang were located on Taman Suropati Street and surrounded with settlements area.

In Menteng Park, the canopy coverage was not massive enough. Menteng Park was built in 2007. It used to be as a stadium of Persija Jakarta. In Suropati Park, the canopy coverage was massive enough because of the vegetation planted for years. In Situ Lembang Park, the vegetation coverage was only on the border side of the park because thereis a lake in the middle of the park.

Menteng Subdistrict classified as tropical climate type [4]. The highest temperature reached 33.4°C in October and the lowest 24°C in February. The average temperature during 2008 in Central Jakarta is 25.2 °C. The average rainfall is 159.1 mm with the highest rainfall in February amounted to 677.6 mm and lowest in July of 9.5 mm with an average humidity of 68% to 79%. The average wind speed 5.3 km/h.

The average of temperature in Menteng Park was 33.6°C. Figure 2 described the distribution of temperature in Menteng Park and figure 3 described the distribution of humidity. With a total area of 24 546 m^2 park, Menteng Park has canopy coverage area of 11 676 m^2. The linear regression equation between the temperature and the canopy coverage in Menteng Park is $y = -0.004x + 34.72$ with $R^2 = 0.801$. For the humidity, the equation is $y = 0.025x + 46.55$ with $R^2 = 0.864$. From these results it can be concluded that 80% of decreased temperature was influenced by the canopy coverage and 86% of increased humidity in the park was affected by canopy coverage.

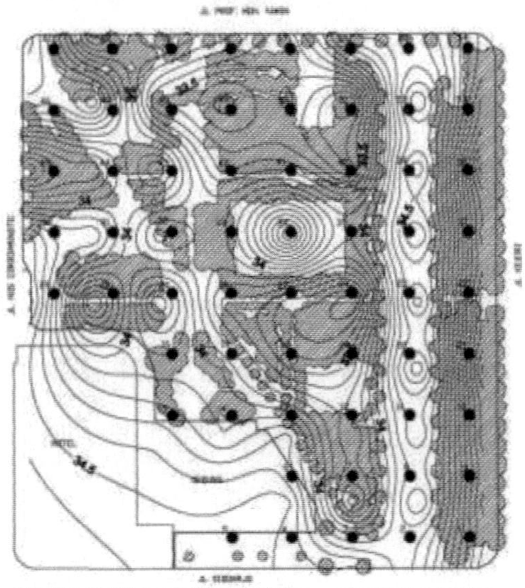

Figure 2. Isotherm map of Menteng Park.

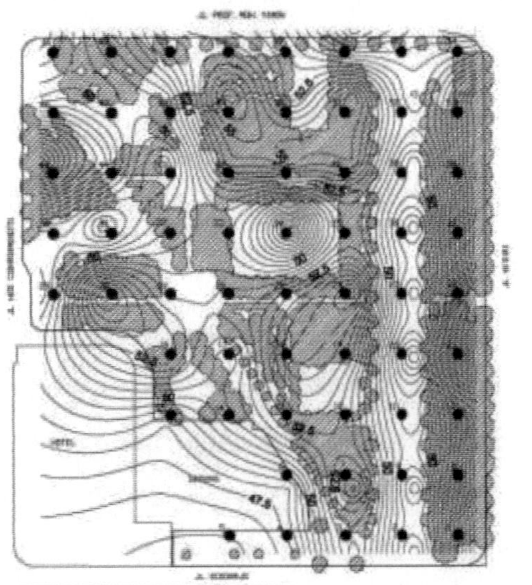

Figure 3. Iso-hygro map of Menteng Park.

Suropati Park has total area of 16 328 m^2. Figure 4 describes the distribution of temperature in Suropati Park and figure 5 describes the distribution of humidity. Suropati Park has canopy coverage area of 3 944 m^2. The linear regression equation between the temperature and the canopy coverage in

Suropati Park is $y = -0.003x + 34.52$ with $R^2 = 0.962$. For the humidity, the equation is $y = 0.032x + 42.21$ with $R^2 = 0.901$. From these results it can be concluded that 96% of decreased temperature was influenced by the canopy coverage and 90% of increased humidity in the park was affected by canopy coverage.

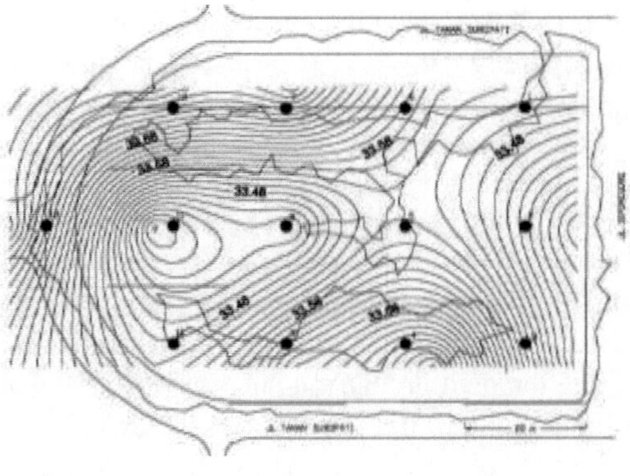

Figure 4. Isotherm map of Suropati Park.

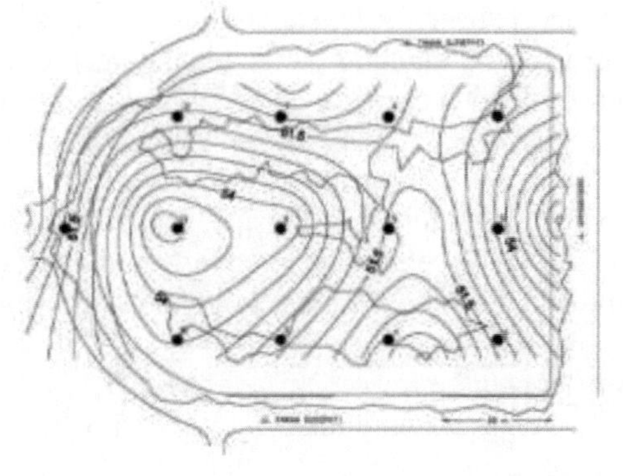

Figure 5. Iso-hygro map of Suropati Park.

Situ Lembang Park has total area of 14 700 m^2. Situ Lembang Park has canopy coverage area of 1 074 m^2. Figure 6 described the distribution of temperature in Suropati Park and figure 7 describes the distribution of humidity. The linear regression equation between the temperature and the canopy coverage in Situ Lembang Park is $y = -0.002x + 34.7$ with $R^2 = 0.936$. For the humidity, the equation is $y = 0.027x + 44.09$ with $R^2 = 0.891$. From these results it can be concluded that 93% of decreased temperature was influenced by the canopy coverage and 89% of increased humidity in the park was affected by canopy coverage.

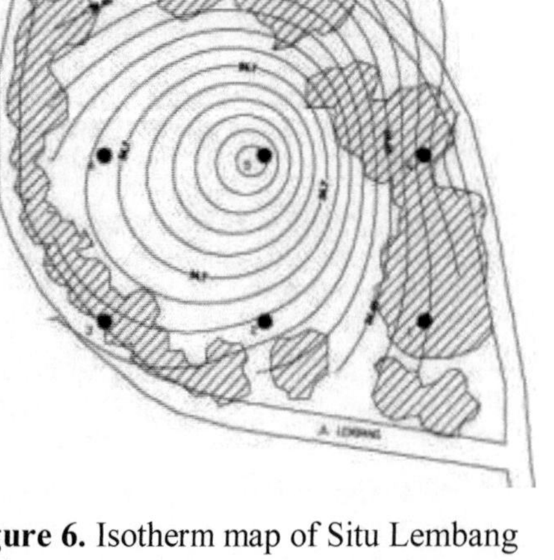

Figure 6. Isotherm map of Situ Lembang Park.

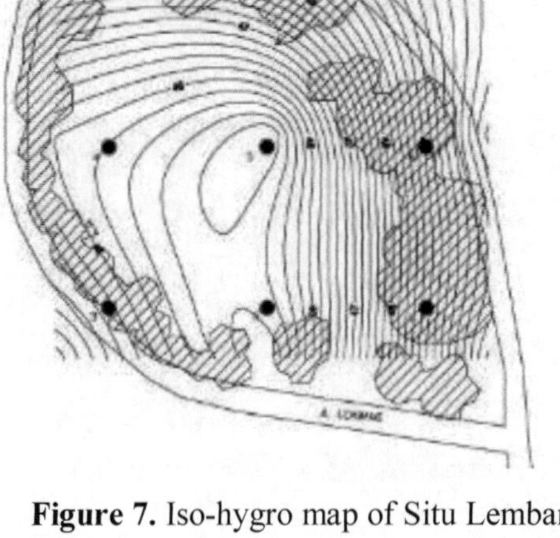

Figure 7. Iso-hygro map of Situ Lembang Park.

From the calculation, the THI value of Situ Lembang Park (30.76) is the highest when compared to Menteng Park (30.64) and Suropati Park (30.3). THI value from the three parks exceeded the THI comfort value which is 21 to 27. The ideal climate for people ranged from 27 to 28°C for the temperature and 40 to 75% for the humidity. Compared with the standard, it could be inferred that the three parks were not comfort enough.

The comfort factors not only from temperature but also from humidity, solar radiation, vegetation, and air movement. The pavement in the park made the temperature at some points in Menteng Park become hot. Beside the pavement, there were sports area, walkways, plazas, office building, and greenhouse area. Some of the existing vegetation in Menteng Park, Suropati Park, and Situ Lembang were *Swietenia macrophylla* and *Samanea saman.* Their canopies intersect each other and created a great shade at some parts in the park. The shaded area ameliorated the microclimate so that decreased the temperature and increased the humidity level.

4. Conclusions

The result of this study showed that from 3 repetitions of collected data, the average temperature and humidity in Menteng Park, Suropati Park, and Situ Lembang Park exceeded the THI comfort value. The grid method helps in order to analyze the distribution of temperature and humidity in the parks connected with the canopy coverage. The area under the canopy coverage mostly has lower temperature and higher humidity value.

References

[1] Carpenter P L, Walker T D and Lanphear F O 1975 *Plants in the Landscape* (San Francisco: WH Freeman and Company)
[2] Mulgiati U 2009 *Pengaruh Penutupan Vegetasi terhadap Kenyamanan Kota* (Bogor: Program Pascasarjana Institut Pertanian Bogor)
[3] Laurie M 1990 *Pengantar kepada Arsitektur Pertamanan* (Bandung: Intermata)
[4] BPS 2009 *Kecamatan Menteng dalam Angka 2009* (Jakarta: Badan Pusat Statistik)

Evaluation of roadside greenbelt trees damage caused by strangler plants in Bogor

Dibyanti Danniswari[1], Nizar Nasrullah[2]

[1]Graduate Student of Department of Landscape Architecture, Faculty of Agriculture, Bogor Agricultural University (IPB), Indonesia;
[2]Department of Landscape Architecture, Faculty of Agriculture, Bogor Agricultural University (IPB), Indonesia
dibyanti@gmail.com

Abstract. Certain plants are called stranglers (hemiepiphyte) because they grow on host trees and slowly choking the host, which often results in the host's death. The existence of strangler plants on roadside greenbelt trees is quite common in Bogor, but they may cause tree's failure and threaten users' safety. To prevent such hazard, evaluation of roadside greenbelt trees damage caused by strangler plants is important. This study was directed to analyse the vegetation of strangler plants in Bogor, to assess the damage caused by stranglers, and to compose strangled trees maintenance recommendations. This study was conducted in March to May 2014 by doing survey at five major roads in Bogor, which were Jalan Ahmad Yani, Jalan Sudirman, Jalan Pemuda, Jalan Semeru, and Jalan Juanda. The results showed that strangler species found in Bogor are *Ficus benjamina*, *Ficus glauca*, *Ficus elastica*, and *Schefflera actinophylla*. The most common species in Bogor is *F. benjamina*. Host trees that tend to be preferred by strangler plants are trees with large trunk, many branches, and medium to high height. The maintenance for every strangled tree is different according to the damage level, mild to severe damage could be treated by strangler root cutting to tree logging, respectively.

Keywords: tree maintenance, roadside greenbelt, strangler plant, hemiepiphyte

1. Introduction

Roadside greenbelt is an area of a road that is designated to be planted with trees, shrubs, or groundcovers. It is an effective type of green space to be used in cities because it does not require large space, but still manage to function as a tool for air and noise pollution mitigation [1]. There are many old and large roadside greenbelt trees in Bogor and it is quite common to find those trees strangled by strangler plants, which are hemiepiphytes. Hemiepiphytes are plants that start their life as epiphytes – plants that grow entirely on other plants – but later live as free-standing individuals by rooting in the ground. They begin their lives as a seed deposited high in a tree by a bird or monkey. They germinate on branches or stems of trees, then grow and develop leaves and roots using nutrient and water resources that accumulate within the crevices of the stem bark [2]. Their roots eventually reach the soil and develop similar root system as their host trees. Usually it takes four or five years until the roots reach the ground, depending on the size of the host tree [3]. Stranglers pass from epiphytic stage in the process and enter a self-supporting stage that function as a normal tree.

Hemiepiphytes are called stranglers because of their roots grow around the host tree and entwine it. As the root diameter increases, they compress the host stem underneath. This process continues until xylem in the host stem can no longer transport nutrient and water upward and therefore dies [2]. Host's nutrient transport is interrupted because of the strangler, but in most cases, the host is eventually killed by the strangler's dense shade [4]. Even though most stranglers end up killing their hosts, they are not parasites because they do not derive nutrients from the host tree.

The existence of strangler plants on roadside greenbelt trees becomes a problem since their roots keep growing larger, but they only have limited space which later become insufficient for their roots. Strangler plant roots are unable to fully break through asphalt so they cannot grow a deep, strong root system. This issue may cause hazards to the surrounding users or properties, moreover when the host tree is dead. Dead host tree will start decaying and forms a humus for the strangler, leaving a hollow trunk [5]. Dead host tree is more likely to fail since the decaying stem and roots will make the wood structurally weak. When the host tree fail, the strangler roots that grows around the stem is not strong enough to hold the weight of the stem. Therefore, both the host tree and strangler result in failure.

Tree failure tends to happen after extreme weather, such as heavy rain, strong wind, and severe storms. Bogor city has high average precipitation, so climate condition is also a possible threat for users' safety. In 2008, a large, old rubber fig tree (*Ficus elastica*) failed and caused hazards to a school property and a vehicle nearby [6]. Fortunately, there was no victim in the accident. Similar case also happened in 2013 even though the damage was relatively smaller. Similar accident in the future should be prevented, therefore this study was conducted. Evaluation of roadside greenbelt trees damage, specifically caused by strangler plants, is needed before we could suggest the right actions for damaged trees maintenance. There are some major roads in Bogor which its greenbelt trees were seemed to be damaged by strangler plants. Those major roads are Jalan Sudirman, Jalan Juanda, Jalan Ahmad Yani, Jalan Pemuda, and Jalan Semeru. The fact that the traffic intensity at those roads is high makes them in dire need of attention. In order to do the evaluation, this study was directed to analyze the vegetation of strangler plants in Bogor, to assess the damage caused by stranglers, and to compose strangled trees maintenance recommendations.

2. Material and Methods

2.1. Study area and sampling technique

This study was conducted at five major roads in Bogor city, West Java, Indonesia, namely Jalan Ahmad Yani, Jalan Juanda, Jalan Pemuda, Jalan Semeru, and Jalan Sudirman. All of these roads are located in Central Bogor district. These roads have similar macroclimatic condition with average monthly temperature was 32.1° C, the minimum was 22.4° C and the maximum was 33.7° C. Average monthly humidity was 92%. Average monthly rainfall was 304-535.3 mm with highest rainfall occurred in November and February [7]. The data was collected in March to May 2014.

To record the presence or absence of strangler plants in the study area, line-intercept method was used, which–instead of quadrats–was based on a single line of only one dimension, namely length [8]. Sampling of strangler plants was done using a number of 100 m long linear plots located at each roadside greenbelt. Each road has different length as well as the greenbelt. The road length of Jalan Ahmad Yani, Jalan Juanda, Jalan Pemuda, Jalan Semeru, and Jalan Sudirman are ±2.2 km, ±1.7 km, ±1.9 km, ±1.4 km, and ±1.4 km, respectively. Ten linear plots were made at each roadside greenbelt, except at Jalan Juanda. There were only five plots made at Jalan Juanda due to insufficient length of greenbelt. In total, samples were taken from 45 linear plots.

2.2. Strangler plant assessment

The species of each strangler plant found in each plot were identified along with the host tree species. There were four aspects measured in this study: dbh (diameter at breast high), height, canopy diameter, and strangler's coverage over the host tree. Measurement of height and canopy diameter were done separately for the host tree and strangler plant. Tree height was measured using hagameter. For strangler plants, the height was measured from the bottom point of the longest root (nearest to the ground) to the top of its canopy. The number of strangler roots was estimated by counting visible roots. To estimate the coverage of the strangler roots and canopy over the host tree, at least six photos were taken from

IOP Conf. Series: Earth and Environmental Science **91** (2017) 012012 doi:10.1088/1755-1315/91/1/012012

different angles; three photos for estimating strangler root density and the other three for strangler canopy density. The coverage was later calculated with digital version of photo grid analysis [9].

2.3. Data analyses

Frequency, dominance, density, and important value index (IVI) of strangler species were calculated following Mueller-Dombois and Ellenberg's method [10]. The IVI were calculated by summing the relative frequency, relative dominance, and relative density for strangler plants. The species diversity index was calculated following Shannon-Wiener index [11], $H = -\sum p_i \ln p_i$; where $p_i = N_i/N$ and H = Shannon-Wiener index of general diversity, N_i = the abundance of the i-th species in the sample.

The measurement data were used for assessing host trees and greenbelts damage. For assessing host tree damage, there were a set of evaluation criteria that consisted of (1) trunk diameter, (2) strangler-host height ratio, (3) density of strangler canopy, (4) density of strangler roots, (5) number of strangler roots, and (6) strangler growth. Those criteria were used to score each strangler plant found in the study area. Measurement data were scored using 5-point scale in each criterion. Table 1-6 presents the score classification for each criterion.

Table 1. Trunk diameter classification.

Score	Qualification	Diameter (cm)
1	Poles	$10 \leq D$
2	Small	$10 < D \leq 30$
3	Medium	$30 < D \leq 60$
4	Large	$60 < D \leq 100$
5	Giant	$D > 100$

Table 2. Strangler-host height ratio.

Score	Qualification	Strangler height (%)
1	Very young	$H \leq 20$
2	Young	$20 < H \leq 40$
3	Medium	$40 < H \leq 60$
4	Nearly adult	$60 < H < 80$
5	Adult	$80 < H \leq 100$

Table 3. Density of strangler canopy.

Score	Qualification	Canopy density (%)
1	Very sparse	$C \leq 20$
2	Sparse	$20 < C \leq 40$
3	Moderate	$40 < C \leq 60$
4	Dense	$60 < C \leq 80$
5	Very dense	$80 < C \leq 100$

Table 4. Density of strangler roots.

Score	Qualification	Root density (%)
1	Very sparse	$R \leq 20$
2	Sparse	$20 < R \leq 40$
3	Moderate	$40 < R \leq 60$
4	Dense	$60 < R \leq 80$
5	Very dense	$80 < R \leq 100$

Table 5. Number of strangler roots.

Score	Qualification	Root (pcs)
1	Few	$1 < R \leq 8{,}25$
2	A few	$8{,}25 < R \leq 15{,}5$
3	Some	$15{,}5 < R \leq 22{,}75$
4	A lot of	$22{,}75 < R \leq 30$
5	Most	$R > 30$

Table 6. Strangler growth classification.

Score	Qualification
1	Very few roots, roots have not reached the ground, canopy just starts growing
2	A few roots, some roots have reached the ground, canopy starts growing
3	More roots, many roots have reached the ground, about half of the host's canopy is covered
4	Many roots, most of the roots have reached the ground,
5	So many roots, most part of the stem have been covered by roots, strangler canopy is dominating

After scoring these criteria, the next step was to assess host tree damage. Host tree damage was calculated using the following equation and then the result was classified according to table 7.

$$Tree\ damage\ (\%) = \frac{Sum\ of\ 6\ criteria\ scores}{Sum\ of\ maximum\ 6\ criteria\ scores} x100\%$$

Table 7. Classification of tree damage

Stadium	Qualification	Description	Tree damage (%)
1	Mild	Host stem is clearly visible; host canopy is not covered by strangler canopy	$D \leq 20$
2	Low	A small part of host stem is covered by strangler roots; strangler canopy coverage is very small	$20 < D \leq 40$
3	Moderate	About half of host stem is covered by strangler roots; some part of host canopy is covered by strangler canopy	$40 < D \leq 60$
4	High	Almost the whole stem is covered by strangler roots; most of the canopy is covered by strangler canopy	$60 < D \leq 80$
5	Severe	Host stem is fully covered by strangler roots; host canopy is fully covered by strangler canopy	$80 < D \leq 100$

Roadside greenbelt damage was calculated using the following equation and then the result was classified according to table 8. The damage results were then used to estimate the effective maintenance recommendation based on each damage level. The recommendation is expected to be an effective solution for dealing with strangler plants in roadside greenbelts.

$$Greenbelt\ damage\ (\%) = \frac{Sum\ of\ damaged\ trees\ scores}{Total\ trees\ in\ greenbelt\ x\ 5} x100\%$$

Table 8. Classification of greenbelt damage

Stadium	Qualification	Greenbelt damage (%)
1	Mild	$D \leq 20$
2	Low	$20 < D \leq 40$
3	Moderate	$40 < D \leq 60$
4	High	$60 < D \leq 80$
5	Severe	$80 < D \leq 100$

3. Results

3.1. Vegetation analysis

A total of 4 strangler species were recorded at the study area, namely weeping fig (*Ficus benjamina*), rubber fig (*Ficus elastica*), bunut (*Ficus glauca*), and umbrella tree (*Schefflera actinophylla*). Among those species, the most frequent one was *Ficus benjamina* with relative frequency 63.6%, and then followed by *F. elastica, F. glauca,* and *S. actinophylla* (table 9). Compared to the other species, *F. benjamina*'s frequency seemed to be significantly different which clearly indicated that it was the most common species found in the study area. From a total of 45 plots, it was found in 28 plots. *F. benjamina* also had the highest density in all study locations with relative density 60.9%, followed by *F. glauca, S. actinophylla,* and *F. elastica* (table 9). There is no certain explanation behind this phenomenon, but it was suspected because *F. benjamina* seeds are easier to be dispersed and/or germinated compared to the other species and it also has a high degree of flexibility that enables them to adapt with a wide range of habitats. Its rather showy fruit might attract wildlife better than the other species. To germinate, the seeds depend on a bacterium that helps crack its seed coat. Wildlife that eats fig fruits cannot digest the hard seeds, but the outer coat of the seed is dissolved in their stomach during digestion, so the 'uncovered' seeds are later being defecated [3]. If *F. benjamina*'s seed coat is somewhat softer than other species, this might be another possible reason why *F. benjamina* dispersed widely.

Table 9. Strangler plant vegetation analysis results for each species. (F: Frequency, RF: Relative Frequency, De: Density (individual/hectare), RDe: Relative Density, D: Dominance, RD: Relative Dominance, IVI: Important Value Index)

Species	Number of trees	Number of plots	Canopy area (m^2)	F	RF (%)	De (i/ ha)	RDe (%)	D	RD (%)	IVI
F. benjamina	39	28	2648.9	0.6	63.6	7.1	60.9	4.8	61.3	185.9
F. elastica	7	6	1214.7	0.1	13.6	1.3	10.9	2.2	28.1	52.7
F. glauca	9	5	390.8	0.1	11.4	1.6	14.1	0.7	9.0	34.5
S. actinophylla	9	5	64.9	0.1	11.4	1.6	14.1	0.1	1.5	26.9
Total	64	45	4352.4	1.0	100.0	11.7	100.0	7.9	100.0	300.0

Table 10. Strangler plant vegetation analysis results in each greenbelt. FB: *Ficus benjamina,* FE: *Ficus elastica,* FG: *Ficus glauca,* SA: *Schefflera actinophylla.* Total in total plot refers to the number of sampling plots in each greenbelt.

Greenbelt location	Species	Number of trees	Number of plots	Canopy Cover (m^2)	F	RF (%)	De (i/ ha)	RDe (%)	D	RD (%)	IVI
Jalan Ahmad Yani	FB	7	6	507.3	0.6	66.7	5.7	70.0	4.2	90.9	227.6
	FE	3	3	50.6	0.3	33.3	2.5	30.0	0.4	9.1	72.4
Total		10	10	557.9	0.9	100.0	8.2	100.0	4.6	100.0	300.0
Jalan Juanda	FB	3	2	201.8	0.4	50.0	4.9	33.3	3.3	14.7	98.0
	FE	2	1	1138.8	0.2	25.0	3.3	22.2	18.7	82.7	129.9
	SA	4	1	36.4	0.2	25.0	6.6	44.4	0.6	2.6	72.1
Total		9	5	1377.0	0.8	100.0	14.8	100.0	22.6	100.0	300.0
Jalan Pemuda	FB	16	9	412.9	0.9	64.3	13.1	64.0	3.4	88.5	216.8
	FE	1	1	10.3	0.1	7.1	0.8	4.0	0.1	2.2	13.3
	FG	4	1	22.1	0.1	7.1	3.3	16.0	0.2	4.7	27.9
	SA	4	3	21.4	0.3	21.4	3.3	16.0	0.2	4.6	42.0
Total		25	10	466.7	1.4	100.0	20.5	100.0	3.8	100.0	300.0
Jalan Semeru	FB	7	7	870.3	0.7	63.6	5.7	58.3	7.1	70.2	192.2
	FG	5	4	368.7	0.4	36.4	4.1	41.7	3.0	29.8	107.8
Total		12	10	1239.0	1.1	100.0	9.8	100.0	10.2	100.0	300.0
Jalan Sudirman	FB	6	4	656.6	0.4	66.7	4.9	75.0	5.4	96.7	238.4
	FE	1	1	15.1	0.1	16.7	0.8	12.5	0.1	2.2	31.4
	SA	1	1	7.1	0.1	16.7	0.8	12.5	0.1	1.1	30.2
Total		8	10	678.8	0.6	100.0	6.6	100.0	5.6	100.0	300.0

The fact that *F. benjamina* had the highest frequency and density in the study area made it had the highest dominance as well. Dominance is related with strangler canopy coverage. Canopy size could be determined by genetic factor and adaptability factor. *F. benjamina* has spreading canopy. When it grows older, it tends to use the nutrient for spreading its canopy rather than growing in height. Then the canopy will spread like an umbrella with hanging branch [12]. That factor supports *F. benjamina* to be the most dominating species in the study area. *F. glauca* and *S. actinophylla* had the smallest relative frequency. This low number means their seeds dispersal is limited. Their relative density and dominance also showed low numbers. This might be caused by their unattractive fruits [13] and their rather low adaptability. Similar case might have happened to *F. elastica* as well, but it showed higher relative dominance because there was a large *F. benjamina* tree at Jalan Juanda (table 10).

Table 10 presents strangler plants frequency, density, and dominance at each road. Among the five roadside greenbelts, greenbelt at Jalan Pemuda had the highest total frequency and density. Strangler plants were found in 9 out of 10 plots. This might be caused by the similarities between each plot. Trees continuity in each plot was similar, there was no plot with very distinctive trait, all trees were lined up consistently. Jalan Sudirman had the lowest total frequency (table 10). In contrast to Jalan Pemuda, plots in Jalan Sudirman had high variability, some plots had a lot of trees and the others only had a few. Plots with only few trees lessened the possibility of suitable host tree availability. It is possible for one host tree to be strangled by more than one stranglers. *F. benjamina* nearly had the highest relative density at every greenbelt, except at Jalan Juanda. The highest relative density at Jalan Juanda belonged to *S. actinophylla*. This might indicate that microclimate condition in Jalan Juanda was advantageous for it.

According to the results, strangler species with highest to lowest IVI are the following: *F. benjamina, F. elastica, F. glauca,* and *S. actinophylla*. Highest IVI value belonged to *F. benjamina* with 185.9 (table 9). IVI describes the most dominant and important species in its ecosystem [14]. In terms of roadside greenbelt maintenance, it means that closer attention should be paid to *F. benjamina* since it has the biggest potential to cause damage to greenbelt trees. *S. actinophylla* ranked the lowest IVI, it means the damage caused by it was the lightest.

3.2. Strangler plants diversity

Based on the results, strangler plants had low diversity. Diversity index (H) for Jalan Ahmad Yani, Jalan Juanda, Jalan Pemuda, Jalan Semeru, and Jalan Sudirman was 0.6, 1.1, 1.0, 0.7, and 0.7, respectively. The average was 0.8. According to Shannon and Wiener [11], vegetation with $H \geq 3$ is considered to have high diversity, $1 \leq H < 3$ is considered moderate, and $H < 1$ is considered low. Three of the greenbelts had low diversity, while the other two had moderate diversity. Relatively low diversity may be caused by two possible factors; namely low species variability and/or dominant density by a certain species. Although H value may increase with increasing number of species, it will not work if there is any very dominant species in the study area, because diversity incorporates species richness, commonness, and rarity [15].

3.3. Relationship between strangler plant and host tree

Based on the survey, there were four host tree species, which were *Canarium commune, Samanea saman, Roystonea regia,* and *Thuja orientalis* (table 11). It was found that 59 of 64 host trees (92.2%) were *C. commune*. It was indeed the most common tree species found in the study area. Previous study by Athreya [16] suggested that dominant host tree species is significantly correlated with strangler plants preference of host tree. Another suggestion is that *C. commune* has many branches which is liked by birds. With birds coming and perching on the branch, the possibility they will drop strangler seeds increases. Male and Roberts [17] suggested that strangler distribution is limited by the availability of, and dispersal to suitable host tree. Mostly, stranglers prefer hosts with rough bark. Host's rough bark let strangler seeds be deposited on it and water and nutrient in the crevices will be advantages for them.

Only few host trees belong to species other than *C. commune*. There were some possible factors that explain this situation, which were (1) the host species was not a dominant species, (2) the bark was rather soft, (3) host tree does not have many branches, or (4) it is still a small tree. Previous studies suggested that strangler plants will not likely live on another strangler. Patel [18] suggested that there hasn't been any evidence about strangler that lives on another strangler. It was also confirmed by Titus

[19] saying there is little possibility that a strangler will live on another strangler since their seeds have low survival rate towards pathogen and seed predators, meanwhile those are commonly found on adult strangler trees. It is not likely for a strangler to be a host for another strangler.

Table 11. Host tree and its strangler plant species.

No	Host / Strangler	*F. benjamina*	*F. elastica*	*F. glauca*	*S. actinophylla*	Total	Percentage (%)
1	*Canarium commune*	37	6	7	9	59	92.2
2	*Samanea saman*	1	-	-	-	1	1.6
3	*Roystonea regia*	1	1	-	-	2	3.1
4	*Thuja orientalis*	-	-	2	-	2	3.1
	Total	39	7	9	9	64	100.0

Stranglers' preference of host tree might be affected by the tree's dbh (table 12). Based on survey results, 33 out of 58 strangled trees (56.9%) have large dbh that is more than 100 cm. This showed that stranglers had the tendency to live on large trees. A previous study also showed a similar result, in which stranglers usually found on host trees with large dbh (>80cm) [18]. Another similar trend also found in other studies [16,17]. Larger tree trunk often means older tree. It is suspected the age of the host tree plays an important role in this case. Old trees usually have more crevices that could be where strangler seeds start germinating and also may be better at providing water and nutrients for stranglers.

Table 12. Strangler host preference based on tree trunk diameter.

Strangler	Trunk diameter (cm)					Total
	$d \leq 10$	$10 < d \leq 30$	$30 < d \leq 60$	$60 < d \leq 100$	$d > 100$	
Absent	41	114	206	75	27	463
Present	2	1	8	14	33	58
Total	43	115	214	89	60	521

Another factor that might as well affected stranglers' preference is host tree height (table 13). According to the result, most stranglers tended to live on host tree with height Class 4, which is between 15 to 28 m. It can be seen that 49 out of 58 strangled trees were in this class. This finding could be explained as that is the dominant height of trees in the study area. If it is observed closely, 4 out of 16 trees in Class 5 ($t \geq 28$), equals to 25%, were strangled. Meanwhile, only 49 out of 294 trees in Class 4 that were strangled, which made the percentage equals to 16.67%. With that finding, it is possible if there were more trees in Class 5 in the study area, most of trees in Class 5 might be strangled.

Table 13. Strangler host preference based on tree height.

Strangler	Tree height (m)					Total
	$t < 1$	$1 \leq t < 6$	$6 \leq t < 15$	$15 \leq t < 28$	$t \geq 28$	
Absent	0	41	165	245	12	463
Present	0	3	2	49	4	58
Total	0	44	167	294	16	521

3.4. Roadside greenbelt damage assessment

Damage assessment was done by considering five indicators; trunk diameter, strangler height, strangler canopy density, strangler root density, and number of strangler roots. The assessment results were then classified into 5 classes: mild, low, moderate, high, and severe. Assessments of each tree were used to assess the roadside greenbelt damage.

Table 14. Greenbelt trees damage in the study area.

Greenbelt location	Mild	Low	Moderate	High	Severe	Total	%
Jalan Ahmad Yani	-	6	-	2	2	10	17.2
Jalan Juanda	-	3	2	-	2	7	12.1
Jalan Pemuda	1	13	7	-	1	22	37.9
Jalan Semeru	-	3	1	3	5	12	20.7
Jalan Sudirman	-	3	1	1	2	7	12.1
Total	1	28	11	6	12	58	100.0
Percentage (%)	1.7	48.3	19.0	10.3	20.7	100.0	

Most of the strangled trees found in the study area were under category low damage (table 14). The fact that there were 12 trees that suffered severe damage makes them should be the top priority for further maintenance. If roadside greenbelts were observed as a whole, rather than individual tree, each greenbelt was still in mild damage. Greenbelt trees with highest to lowest damage were the following Jalan Pemuda, Jalan Juanda, Jalan Semeru, Jalan Ahmad Yani, and Jalan Sudirman with damage percentage 5.0%, 7.3%, 11.0%, 6.8%, and 4.3%, respectively. Regardless of the greenbelt damage level as a whole, the attention must be paid to trees individually.

3.5. Maintenance recommendation

Greenbelt trees maintenance should be started with taking care of the stranglers. Stranglers may be treated physically, chemically, and biologically. Since there is not much known about chemical and biological treatment for stranglers, physical treatment seemed to be the most practical way. Physical treatment for each damage level should be differentiated. Trees with mild damage (Stadium 1) could be treated with cutting its strangler roots. Strangler roots should be prevented from reaching the ground since their growth rate will speed up once they get into the ground. Treatment for low damage tree (Stadium 2) should be separating strangler roots that are already attached to stem and cutting all its hanging roots. Any strangler canopy should be cut too. Similar treatment could also be used for tree with moderate damage (Stadium 3), but it needed to be done more carefully since stranglers in Stadium 3 usually have thickened roots and denser canopy. Heavy pruning and cutting should be done on strangler plant in trees with high damage (Stadium 4). Host trees at this stage usually still have strong stem, therefore heavy pruning will still able to save the host tree. Heavy pruning could be done by cutting lateral branches so the stranglers only have one straight stem. While treatment for Stadium 1 to 4 suggests to remove the stranglers, then save the host, treatment for severe damage (Stadium 5) should be done with either logging the tree or leave it be instead, because the host tree probably has died at this stadium. The tree should be logged when it seems to be leaning, but may still be preserved if it has enough space to grow its roots. If the branches are too wide, it will be good to prune them as well. There are several reasons why it should be preserved, strangler tree has aesthetic, architectural, and also ecological value. Strangler could be an element of users' mental map or road identity since it has a unique and remarkable visual. It can also be a nice shade tree and ornamental tree. Ecologically, strangler plants provide food and habitat for wildlife. These physical treatments may not be able to remove the strangler completely, therefore it is needed to do the treatment periodically. In order to maintain every tree in roadside greenbelts, it would be helpful if there were a regular tree 'check-up' so stranglers' growth could be controlled. The survey will might be easier if the tree is numbered.

4. Conclusion

Based on the results, there were four species found in the study area, which were *F. benjamina, F. elastica, S. actinophylla,* and *F. glauca*. Strangler population in the study area was 64 plants and some of them were living on the same host. The most frequent species found was *F. benjamina*. It had the highest Important Value Index compared to the other species. Strangler diversity in Bogor city was considered low. The most preferred host tree species was *C. commune.* Strangler plants tended to live on host tree with large trunk diameter (>100 cm) and high height ($15 \leq t < 28$). Every greenbelt trees in the study area suffered damage from stranglers. Among 5 greenbelts, Jalan Pemuda had the highest damage and Jalan Sudirman had the lowest damage, but the greenbelt that has severely damaged trees the most is Jalan Semeru. Attention should be paid to each tree individually, specifically trees that suffered severe damage.

Damaged trees should receive different treatment according to their damage level, which were classified to 5 stadiums; mild, low, moderate, high, and severe, with mild is Stadium 1 to severe is Stadium 5. Treatment for damage level 1 to 4 should be focusing on saving the host by pruning and cutting strangler canopy and roots, respectively. If it is not possible to remove the strangler completely, trees that suffered Stadium 4 damage should be heavily pruned so they don't have lateral branches anymore. Treatment for damage level 5 should be done with either logging the tree or leave it be instead. The tree should be logged when it seems to be leaning, but may still be preserved if it has enough space to grow its roots. Trees maintenance should be accompanied with regular tree 'check-up'.

References

[1] Islam MN, Rahman KS, Bahar MM, Habib, MA, Ando K and Hattori N 2012 Pollution attenuation by roadside greenbelt in and around urban areas *Urban For. Urban Gree.* **11** 460-4 (http://dx.doi.org/10.1016/j.ufug.2012.06.004)
[2] Medina E 2007 *Functional Plant Ecology 2nd*, ed Pugnaire F and Valadares F (Boca Raton: Taylor & Francis Group, LLC) p 333
[3] Hood MA 2004 *The Strangler Fig and Other Tales* (Walnut Creek: Altamira Press) p 61
[4] Lodge TE 2010 *The Everglades Handbook: Understanding the Ecosystem 3rd* (Boca Raton: Taylor & Francis Group,LLC) p 72
[5] Mattheck C 1998 *Design in Nature: Learning from Trees* (New York: Springer) p 112
[6] Sulistyantara B 2014 Upaya menurunkan resiko pohon tumbang *Risalah Kebijakan Pertanian dan Lingkungan* **1** 7-11
[7] BPS Kota Bogor 2013 *Bogor Municipality Figures* (Bogor: BPS Kota Bogor) pxxv
[8] Mueller-Dombois D and Ellenberg H 1974 *Aims & Methods of Vegetation Ecology* (New York: Wiley& Sons) p 100
[9] Hall FC 2002 *Photo Point Monitoring Handbook: Part B – Concepts and Analysis* (Pacific Northwest: USDA) pp 59-80
[10] Mueller-Dombois D and Ellenberg H 1974 *Aims & Methods of Vegetation Ecology* (New York: Wiley& Sons) pp 107-119
[11] Shannon CE, Wiener W 1963 *The Mathematical Theory of Communication* (Illinois: Urbana)
[12] Starr F, Starr K, Loope L 2003 *Ficus benjamina* United States Geological Survey, Biological Resources Division, Haleakala Field Station, Maui, Hawai'i
[13] Gilman EF dan Watson DG 1994 *Schefflera actinophylla: Schefflera* Institute of Food and Agricultural Service, University of Florida
[14] Fachrul MF 2007 *Metode Sampling Bioekologi* (Jakarta: Bumi Aksara)
[15] Begon M, Townsend CR and Harper JL 2006 *Ecology: From Indviduals to Ecosystem* (Oxford: Blackwell Publising) p 471
[16] Athreya VR 1999 Light or presence of host trees: which is more important for strangler fig? *Journal of Tropical Ecology* **15** 589-602
[17] Male TD and Roberts GE 2009 Host associations of the strangler fig *Ficus watkinsiana* in a subtropical Queensland rain forest *Austral Ecology* **30** 229-36
[18] Patel A 1996 Strangler fig-host associations in roadside and deciduous forest sites, South India *Journal of Biogeography* **23** 409-14
[19] Titus JH 2000 *Monteverde: Ecology and Conservation of a Tropical Cloud Forest*, ed Nadkarni NM and Wheelwright NT (New York: Oxford University Press, Inc.) p 71

Converging social classes through humanized urban edges

M V Abuan[1] and Z D Galingan[2]

[1,2] University of the Philippines - Diliman, Philippines

Email: abuan.miko@gmail.com

Abstract. Urban open spaces are created to be used by people. It is a place of convergence and social activity. However, these places have transformed into places of divergence. When spaces become dehumanized, it separates social classes. As a result, underused spaces contribute to urban decay. Particularly an urban edge, the JP Rizal Makati Waterfront Area is the center of this paper. The JP Rizal Makati Waterfront Area is a waterfront development situated along the banks of one of Metro Manila's major water thoroughfare --- Pasig River. The park and its physical form, urban design and landscape tend to deteriorate over time --- creating a further division of social convergence. Social hostility, crime, negligent maintenance and poor urban design are contributing factors to this sprawling decay in what used to be spaces of bringing people together. Amidst attempts to beautify and renew this portion of Makati City's edge, the urban area still remains misspent.This paper attempts to re-humanize the waterfront development. It uses the responsive environment design principles to be able to achieve this goal.

1. Introduction

1.1 Makati City and the Study Area - JP Rizal Waterfront Area

Makati City is known as one of the highly urbanized cities in Metro Manila. With a population of 529.039, it ranks 32nd in the most populated cities in the world. Aside from this number, Makati, being a CBD, attracts transient visitors that increase daytime population. The Makati Transportation Study in 2012, projected the maximum daytime population of Makati at 4,241,105. With the continued growth of urbanization, population growth is also imminent.

The vast movement of people going in and out of the city implicates the movement of social classes. The amount of jobs available in the city, from white to blue collar categories, attracted different workers depending on offered industries. Professionals, utility workers and other job workers enter the city and bring about a mix of high, mid and low classes in the socio-economic scale.

The Profile of Makati included in its development plan an Urban Greening program. It listed down 12 urban parks in Makati including the Guadalupe Nuevo Linear Park which is included in this paper's stud area.

The study area is called the JP Rizal Waterfront Area. It is composed of several linear parks and has the potential to become an urban park system. It is terminated at the east by the Makati Park and Garden and the at west by the Casa Hacienda Park. It is good to note that the Makati City Local Government spent efforts in beautifying this stretch by renovating existing infrastructures and providing landscaping. However, the area remains underused for various reasons that this paper will tackle.

The JP Rizal Waterfront Area is comprised of three major areas spread in three different barangays --- Poblacion, Guadalupe Viejo and Guadalupe Nuevo. The three barangays outline the northern border of Makati City and is bounded by the Pasig River. It is considered as an edge of Makati City. The site is 43,151 square meters in land area with a waterfront length of 2,236 meters (Figure 1).

The site consists of boardwalks and several establishments. Residential, industrial and institutional uses populate the central area of the area coverage.

Figure 1. Scope of the JP Rizal Waterfront Area (Source: Google Earth, 2016)

1.2. Urban Edges

Edges are linear elements which are not used as paths. Although not dominant as paths, these edge elements have important organizing features, particularly holding together generalized areas, as in the outline of a city [5].

The urban edge is one of the five mental mapping ideas as introduced by Lynch. Together with the other four (nodes, paths, districts and landmarks), these elements of the city help in understanding how and why people perceive a place in the way that they do.

For the purposes of this study, the edge is defined and considered as an urban space. Due to the fact that edges are attached to certain paths, districts or nodes, as well as, it can hold special landmarks, the edge is considerably a place in the city where activity happens. And where activity happens, it becomes a space for convergence. Provided with the correct design considerations and strategies, the urban edge should bring people together in social activities. However, with the current situation and case of the JP Rizal Waterfront Area, this is not the case

1.3. Social Classes

The JP Rizal Waterfront Area is home to different social classes. Ranging from the upper class A-B (residents of nearby high end residential enclave) to lower class C-D (vagrants and informal settlers) that frequent the area, the different social classes change the scene of the area. In the morning, joggers (Class A and B) frequent the place, while in the afternoon and late night, the space turns into a camp ground for Class C and D individuals (or groups).

The disparity of social classes depend on two forces: the pull of the place and the push of each class. The place, as a an area of activity, would naturally pull users. However, if there is no strong pull, the place becomes weak. The push of social classes is seen when one class is turned off or intimidated by the other. A combination (or lack) of the push and pull factors greatly enhances or disrupts the usability of an area.

2. Problem Setting

Kelman [4] discussed two features of humanness, namely: identity and belongingness in a community. Reversal of these roles result to loss of identity and exclusion from a community. It then results to dehumanization. Dehumanization occurs when people fail to appreciate fully the needs, desires, feelings and hopes of others [11].

In the urban space, dehumanization is manifested in several apects. The physical or visual factor is one aspect. Spaces tend to be fragmented because of the overwhelming scale of transport routes [3]. Furthermore, Enjuto [3] discussed that environmental conditions such as sounds, air quality, humidity, sunlight, shading, moisture etc., fall under another aspect that affects the city. The third factor is the socio-cultural aspect. Spaces and places are imbued with powerful meanings by different groups of people [7]. The meanings can vary between social groups and individuals, and be fragmented and contradictory [7]. With these three aspects (physical, environmental and socio-cultural), the manifestation of dehumanization results to misused and underused spaces.

Dehumanization of spaces can be traced back to earlier statements regarding the push and pull factors. Placelessness is a factor that weakens the pulling force. The "melting pot" experience [8] forces people to integrate in a society void of individuality. The requirement to conform strengthens the pushing force that segregates social classes.

2.1. The Problem
In 2015, the UN-Habitat launched in its annual World Habitat Day the advocacy on "Public Spaces for All, Designed to Live Together". It promotes certain points about public spaces, which includes open spaces. The advocacy states that public spaces: 1.) are crucial for the urban poor, 2.) contribute to building social cohesion, 3.) promote gender equality, 4.) enhance safety, 5.) support economic development, 6.) improve public health; and 7.) improve the environment [10]. For these points to be satisfied, the convergence or bringing together of people is key.

However, the urban areas intended to bring people together, more often than not, create social and cultural segregation. Dehumanized urban elements contribute to this phenomena. As a result, segregation creates constructs that lead to these urban spaces as either misused or underused. The urban space improperly used and left unmanaged will then lead to urban blight.

2.2. Hypothesis
The hypothesis of this paper is based heavily on the assumption that the re-humanization of these dehumanized spaces will reverse its current effects. Humanizing these urban spaces will solve socio-cultural segregation --- allowing people to come together in social activities. And without segregation, social classes will converge. And with the creation of convivial spaces, sustainability of the spaces are achieved. For redeveloped areas, this can be a catalyst for urban renewal.

2.3. Goals and Objectives
The goal of this paper is to establish a framework in humanizing urban spaces. To arrive at this goal, the following objectives are drawn:

- to recommend design strategies to address dehumanization of the JP Rizal Waterfront area
- to study correlation of social classes with urban landscape design

2.4. Significance of the Study
This paper can aid local government units and communities in re-evaluating their programs on urban renewal programs. The in-depth socio-cultural research entailed in developing open spaces is important so as not to result in dehumanized spaces.The paper can also assist students and the academic professionals in furthering the study of humanely designed spaces. This provides basic information and theories on the subject.This study can be further developed into a more in-depth research that can support stakeholders, especially, law makers in creating decisions or ratifying existing laws and guidelines.

3. Methods, Findings and Recommendations
The theoretical framework of the study relies on the application of humanizing strategies in urban spaces. This is achieved by responding to the physical, environmental and socio-cultural needs of such spaces.

In studying the needs of human beings, we must look at the basics. Maslow introduced the concept of hierarchy of needs, which are intrinsic to each individual. In relation to these needs, the response to the physical, environmental and socio-cultural aspects are satisfied. (Figure 2) A similar study [9] posits that depending on the space, physical needs are fulfilled by cultural advancement.

This framework is used to correlate the connection between the manifestation of humanization in a physical space. With each aspect, responsive design principles should be engaged to come up with humane spaces.

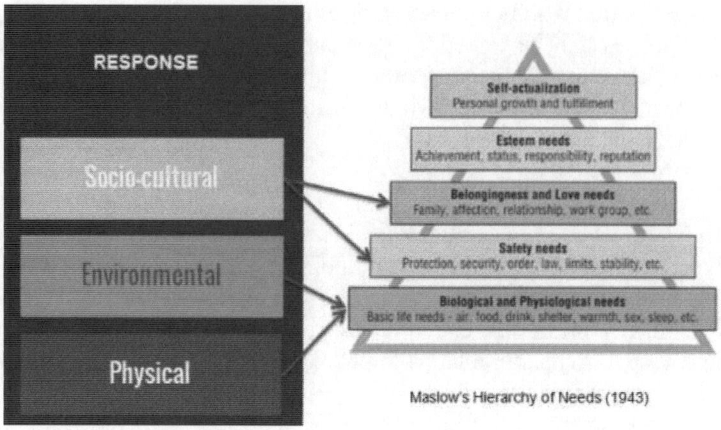

Figure 2. Response to Maslow's Hierarchy of Needs

3.1. Site Inventory and Analysis

During site inventory and analysis, it is noted that due to the pre-existing condition that the entire stretch of the site is spread across three towns of Makati, the visual character differs. Site elements such as planting, hardscape patterns and materials, street lights, sheds and street furniture are not uniform from one portion to another, particulary the western side from the eastern side. In terms of continuity, the length of the site can be divided into three portions: west, center and east areas. It is divided, not physically, but rather implicitly by the presence of two overhead roads (Pantaleon Bridge and EDSA) which traverses perpendicularly above the linear park.

It is important to cite that the linear park sits along the bank of the Pasig River --- making it a waterfront development --- and along JP Rizal Avenue, a major thoroughfare at the edge of northern Makati. The road traffic is busier than that of the water traffic, although presence of informal and formal ferry boat terminals integrates with the site. As such, the site has numerous public utility vehicle stops, but no distinct drop-offareas are designated.

It should also be noted that there are existing vegetation across the stretch of the study area. Some of these are potted plants while majority are in-ground street plants. Existing inventory includes: Palawan cherry (*Cassia nodosa*), rain tree (*Albizia saman*), MacArthur palms (*Ptychosperma macarthurii*), golden shower (*Cassia fistula*) and yellow oleander (*Thevetia peruviana*). These major urban trees provide shade in interspersed areas. Only certain spaces are shaded due to the distance between one tree to the next.

These urban trees also have push and pull factors. At late morning and high noon, the shade offered by these trees pull users to conglomerate underneath the canopies. However, at night time, the trees push away people as these are possible spaces where safety and security are questionable.

In the course of the evaluation, certain major trees are important to be maintained in the redevelopment proposal. Some needs to be transplanted in order to achieve a more humanized area.

3.2. Cultural Analysis

Cultural mapping results show that the concentration of convergence happens at the terminals of the site (western most and eastern most areas). This is due to the presence of activity generating areas such as a terminal located at the western end and a park at the eastern end. People can be seen interacting in the park spaces (i.e. children playing, people cooling off from the heat). Towards the center of the site, the activities change into commercial transactions that involve buying and selling of food items, pop-up eateries and the like.

A cultural analysis on the various activities at the site was done. Results are based on actual count done in a 3-day observation. Its frequency is derived on the average of a 2 hour count per timeframe, spread in 3 days (Figure 3).

Activity Inventory	Morning	Noon	Evening
Transitory Use i.e. waiting area for public transportation, river ferry terminals	High	Medium	High
Commercial Activities i.e. street vending, eating area	Low	High	High
Leisure Use i.e. biking, jogging, walking	Medium	Low	Medium
Unwanted Activities i.e. vagrants, theft	Medium	Low	High

Frequency:
High > 25
Medium 15 – 25
Low < 15

Figure 3. Activity inventory results based on a 3-day observation of the site

It should be noted that the different activities have been grouped. In Figure 4, the cultural mapping of the grouped activities is shown.

Figure 4. Activity mapping (Source for Base Map: Google Earth, 2016)

Activity patterns on site also vary on the site elements provided as well as the location. For example, a mini-playground was provided, located away from the western terminus, which resulted in the underuse of the space. In addition to that, the playground has little shade and protection from the elements. Safety is also a concern.

The pop-up stores are usually situated near office developments and on-going construction sites found along the stretch of the site. These transient activities rise during late afternoon to evenings.

Commuter activities are high in the morning and the evening, due to the fact that the JP Rizal Avenue is a major road where plenty of transients pass by. Traffic congestion happens especially at choke points where make shift drop off areas occur.

Based on the perception of the place, inferred qualities of the space dictate issues on 1) sense of place; 2) visual character; 3) need for multifunctional spaces; 4) maintenance; 5) walkability, and; 6) safety.

Correlating these findings into open space planning, there is a need to respond to the user needs. Thompson [8] theorized that open spaces in cities are places that celebrate cultural diversity. She described that although public parks are places for meeting strangers, these are also places where people can be intimate, anonymous and therefore, private [8]. The findings on the users of the JP Rizal Waterfront Area clearly displays the lack of theoretical basis of its design. These dehumanized factors

IOP Conf. Series: Earth and Environmental Science **91** (2017) 012013 doi:10.1088/1755-1315/91/1/012013

are manifested in the space as in poor design, the misuse or underuse of spaces, maintenance, barriers and discontinuity to name a few.

In analyzing the theories of Thompson and Tiwari, the connection between open space and users are clearly established. It is therefore a pliable next step to respond to the needs of the users through the environment in which they belong. Borrowing heavily from the responsive design principles of Bentley et al [1], this paper correlates responsive environments as a solution in re-humanizing urban open spaces.

Responsive Design Principles	Physical	Environmental	Socio-Cultural
Permeability	✓		✓
Variety	✓		✓
Legibility	✓		✓
Robustness	✓	✓	
Visual Appropriateness	✓		✓
Richness	✓	✓	
Personalization	✓		✓
Maslow's Pyramid Level	1	1	2 and 3

Figure 5. Responsive design principles correlated to the three aspects of human needs derived from Maslow's pyramid

3.3. Recommendation

Based on the responsive design principles and the site conditions, the following design strategies are formulated:

3.3.1. Permeability. 1) Improve walkability through covered walkways; 2) remove makeshift ferry terminals; 3) enhance pedestrian access through crosswalks and better vehicular drop-offs; 4) propose reblocking / transferring of some dwellings very near to the river, and; 5) connect east and west side through masking of psychological barriers (i.e. provision of lights, continuity of character, induce movement through curvilinear paths).

3.3.2. Variety. 1) Create 'loose fit' spaces for multi-purpose functions; 2) provide spaces for existing major uses (waiting sheds, vending areas, jog paths), and; 3) provide spaces with blurred distinctions as private or public realm (i.e. benches with interspersed armrests).

3.3.3. Legibility. 1) Create strong terminus at both ends to facilitate movement (i.e plazas, activity areas); 2) provide way finding elements (i.e. signage); 3) establish monuments (i.e. sculptures, etc.) 4) establish clearly demarcation of hardscape and softscape areas, and; 4) preserve major tree locations as these have strong pulling effects to users.

3.3.4. Robustness. 1) Enhance the river experience (i.e. capitalize on vistas, river rehabilitation); 2) reuse existing warehouses and convert to spaces with multiple uses; 3) introduce indoor-outdoor activities near the waterfront, and; 4) introduce water related activities (i.e. dragonboat activity, clean-up drive)

3.3.5. Visual Appropriateness. 1) Establish unifying character (i.e. plants, floor patterns, urban design elements, furniture, materials); 2) maintain one color scheme across the stretch of the site regardless of barangay jurisdiction, and; 3) strengthen sense of place by creating unique spaces across the area, but have unifying elements (i.e. old trees with seating space, view decks with seating space)

IOP Conf. Series: Earth and Environmental Science **91** (2017) 012013 doi:10.1088/1755-1315/91/1/012013

3.3.6. Richness. 1) Improve softscape elements by adding more vegetation (i.e. shrubs, take off from existing urban trees on site: *Albizia saman, Cassia nodosa, Cassia fistula, Pterocarpus indicus*), and; 2) integrate sensory elements (i.e. fragrant trees such as *Azadirachta indica*, play around colors, provide various textures)

3.3.7. Personalization. 1) Create spaces that interface private and public use (i.e. view decks, seating spaces); 2) introduce a light park under the EDSA bridge to connect east and west sides; 3) provide ample lighting to deter unwanted activities; 4) preserve areas of high socio-cultural values (i.e. identified spots where people converge, shade trees, resting places, etc.); 5) address safety concerns through defensive site elements (i.e. choice of lighting fixtures, provision of bollards along drop offs and railings along riverfront areas); 6) correct the distances of tree planting locations in order to address issues on defensible space but also integrating proper shading effects, and; 7) increase security through avoidance of dark areas, provision of paths with unblocked visual corridors, etc.

4. Conclusion

In creating humanized spaces, there is no set template. Designing urban edges that brings people into convergence will always entail an in-depth study of the project area. It is important that during the design process, the users are already considered in a participative manner. One site will always be unique from another site due to the physical, environmental and socio-cultural aspects of that place.

There are however, base guidelines or considerations that can be used as a checklist in studying and designing urban spaces. Responsive design principles are some of the tools that a designer can use in order to achieve an effective and convivial output. These can be applied to any urban design study as a constant, while the site and users are the variables.

On the subject of the site as waterfront development along Pasig River, there is a need to revisit the easement requirements for urban areas (3 m) in order to appropriate the correct distances of built structures from water bodies in the urban setting. A stricter enforcement of the Easement of Servitude provisions found in the Civil Code of the Philippines should also be considered. Furthermore, the Pasig River Rehabilitation Commission (PRRC) has masterplanned the banks of the river including transference of Informal Settler Families (ISF's), recovery of easements and establishment of Environmental Protection Areas (EPA's). These should all be considered in similar waterfront revitalization projects like the JP Rizal Waterfront Area development.

Converging social classes is achievable given that the response to human needs are met. Design should not focus solely on form but rather a big portion should be spent on function. More importantly, design should address the needs of its target users.

References

[1] Bentley I, Alcock A, Murrain P, McGlynn S, and Smith G 1985 *Responsive environments: A manual for designers* (London: The Architectural Press Ltd)

[2] Brown L J, Dixon D, and Gillham O 2009 *Urban Design for an Urban Century: Placemaking for People* (New Jersey: John Wiley & Sons)

[3] Enjuto R 2013 The dehumanization of the city: the car and the invasion of the public space *Researchgate* pp 1679-1680

[4] Kelman H C 1976 Violence without restraint: Reflections on the dehumanization of victims and victimizers G M Kren and L. H. Rappoport (Eds.) *Varieties of Psychohistory* (New York: Springer) pp 282-314

[5] Lynch K 1960 *Image of a City* (Cambridge: The Technology Press & Hardvard University Press)

[6] Pasig River Rehabilitation Commission 2014 *Annual Report*

[7] Roberts M and Greed C 2001 *Approaching Urban Design: The Design Process* (London: Pearson Educated Limited)

[8] Thompson C W 2002 Urban open space in the 21st century *Landscape and Urban Planning* pp 59-72

[9] Tiwari S 2013 Developing a sense of place by humanizing public pedestrian precincts *International Journal of Architecture and Urban Development* pp 21-26

[10] UN-Habitat 2015 *Public spaces for all designed to live together* (Accessed: May 1, 2016)

[11] Waytz, Schoroeder and Epley 2014 The Lesser Minds Problem in Humans and dehumanization Presidential Decree 1067 (1976) *Water Code of the Philippines* Chapter IV Article 51 pp 61-62

IOP Conf. Series: Earth and Environmental Science **91** (2017) 012014 doi:10.1088/1755-1315/91/1/012014

Stakeholders analysis on criteria for protected areas management categories in Peninsular Malaysia

Z Hashim[1,2], S A Abdullah[1] and S Md. Nor[1]

[1]Institute for Environment and Development (LESTARI)
Universiti Kebangsaan Malaysia (UKM)
43600 UKM Bangi, Selangor, Malaysia

[2]Centre of Studies for Parks & Amenity Management
Faculty of Architecture, Planning & Surveying
Universiti Teknologi MARA
40450 Shah Alam, Selangor, Malaysia

E-mail: zackhashim1@gmail.com

Abstract. The establishment of protected areas has always been associated with a strategy to conserve biodiversity. A well-managed protected areas not only protect the ecosystem and threatened species but also provides benefits to the public. These indeed require sound management practices through the application of protected areas management categories which can be is seen as tools for planning, establishment and administration of protected areas as well as to regulate the activities in the protected areas. However, in Peninsular Malaysia the implementation of the protected areas management categories was carried out based on the *'ad-hoc'* basis without realising the important of the criteria based on the local values. Thus, an investigation has been sought to establish the criteria used in application to the protected areas management categories in Peninsular Malaysia. The outcomes revealed the significant of social, environment and economic criteria in establishing the protected area management categories in Peninsular Malaysia.

1. Introduction

Protected areas refer to an area which receives protection due to its natural, ecological and cultural values. Protected areas also involved marine areas involving boundaries of which will include some parts of the ocean [1] [2] [3]. Protected areas have often been associated its primary purpose of existence is to conserve biodiversity and providing habitat as well as safeguard the areas from hunting for threatened and endangered species. This kind of protection helps maintain the ecological processes that cannot survive in most intensely managed landscape and seascapes.

The International Union for Conservation of Nature (IUCN) defined protected areas as "land or sea especially dedicated to the protection and maintenance of biological diversity, and of natural and associated cultural resources, and managed through legal or other effective means". The Convention on Biological Diversity (CBD), however, uses a different definition "a geographically defined area which is designated and managed to achieve specific conservation objectives" [4]. Both the IUCN and the CBD agree that protected areas are used as environmental conservation units and use different terminologies such as national parks, wildlife sanctuary, nature reserve, game reserve, game ranch and nature conservancy [5] [6].

2. Protected Area and Management Categories

The classical concept of protected area movement began with the establishment of Yellowstone National Park in the United States of America in 1872 [7] [8] [9]. It was considered as a pioneered to the protected areas in their classic form, as government owned, government run areas set aside for protection and enjoyment[4]. This model was and remains a simple but powerful expression of a peoples' concern to protect their heritage for all time. In the early1870s protected areas expanded and concerns of environmental conservation were enhanced with the emergence of the concept of managing protected areas [10]. Since then, the establishment of protected areas has witnessed various size, shape, management system, ownership and governance patterns employed to the protected areas [4] [11]. However, there is no uniform standard or understanding on the protected area. During this period protected area often referred to as an area with beautiful scenery, wildlife and recreation [11]. Thus, the introduction of protected areas management categories by IUCN in 1994 is seen as an important global standard for managing protected areas. The IUCN clarified the implication of different types of protected areas under six management categories as indicated in Table 1[12].

Table 1. Protected areas management categories

	Type	Primarily management objective
Ia	Strict Nature Reserve	Scientific purposes
Ib	Wilderness Area	Wilderness protection
II	National Park	Ecosystem protection and recreation
III	Natural Monument or feature	Conservation of specific natural feature
IV	Habitat/species Management Areas	Conservation through management intervention
V	Protected landscape/seascape	Landscape/seascape conservation or recreation
VI	Protected areas with sustainable use of natural resources	Sustainable use of natural resources

The first three of the protected areas management categories (1a, Ib, II and III) were designed mainly to protect biodiversity and natural features with strictly controlled human intervention whereas the last three of the management categories (IV, V and VI) allow the human intervention to the protected areas. Most countries have established protected areas without taking into account the purpose for which various terms are used to refer to protected areas such as the historic area, city parks, game reserves, wildlife sanctuaries, archaeological sites, landmarks and land for forestry use.

In most Asian countries including Malaysia, the protected areas were established during the colonial period with the main purpose of a game reserved mainly for hunting and recreation with less intention and consideration to the interest of local communities [13] [14] 15]. The negligence of the local communities participation has led to the problem on social, economic and environmental effects in many countries [10]. For instance, in Peninsular Malaysia overall protected areas coverage is 1,801,792ha (13.7%) of the total land of 13,189,061ha includes national parks, state parks, wildlife reserves and sanctuaries [16]. These led to the foundation of the Malaysian National Biodiversity Policy [17]. Principally, the conservation strategy is an integral part of sustainable development of the country in providing protection of sources for critical ecosystem services [18]. All of the protected areas gazetted were based on the various provision of the law, but it does not clearly spell out the definition and objectives on the designation of the management categories towards the protected areas [19]. Until now, protected areas in Peninsular Malaysia are classified as national parks, state parks, wildlife reserved and sanctuaries [19] and mostly being managed by the Department of Wildlife and National Parks (DWNP).

The application of protected areas management categories in Peninsular Malaysia began in 1996 however, the determination and the application do not have a definition and clear objectives. This is

IOP Conf. Series: Earth and Environmental Science **91** (2017) 012014 doi:10.1088/1755-1315/91/1/012014

because it only focused on the scientific evidence with less concern on the interest of the local communities surrounding the protected areas. Indirectly, this casts doubt on the suitability and appropriateness of the use of category management in existing protected areas. This situation coupled with the application of the management categories in Peninsular Malaysia was based on *'ad-hoc'* basis with the absence of details in respect to the criteria used in determining the management categories [16]. Due to at the earlier application of the categories were based on its needs at that moment without any detailed consideration were made in assigning the protected areas management category [20]. As a result, the country's immense biodiversity potential is underutilised due to lack of awareness, lack of integration between the local communities and the management.

Over the years, the management of protected areas involved in the discussion on the local empowerment, popular participation, democratisation and devolution of power [21]. These have led the *'paradigm shift'* in the protected areas management from an emphasis on the 'centralization' under the federal government governance to the interest of the local stakeholders [4] [22]. Although protected areas often been associated with the main purpose of biodiversity conservation, these areas also provide essential benefits particularly through recreation, tourism, agriculture and other natural resource product [9] [23] [24]. The conservation of resources in protected areas began to be evaluated regarding social, economic, environment [21].

3. Criteria for environment, economy and social

A set of criteria comprises of the environment, economy and social [25] [26] [27] [28] [29] were employed in determining the criteria associated with the application of protected areas management categories. All of these criteria were selected based on the literature search and had been confirmed by the expert from the locations chosen for this study. The selection of the criteria has been modified to reflect the aims of this research as illustrated in Table 2:

Table 2. Criteria for social, economy and environment

Criteria	Sub-criteria	Description
Environment	Biodiversity	Species and habitat conservation
	Landscape	Landscape protection
	Ecosystem	Water quality
Economy	Employment/Income	Provide job opportunity
	Business opportunity	Product branding
		Rural diversification (agriculture)
		Tourism value and e-business
Social	Social inclusion	Public participation
		Demographic trends/infrastructure
	Health and well-being	Recreation
	Learning and education	Research/volunteering/education

4. Stakeholders Analysis

Stakeholders refer to any individual, group or institution who has vested interest in the natural resources of the project areas and who potentially will be affected by the activities and have something to gain or lose if conditions change or stay the same [30]. The aims of stakeholder analysis are to develop a strategic view of the human and institutional landscape and the relationships between the different interested parties and the issues they care about most [31]. Protected areas often affected by stakeholders participation either directly or indirectly involved in the utilisation of the resources or management duties [9] [32]. Successful protected areas management depends on the interaction between practitioners, policy makers and various stakeholders including local communities [9] 33]. The lack of support from stakeholders, particularly from local communities together with the management make it elusive the goals of integrated management strategies for the sustainable utilisation of resources [34].

5. Methodology

The questionnaire survey was conducted to 728 respondents in 10 locations, i.e., RHL Wang Pinang; Pusat Konservasi Bukit Pinang; Taman Negara Pulau Pinang; RHL Sungkai; Pusat Konservasi Bota Kanan; Santuari Burung Batu Gajah; Taman Negara Pahang; Taman Negara Kelantan; Taman Negara Terengganu and Pusat Konservasi Bukit Paloh comprises of three main group of management, local community and visitors. All of the locations are in Peninsular Malaysia under the management of DWNP. The survey took over a period of more than six months. The main aim is to establish and understands the social, economic and environmental criteria's of protected areas management categories. An expert judgement has been sought before the survey comprises those who directly and indirectly involved in protected areas management. The survey questionnaire comprised of 55 questions that were divided into some sections focusing on basic socio-demographic information and issues on benefits of social, economic and environment from protected areas. Before beginning the potential survey respondents in the study were informed of the goals of the interviews through the statement read by the interviewer and assured the data were analysed anonymously. Interviews were conducted upon verbal consent of the potential respondents to participate.

6. Results

6.1 Respondents Demographics

Based on valid survey responses of 728, the respondents were 75% male and 25% female of Malay, Chinese, Indian and indigenous ethnic groups mostly (Semai, Semelai, Che Wong Jah Hut and Batek). The gender biased was caused by the readiness of the respondents to be part of the interview conducted. The age of the respondents comprises of 62% (25-39 years old); 27% (40-55 years old); 10%(18-24 years old and 1% (>55 years old). The compositions on a group of respondents were 20% management; 36% visitors and 44% local communities. The distribution of the respondents were Wang Pinang Wildlife Reserve 4.8%;Bukit Pinang Conservation Centre 4.9% ;Penang National Park 24.2%; Sungkai Wildlife Reserve 6.9%;Bota Kanan Conservation Centre 6.2%; Batu Gajah Birds Sanctuary 5.5%; Pahang National Park 24.6%;Kelantan National Park 8.0%; Terengganu National Park 7.7% and Bukit Paloh Conservation Centre 7.3%. Based on the survey conducted the location of the protected areas were located far from the city and surrounded by traditional Malay, Chinese and indigenous village.

6.2 Criteria for Protected Areas Management Categories

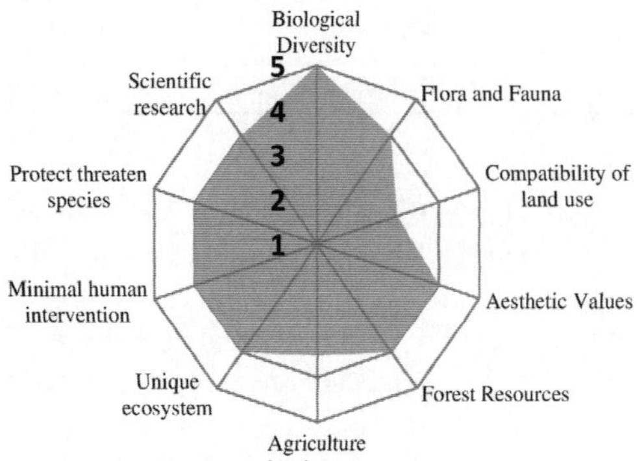

Figure 1. Criteria for Environment

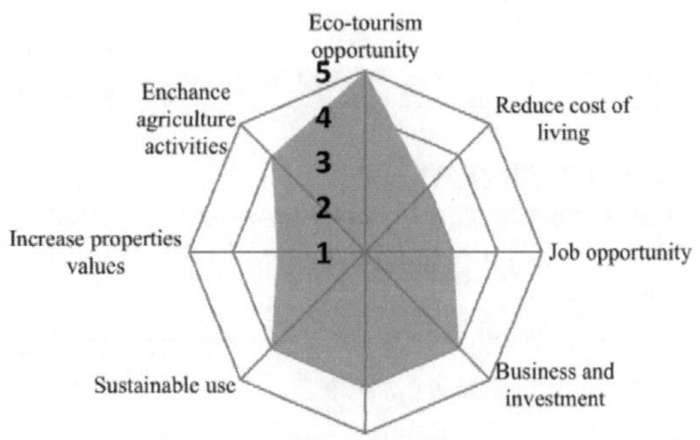

Figure 2. Criteria for Economy

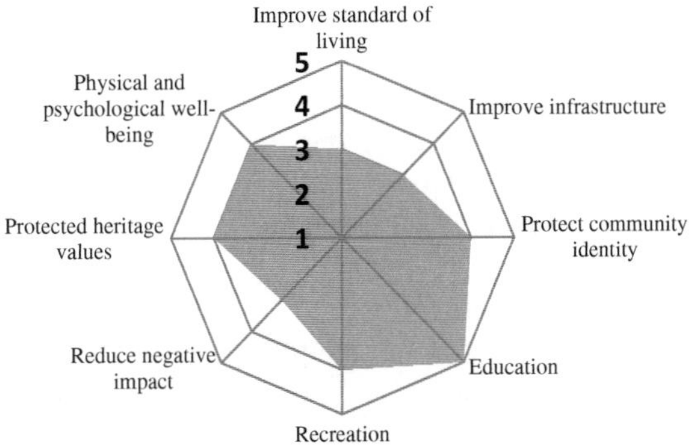

Figure 3. Criteria for Social

In light with the aim of this research and the literature search, a set of criteria has been identified based on three elements, i.e., environment, economy and social as indicated above. The criteria were developed from the general set used in establishing the protected areas [9] [35] [36]. All of the criteria's have been modified to reflect on this study and the policies of National Policy on Biodiversity [17]. A total of 26 criteria has been tested to the respondents comprises of 10 criteria for the environment, eight criteria for the economy and eight criteria for social. The 'like types' questions by using 1= Strongly agree; 2= Disagree; 3=Neither agree or disagree;4= Agree and 5= Strongly agree been used to measure a different kind of variables. Because of this observation the median (*Md*) was used as the measure of a central tendency [37] [38].

Figure 1, shows most of the respondents 'strongly agree' on the environmental criteria associated with the protected areas should possess biological diversity, 'agree' on unique eco-system, biological diversity, flora and fauna, minimal human intervention, protect threaten species, scientific research, enhance agriculture development and protect aesthetic value. Whereas, the respondents 'neither agree or disagree' towards the compatibility of the land use. It indicates, protected areas should maintain its main purpose for the protection and conservation of the biological diversity of the areas. This should be translated as the prime function for the establishment of the protected management categories.

Meanwhile, as illustrated in Figure 2 respondents were asked on the importance of protected areas to economic wellbeing. It seems that they 'strongly agree' on the opportunity for eco-tourism, 'agree' on enhancing the agriculture activities, business and investment, increase income and sustainable use of natural resources but 'neither agree or disagree' to the increase in properties value, reduce cost of living and providing job opportunities. The mix responses among the respondents were due to the facts that not all of the protected areas provides opportunities to generate income or job opportunities as some of the location was established mainly for conservation and education.

Figure 3, indicates the response by the respondents to the importance of social criteria. Most of the respondents suggested 'strongly agree' towards opportunity for education, 'agree' on protecting the culture, protect heritage values, physical and psychological well-being and opportunity for recreation. These shows the dependency of the stakeholders to the protected areas is high. This pattern suggests that the important of protected areas should be extended to the needs of the stakeholders. Meanwhile, they were 'neither agree or disagree' towards reducing negative impact, improve the standard of living and improve the infrastructure.

7. Conclusion

To our knowledge, this is the study that describes the criteria that influence stakeholders towards values of protected areas in Peninsular Malaysia. The design of this research allowed us to address a range of protected areas values and stakeholders that were often omitted in the process of determining the protected areas management categories. Interestingly this survey suggests that most of the stakeholders surrounding the protected areas are willing to involve and participate in the process of

IOP Conf. Series: Earth and Environmental Science **91** (2017) 012014 doi:10.1088/1755-1315/91/1/012014

designing the protected areas management. This indicates the importance of their views in line with the *'paradigm shift'* which emphasis on the stakeholders surrounding the protected areas.

On the other hand, the implication on the primary objectives of protected remains as the main agenda, while at the same time the function of economic and social has become importance in managing protected areas increasingly and it should be translated as secondary objectives to the existence and future of protected areas establishment. This study should further facilitate the government agency particularly DWNP in reviewing their protected areas management categories to the protected areas under their jurisdiction which until now due to be made. Further studies such as this are importance if we firmly believe that the people, particularly the local communities, have a voice in the future development of protected areas in Peninsular Malaysia.

References

[1] Salafsky N, Margoluis R, Redford KH, et al. Improving the Practice of Conservation: A Conceptual Framework and Research Agenda for Conservation Science. *Conserv Biol* 2002; 16: 1469–1479.
[2] Eagles P, McCool S, Haynes C. *Sustainable Tourism in Protected Areas: Guidelines for Planning and Management*. Gland, Switzerland and Cambridge, UK: IUCNhttp://cmsdata.iucn.org/downloads/pag_008.pdf (2002, accessed 10 July 2014).
[3] Hinch PR, De Santo EM. Factors to consider in evaluating the management and conservation effectiveness of a whale sanctuary to protect and conserve the North Atlantic right whale (Eubalaena glacialis). *Mar Policy* 2011; 35: 163–180.
[4] Philips A. *Turning Ideas on Their Head - The New Paradigm for Protected Areas*. 2003.
[5] Timko J, Innes J. Evaluating ecological integrity in national parks: Case studies from Canada and South Africa. *Biol Conserv* 2009; 142: 676–688.
[6] Teh L, Cabanban AS. Planning for sustainable tourism in southern Pulau Banggi: an assessment of biophysical conditions and their implications for future tourism development. *J Environ Manage* 2007; 85: 999–1008.
[7] Chape S, Harrison J, Spalding M, et al. Measuring the extent and effectiveness of protected areas as an indicator for meeting global biodiversity targets. *Philos Trans R Soc Lond B Biol Sci* 2005; 360: 443–55.
[8] Worboys GL, Lockwood M. Issues in Australian Protected Area Management. *George Wright Forum*.
[9] Lockwood M, Worboys G, Kothari A. Managing Protected Areas: A Global Guide. *Ann Tour Res* 2008; 35: 278–279.
[10] Pimbert M, Pretty J. Parks, people and professionals: Putting 'participation' into protected area management. *Soc Chang Conserv* 1995; 1–15.
[11] Bishop K. *Speaking a Common Language: the uses and performance of the IUCN System of Management Categories for Protected Areas*. Cardiff: Cardiff University, IUCN-The World Conservation Union and UNEP-World Conservation Monitoring Centre, 2004.
[12] Dudley N (ed). *Guidelines for Applying Protected Area Management Categories*. IUCN Publication Services Unit. Epub ahead of print 2008. DOI: 10.2305/IUCN.CH.2008.PAPS.2.en.
[13] Aiken SR, Leigh CH. Environment and the federal government in Malaysia. *Appl Geogr* 1988; 8: 291–314.
[14] Véron R, Fehr G. State power and protected areas : Dynamics and contradictions of forest conservation in Madhya Pradesh , India. *Polit Geogr* 2011; 30: 282–293.
[15] Sladonja B. *PROTECTED AREA*. Rijeka, Crotia: InTech, 2012.
[16] PERHILITAN, EPU, DANCED. Capacity Building and Strengthening of the Protected Areas System in Peninsular Malaysia. 1996; 305.
[17] Ministry of Science Technology and The Environment Malaysia. *National Policy on Biological Diversity (Malaysia)*. 1998.

[18] Wern H, Weng C. The potentials, threats and challenges in sustainable development of Penang National Park. *Malaysian J Env Manag*.
[19] DWNP/DANCED. *Protected Area Management Guidelines 1*. Kuala Lumpur: DWNP, 1994. Epub ahead of print March 1994. DOI: 10.1017/S0376892900024073.
[20] Minsitry of Natural Resources and Environment. *Management Effectiveness Assessment Of National And State Parks In Malaysia*. Petaling Jaya, Selangor: Sasyaz Holding Sdn. Bhd., 2006.
[21] Naughton-Treves L, Holland MB, Brandon K. the Role of Protected Areas in Conserving Biodiversity and Sustaining Local Livelihoods. *Annu Rev Environ Resour* 2005; 30: 219–252.
[22] Hamzah A, Ong DJ, Dario P. Asian philosophy of protected areas.
[23] Buckley R, Robinson J, Carmody J, et al. Monitoring for management of conservation and recreation in Australian protected areas. *Biodivers Conserv* 2008; 17: 3589–3606.
[24] Buckley R. Parks and Tourism. *PLoS Biol* 2009; 7: e1000143.
[25] Phillips A. *Economic values of protected areas: Guidelines for protected area managers*. Gland, Switzerland and Cambridge, UK: IUCN Gland, Switzerland and Cambridge, UKhttp://library.wur.nl/WebQuery/clc/1681593 (1998, accessed 23 January 2014).
[26] Harmon D. Intangible values of protected areas: What are they? Why do they matter. *George Wright Forum* 2004; 21: 9–22.
[27] Dudley N, Stolton S. *Defining Protected Areas*. Gland, Switzerland: IUCN, 2008.
[28] Hein L. Economic Benefits Generated by Protected Areas : the Case of the Hoge Veluwe Forest , the Netherlands. *Ecol Soc*.
[29] Tomićević J, Shannon M a., Milovanović M. Socio-economic impacts on the attitudes towards conservation of natural resources: Case study from Serbia. *For Policy Econ* 2010; 12: 157–162.
[30] Golder B, WWF-US, Gawler M. Cross-Cutting Tool Stakeholder Analysis. *Sourceb WWF Stand*http://heapol.oxfordjournals.org/content/15/3/338.short (2005).
[31] IUCN. Stakeholders analysis in the Azraq Oasis Restoration Project.
[32] Lockwood M. Good governance for terrestrial protected areas: A framework, principles and performance outcomes. *J Environ Manage* 2010; 91: 754–66.
[33] Kothari A. Protected areas and people : the future of the past 1. *Parks* 2008; 17: 2–5.
[34] Coad L, Burgess ND, Loucks C, et al. The ecological representativeness of the global protected areas estate in 2009: progress towards the CBD 2010 target. *Analysis* 2010; 1–35.
[35] Lockwood M, Bos D, Glazebrook H. Integrated Protected Area Selection in Australian Biogeographic Regions. *Environ Manage* 1997; 21: 395–404.
[36] Dearden P. Progress and problems in Canada's protected areas: overview of progress, chronic issues and emerging challenges in the early 21st Centuryhttp://dspace.ucalgary.ca/handle/1880/46957 (2008, accessed 23 January 2014).
[37] Garland R. The mid-point on a rating scale: Is it desirable. *Mark Bull* 1991; 2: 3–6.
[38] Sullivan GM, Artino AR. Analyzing and Interpreting Data From Likert-Type Scales. *J Grad Med Educ* 2013; 541–542.

User Satisfaction Assessment To Edu-Eco Tourism Services Of Cibodas Botanical Garden

I W Hidayat [1] and Winarni[2]

[1] Researcher, Cibodas Botanical Garden-Indonesian Institute of Sciences, PO Box 19 SDL, Cipanas-Cianjur, West Java Province, Indonesia 43253

[2] Public relations, Cibodas Botanical Garden-Indonesian Institute of Sciences, PO Box 19 SDL, Cipanas-Cianjur, West Java Province, Indonesia 43253

E-mail: imawan.wahyu.hidayat@lipi.go.id

Abstract. Cibodas Botanical Garden (CBG) is a government institution which has principal duties and functions as area of conservation *ex situ* of wet highland plants, research, education and tourism, it very closely related to aspect of the services to user. Good services will support the sustainability and existence of CBG as a world class edu-eco tourism destination. The purpose of this study was to measure the quality of services which delivered and improvement which necessary at the future. Assessments were made based on 14 criteria of services aspect for user which need research-education services and regular tourism services activities. The study was conducted by distributing questionnaires to users of these services. Questionnaires distribution was conducted in early August 2015 and August 2016, the respondents were 124 and 207. The results were showed the user satisfaction at good level, there were 77.685 in 2015 and 72.08 in 2016. Although still at a good level, there were a decline in satisfaction levels based on the value. In the future, the managerial needs to continuously to improve it, in order to get a good or very good valuation.

1. Introduction

Cibodas Botanical Garden (CBG) was established on 11 April 1852 by Johannes Ellias Teijsmann, a curator of the Garden at that time, with the name of *Bergtuin te Tjibodas* (*Kebun Pegunungan Cibodas*). CBG was initially intended as a place of acclimatization plant species of foreign origin which have value and high economic importance, one of which is Quinine or *Kina* (*Cinchona calisaya*). Then CBG was developed into a part of the Bogor Botanical Gardens with name as Branch Office of Cibodas Botanical Garden. Starting in 2003, the status of the CBG become more independent as a Technical Implementation Unit of Plant Conservation Office of Cibodas Botanical Garden under the Center for Plant Conservation of the Bogor Botanical Gardens in Deputies of Life Science of Indonesian Institute of Sciences (*LIPI*). And in February 2016, CBG was re-determined by regulation of the Head of Indonesian Institute of Sciences (*LIPI*) by eliminating the Technical Implementation Unit, and adds structural position in it which is Services and Information Unit, in the expectancy of CBG can be more engaged and improve of public services.

CBG has principal duties and functions as area of conservation *ex situ* of wet highland plants and developing research in tropical plant conservation and utilization of plant particularly wet tropical highlands. It is also developing environmental education to increase public knowledge and appreciation of the importance of plants and the environment for life, and improving the quality of public services in edu-eco tourism. CBG has also to conduct an inventory, exploration, plants collection, planting and maintenance of plants of the mountains, especially the western part of Indonesia which has a high value of science and economic potential and to be collected in the form of botanical garden.

Besides as ordinary tourism destination which has been famously known regionally even international, in order to improve public knowledge and awareness in the field of conservation, CBG has also developed edu-eco tourism which called "*Wisata Flora*". It was an CBG's attempt to answer the challenge the social changes taking place in contemporary society, which is considering leisure and "back to nature" activities as one of the necessities of life dominated by technology, and incorporate elements of conservation and science. In this case, public services has the most importance role because good services will support the sustainability and existence of CBG as a world class edu-eco tourism destination. Destinations invest considerable amounts of resources into planning and marketing tourism products and services. The assessment of the success of these efforts is thought to be best evaluated based on the satisfaction of tourists who visit the destination [1]. Tourist satisfaction is critical to destination competiveness and considerably affects destination choice, revisit intention and loyalty [1].

High levels of tourist satisfaction with a tourism destination is seen to result in increased revenues for service providers [2]. The critical implication is that the ability to understand the reasons for tourist satisfaction and the consequences of satisfied tourists is crucial [3] [4]. Establishing a system for measuring tourist satisfaction allows suppliers in a destination to understand and respond effectively to visitor needs [2] [5]. Therefore, the level of user satisfaction is one of important parameter for CBG to evaluate whether the services is good or not. The purpose of this study was to measure the quality of services of edu-eco tourism of CBG which delivered to user and improvement which necessary at the future, and for public, would be used as a description of the performance of the concerned service unit.

2. Edu-eco tourism of CBG

Edu-eco tourism of CBG or *Wisata Flora* is adopting the principles of smart tourism. Its service objects includes tourists, tourism service organizations and local governments in tourism destination [6]. Smart tourism provides varying forms of tourism information to all the entities within the tourism industry [7]. Smart tourism is the typical example of integrated development by combining tourism industry with technological innovations. It can be concluded that smart tourism is the ubiquitous tour information service received by tourists during a touring process. Tour information service is the summation of the common attributes of smart tourism [8].

Edu-eco tourism of CBG is a package of educational tour which offers an introduction to the plants collection in CBG, from morphology (e.g. the shape of leaves, stems, roots, fruits, flowers), seed dispersion, the process of propagation (this introduction is conducted inside and outside the room). The process of propagation learning is how to reproduce the plant by vegetative and generative procedures. Propagation of plants which studied are grafting, inosculating, cutting and seeding. Furthermore, participants are guided to recognized of CBG with a direct view to leisure or attractive spots and theme parks which exist.

Other activities are offers an introduction to the gardens (e.g. function, designation, plant collections, and points of interest) and the tropical rain forest (e.g. forest function description, characteristics and functions of canopy and plants typical of the tropical rain forest). In addition, user also guided into remnant forest at CBG, so they will feel the difference between the garden and forest.

Other package activity is learned to make dried herbarium which guided by a facilitator. Activities are began to introduction of the herbarium and the equipment which needed to make it. Then, the material collected from making in the garden and the practice of making herbarium to complete. Participants can bring home the results of the practice as an instrument of learning in their school. Each group who desires these services can choose what activities they want in a edu-eco tourism package at CBG.

3. Methodology

The study was conducted at CBG which administratively located in Cipanas District, Cianjur Regency, West Java Province (see figure 1). CBG is located at mountainside of Mount Gede and Mount Pangrango at an altitude of approximately 1,300-1,425 meters above sea level, with an area of 84.99 hectares. The average temperature of 20 °C, humidity of 80.82% and an average rainfall of 2,950 mm per year. CBG is a comfortable place to rest while enjoying the beauty of the various types of plants that originated from Indonesia, especially from tropical wet highland, and other foreign countries. CBG ± 100 km from Jakarta and ± 80 Km from Bandung.

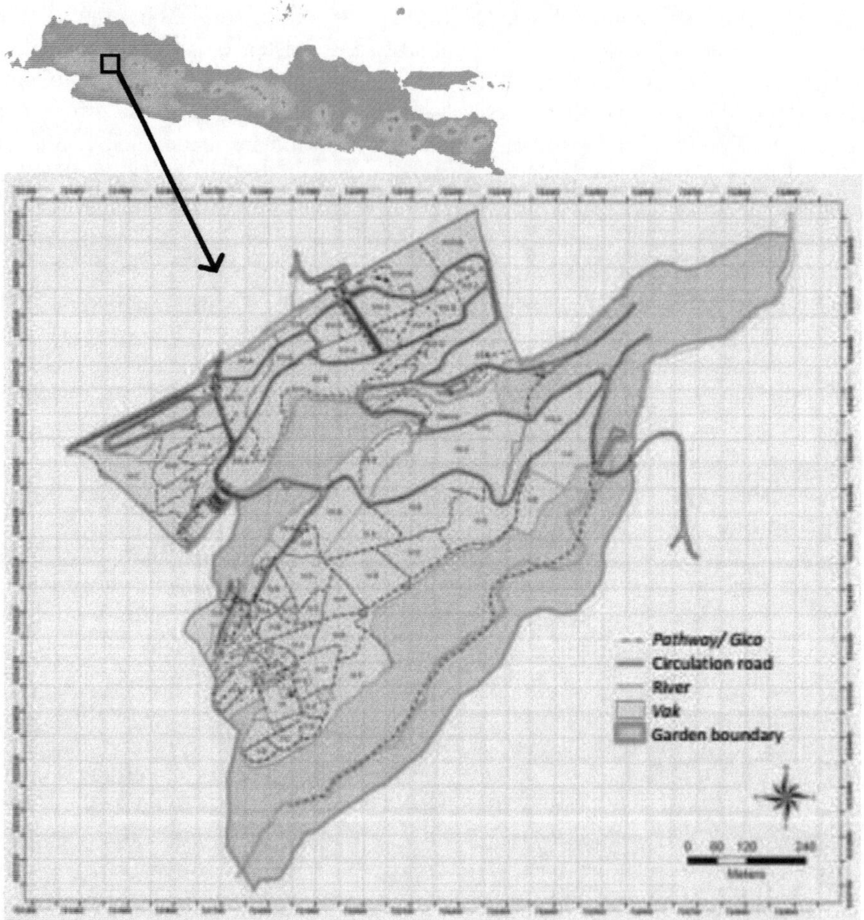

Figure 1. Study area at Cibodas Botanical Garden, Cipanas-Cianjur, West Java. (*Courtesy of* Agus Darmawan).

The subject of study was the user of edu-eco tourism services which provided by CBG as *Wisata Flora* program. The methods of study was conducted by distributing questionnaires to be filled out by respondents. Questionnaires distribution was conducted in early August 2015 and early August 2016. These both time selected because the user services at CBG were very high, especially for *Wisata Flora* program.

3.1. Questionnaires design and respondents

Questionnaires which compiled was a set of questions that indicate the quality of edu-eco tourism services of CBG. There were 14 questions which part of a service quality, (1) service procedures;

(2) service requirements; (3) a distinct operator; (4) discipline of the operator; (5) responsibility of the operator; (6) operator's capability; (7) fast response; (8) fairness of service; (9) hospitality; (10) charges fairness; (11) assurance of service charges; (12) assurance of schedule; (13) environment amenity; (14) services comfort.

Each respondents must to provide a value for each indicators. Each indicators has a weight rating from value "1" to "4", value "1" for services quality at "not good" level, value "2" for "less good" or "poor" level, value "3" for services at "good" level and value "4" for "very good" level of services quality. Respondents in this study were selected randomly from any user who has received the services. In 2015, the number of respondents was 124 and 207 in 2016 (see table 1).

Table 1. Characteristic of respondents in this study.

Respondents characteristic	2015 (*N*= 124)	2016 (*N*= 207)
Gender		
Male	39	84
Female	85	123
Age (years old)		
0-17	40	89
18-25	52	62
26-30	11	13
31-35	5	19
36-40	7	20
41-45	2	3
46-50	4	-
> 50	3	1
Education		
Elementary school or under	19	26
Junior high school	18	65
High school	63	85
Diploma	4	7
Bachelor	14	23
Post graduated	6	1
Occupation		
Government employee	6	12
Private employee	16	34
Entrepreneurial	2	24
Student	89	87
Others	11	50

3.2. Data collection and analysis

Each questionnaires which compiled from respondents was a set of values to each services indicators. Average value of each indicators formulated as:

$$\overline{Is_i} = \frac{\sum Is_i}{N} \quad (1)$$

$\overline{Is_i}$ is an average value of indicator of services of *i* (the value is between 1 to 4); $\Sigma\, Is_i$ is total value number of indicator of services of *i*; N is total number of respondents at the period; *i* is the number of indicators (1,2,3,... 14). Then, each indicators has a weight value, 1/14 or 0.071, so that average weight value of indicator of services *i* was formulated as $\overline{Is_i}' = \overline{Is_i}$ x 0.071. In order to convenient of understanding that the value was expected in range 25-100, each indicators was multiplied with based value or constants (*c*) of "25". Then, user satisfaction index (*USI*) can be formulated as:

$$USI = \sum \overline{Is_i}' \times c \quad (2)$$

User satisfaction index was a quantitative method and an objective parameter to assess the quality of edu-eco tourism services of CBG. Classification of USI based on range 25-100 value, which value 25 to 43.75 has a "D" qualification or quality of services was "not good", value 43.76 to 62.5 has a "C" qualification or quality was "less good" or "poor", value 62.51 to 81.25 has a "B" qualification or "good" quality and value 81.26 to 100 has an "A" qualification or "very good" quality. The assessment was also can be referenced as the level of performance of the related service unit and for further improvement at the future, and it can be as medium of public information. These assessment procedures were consistent with The Decree of the Indonesian Minister of Empowering the State Apparatus which about general guidelines for preparing the public satisfaction index at government services unit [9].

4. Result

The details of 2015 and 2016 questionnaires data which delivered to respondents about their opinions to the edu-eco tourism services of CBG can be seen in table 2.

Table 2. Data analysis of user satisfaction questionnaires of 2015 and 2016.

Indicator of services (Is_i)	2015 (N=124)			2016 (N=207)		
	Total value (ΣIs_i)	Average ($\overline{Is_i}$)	Average weight ($\overline{Is_i}'$)	Total value (ΣIs_i)	Average ($\overline{Is_i}$)	Average weight ($\overline{Is_i}'$)
Service procedures	375	3.024	0.215	597	2.884	0.205
Service requirements	375	3.024	0.215	592	2.860	0.203
A distinct operator	392	3.161	0.224	569	2.749	0.195
Discipline of the operator	393	3.169	0.225	616	2.976	0.211
Responsibility of the operator	400	3.226	0.229	619	2.990	0.212
Operator's capability	403	3.250	0.231	618	2.986	0.212
Fast response	373	3.008	0.214	568	2.744	0.195
Fairness of service	377	3.040	0.216	606	2.928	0.208
Hospitality	411	3.315	0.235	635	3.068	0.218
Charges fairness	383	3.089	0.219	610	2.947	0.209
Assurance of service charges	375	3.024	0.215	569	2.749	0.195
Assurance of schedule	369	2.976	0.211	573	2.768	0.197
Environment amenity	411	3.315	0.235	625	3.019	0.214
Services comfort	390	3.145	0.223	609	2.942	0.209

Based on table 2, the user satisfaction index in 2015 and 2016 were 77.685 and 72.08. These values were indicated that user satisfaction at the "good" level, or it can generated that the performance of services unit at grade "B" [9]. Although it has the same value range, but the difference in the number of respondents and the characteristics in it on which the survey in both years was affected the difference in result. The highest value of the indicators in the both years were "hospitality" (average values were 3.315 and 3.068) besides environment amenity indicator in 2015 was also reached at the same value. In 2015, the lowest value was the assurance of schedule (2.976) and 2016 was fast response with an average 2.744. Nevertheless, on the indicator with the lowest values were only slightly to the value of "3" as limit ratings of good category.

In 2015, except the assurance of schedule, the other indicators has a higher value than 3 (3.008 to 3.315). Furthermore, in 2016, except hospitality and environment amenity indicators, the others were slightly lower than the value of "3" (2.744 to 2.99). The results were showed that the performance of the services unit of CBG to delivered of edu-eco tourism program has been good.

5. Discussion

The results indicated that user satisfaction were declined between 2015 to 2016. Though still in good category [9], it showed a decline in the performance of the related units in satisfying the users. All indicator in 2016 have a lower average value than 2016 (see figure 2). It needs to get the attention of management in order to improve the quality of service in the future, if not, then CBG will be left by their user.

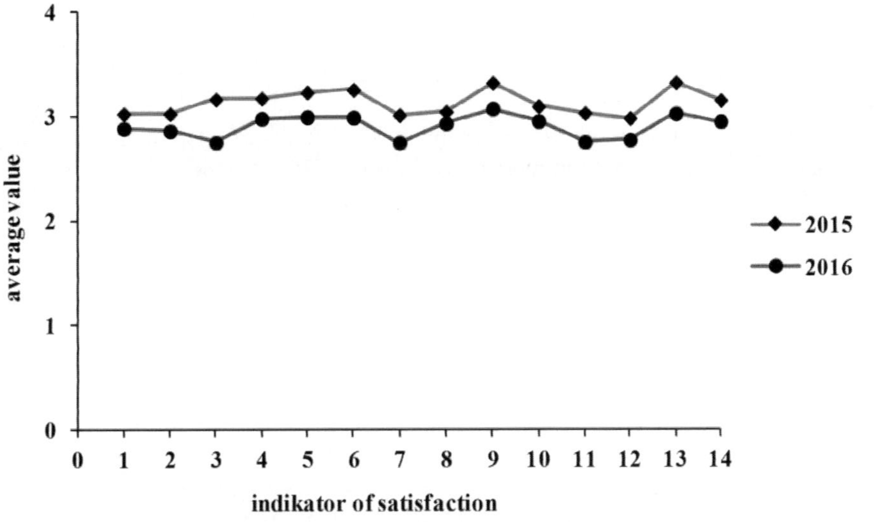

Figure 2. The comparisons of average value per-indicator of user satisfaction between 2015 and 2016.

The biggest gap of average value was at a distinct operator indicator (0.412). This occurs because the users feel operators in the field is limited. When edu-eco tourism activities were involved many participants, while the availability of operator guides just a few, then users will not or just have received less attention, for example, the discussion about plants or surrounding condition is not answered or less satisfactory. The total operators of CBG which run the edu-eco tourism program less than 10 persons. So that when a surge in demand for this service, the policy which taken is recruiting person from other units, which sometimes constrained by availability or other tasks.

This less operator was also impacted to the fast response and assurance of the schedule program, so at the both indicators were also reached a lower value (see table 2). The ideal condition which user can be delivered a satisfaction information, discussion and other needs in edu-eco tourism program of

CBG is 1 operator or guide per-20 persons of user or less. Therefore, this needs to add a sufficient of regular and capable operator to anticipate a surge demand of this service, especially at the peak time of visit like July and August and at the end of year.

Furthermore, both hospitality and environment amenity has a highest value from the user. These indicates that the performance of related unit has been satisfying the user. This was a positive relationship between perceived performance and tourist satisfaction [10]. It also indicates that pre-travel expectations and post-travel experiences of the user has been consistent [11]. This experiences was also evoked positive feelings [11], so both indicators show a highest value. CBG which offers views of the tropical mountains and a fine landscape gives a new experience to visitors. Where most of population lives and works in urban places, landscape values and services take on new meanings, and the economic dimension of landscape becomes increasingly important [12]. These good quality need to be managed and in the future CBG has to increase the quality of service in order to get long-term relationships with the customer. It affects not only immediate repeat purchases but also the reputation and the trust. Enhanced tourist satisfaction may lead to increased revenues and profits for service providers. Thus, recognizing and understanding the cognitive and behavioural consequences of satisfaction has important implications for destination management [2].

6. Conclusion

Generally, user satisfaction on edu-eco tourism of CBG at a good level. This assessment can be generated as early based-value and an objective valuation for the user satisfaction and the performance level of the related unit. In order to increase the appreciation of users, then CBG need to continuously improve the quality of services, since there were indications of a decline, although slightly. It is important to improve this study in the future with involved more respondents and periodically.

References

[1] Yoon Y and Uysal M 2005 An examination of the effects of motivation and satisfaction on destination loyalty: a structural model *Tour. Manag.* **26** pp 45-56

[2] Dmitrović T, Cvelbar L K, Kolar T, Brenčič M M, Ograjenšek I and Žabkar V 2009 Conceptualizing tourist satisfaction at the destination level *Int. J. Cult. Tour. Hospitality Res.* **3** pp 116-26

[3] del Bosque I and San Martin H 2008 Tourist satisfaction: a cognitive-affective model *Ann. Tour. Res.* **35** pp 551-73

[4] Song H, Li G, van der Veen R and Chen J L 2011 Assessing mainland Chinese tourists' satisfaction with Hong Kong using tourist satisfaction index *Int. J. Tour. Res.* **13** pp 82-96

[5] Myers J 1999 *Measuring customer satisfaction* (Chicago: American Marketing Association)

[6] Fu Y and Zheng X 2013 China smart tourism development status and countermeasures *Dev. Res.* **4** pp 62-5

[7] Xu B, Li D, Qian Y, *et al* 2013 Smart tourism: a new trend for tourism industry based on existed research *Resour. Dev.Mark.* **7** pp 781-4

[8] Li Y, Hu C, Huang C and Duan L 2016 The concept of smart tourism in the context of tourism information services *Tour. Manag.*[*in press*]

[9] The Decree of the Indonesian Minister of Empowering the State Apparatus *KEP/25/M.PAN/2/2004* about The General guidelines for preparing the public satisfaction index at government services unit (Indonesian)

[10] Song H, van der Veen R, Li G and Chen J L 2012 The Hong Kong tourist satisfaction index *Ann. Tour. Res.* **39** pp 459-79.

[11] Chen C F and Chen F S 2010 Experience quality, perceived value, satisfaction and behavioral intentions for heritage tourists *Tour. Manag.* **31** pp 29-35

[12] van der Heide M and Heijman W J M. (Eds.) 2013 *The Economic Value of Landscapes* (London/New York: Routledge Studies in Ecological Economics)

IOP Conf. Series: Earth and Environmental Science **91** (2017) 012016 doi:10.1088/1755-1315/91/1/012016

Identifying the characteristic of *SundaParahiyangan* landscape for a model of sustainable agricultural landscape

M Z Dahlan[1], H S A Nurhayati[2] and W Q Mugnisjah[2]

[1]Laboratory of Landscape Ecology and Planning, Department of Technology and Ecology, Graduate School of Global Environmental Studies, Kyoto University, Japan
[2]Laboratory of Landscape Management, Department of Landscape Architecture, Bogor Agricultural University, Indonesia

E-mail: zai_ni@ymail.com

Abstract. This study was an explorative study of the various forms of traditional ecological knowledge (TEK) of Sundanese people in the context of sustainable agriculture. The qualitative method was used to identify *SundaParahiyangan* landscape by using Rapid Participatory Rural Appraisal throughsemi-structured interviews, focus group discussions, and field survey. The Landscape Characteristic Assessment and Community Sustainability Assessment were used to analyze the characteristic of landscape to achieve the sustainable agricultural landscape criteria proposed by US Department of Agriculture. The results revealed that the *SundaParahiyangan*agricultural landscape has a unique characteristic as a result of the long-term adaptation of agricultural society to theirlandscape through a learning process for generations. In general, this character was reflected in the typical of Sundanese's agroecosystems such as forest garden, mixed garden, paddy field, and home garden. In addition, concept of *kabuyutan* is one of the TEKs related to understanding and utilization of landscape has been adapted on revitalizing the role of landscape surrounding the agroecosystem as the buffer zone by calculating and designating protected areas. To support the sustainability of production area, integrated practices of agroforestry with low-external-input and sustainable agriculture (LEISA) system can be applied in utilizing and managing agricultural resources.

1. Introduction

The issue of sustainable development in agriculture requires a change in direction, systems, strategies, and policies to establish agriculture highly competitive, productive, profitable, environmentally friendly, as well as improving health, food quality and safety in the present and future. Thus, agricultural development should be able to conserve soil, water, maintain biodiversity, eco-friendly, technically appropriate, economically viable, and socially and culturally acceptable [1]. As an agriculture-based development country, Indonesia has a strong linkage with sustainable agricultural development. This is proved by the condition of most Indonesian people (35% of Indonesia's total population) are dependent on agriculture and the majority are living in rural areas [2].

Agricultural society in the *SundaParahiyangan* region is an integral part of the agricultural landscape who utilize natural resources and the environment into various forms of agroecosystems [3][4]. This integrated system is formed by the union of ecological systems, socio-economic and spiritual-cultural society. Aside from being a source of life, with their wisdom, the agricultural society plays an important role in managing natural resources and the environment [5]. Thus, the local wisdom that reflected in the traditional ecological knowledge (TEK) can be a trigger in achieving the sustainable agriculture development in rural areas, respectively. The existence of TEK of Sundanese people is expected to filter the negative effects as well as to improve the positive effects of modernization for both socio-cultural life of the community and their environment. Accordance with the important of TEK [6][7][8], the purposes ofthis study was to (1) identify the characteristics of the *SundaParahiyangan* agricultural landscape along with a variety of manifestation of TEK of *SundaParahyangan* society in the context of sustainable agricultural landscape, and further to (2) develop a model of sustainable agriculture landscape management.

2. Methodology

2.1. Study site

This study was conducted in Ciomas Village,Panjalu District, Ciamis Regency, West Java Province, Indonesia (latitude 07^{0}07'00" to 07^{0}12'00" S, longitude 108^{0}15'00" to 108^{0}19'00" E). The study sites were determined based on etymology of *SundaParahiyangan*which defined as the mountainous area with the elevation $\geq$ 600 m above sea level (hereafter: m asl) [9][10].The area of this villageis about 1350 ha where located between 512 and 1762 m asl with the average daily temperature is about 25 ^{0}C, humidity is 80%, the average annual precipitation is about 2000 mm, and the rainy season lasting almost all year long except June, July, and August. The dominant of landscape (36.6%)is used forproduction forest that managed mainly by resident. Forest upland has 31.7% in total area that used for conservation forest designated by national law as a game reserve of Sawal Mountain as well as for production forest that managed by IndonesianForest Enterprises (*PerumPerhutani*).

2.2. Data collection and analysis

Rapid Partisipatory Rural Appraisal (rPRA) method through semi-structured interview, focus group discussion, and field survey was conductedduring 2012-2014 to identify the characteristic of *SundaParahiyangan* landscape and its society by using TEK's approach[6][8][11][12].Representative respondents were selected through purposive sampling such as community leader, spiritual leader, local administrative government, and professional in agricultural activity. A set of question related to sustainable agricultural landscape management was conducted in relevant to Landscape Characteristic Assesment (LCA)[13]andCommunity Sustainability Assessment (CSA)[14] to describe the characteristic of Sunda Parahiyangan both landscape and its society. Sustainability analysis was conducted by combining LCA and CSA method which focussed on three main factors (ecology, social-economy, and spiritual-culture) to achieve the USDA's sustainability criteria [15].

3. Results and Discussion

3.1. Characteristic of SundaParahiyangan landscape

As part of *SundaParahiyangan* mountainous landscape, Ciomas Village has typical agroecosystemsthat reflect the Sundanese philosophy of life in interacting with nature. *SundaParahiyangan* landscapewas generally used for several types of agricultural activity.This condition wasfully supported by the physical environment factors which are ideal for farming such as domination of latosol soil type with sloping degree from flat to very steep, the average of daily temperature about 25 ^{0}C,the average of relative humidity of 80%, and an abundant of water resource throughthe average annual rainfall of 2,000 mm.Those factors contribute to the optimal developmentfor various types of agroecosystem as well as providing habitat for a variety of vegetation and wildlife in accordance with the physical environment [16]. The agroecosystem formed in the study area consists of forest garden (*kebun-talun*), mixed garden (*kebuncampuran*), paddy field (*sawah*), and home garden (*pekarangan*), while shifting cultivation system (*huma*)has been discontinued (Figure 1).

IOP Conf. Series: Earth and Environmental Science **91** (2017) 012016 doi:10.1088/1755-1315/91/1/012016

Figure 1. Landscape structure of agroecosystems in Ciomas Village: (a) forest garden/*kebun-talun*, (b) mixed garden/*kebun campuran*, (c) paddy field/*sawah*, and (d) home garden/*pekarangan*

This study revealed that these agroecosystemshave been formed through a mutual interaction and deep understanding between human and their landscape. Agroecosystem formation begun with the opening of forest areas into agricultural areas in general. The main factor that influence the type of agroecosystem was the availability of water resources. Land that was not supported by the availability of abundant water resources, functioned as a dry land farming system (*kebun-talun* or *kebuncampuran*). While the water resources available, land was clearedand watered to be functioned as a wetland farming systems (*sawah*). The next development was influenced by people's demand for a permanent residence. People usedtheir land in each agroecosystem to build a permanent house at the time of fallow land. The remaining land after optimized for buildings was used for agriculture called *pekarangan* (Figure 2). In general, the utilization of land resources to be various types of agroecosystem was more adapted to the physical condition of the land as well as the perceptions and preferences of local people. An example, mixed garden was dominant (26.5%) in this village due to local people's preference to utilize their land efficiently by planting various type of plant species [3].

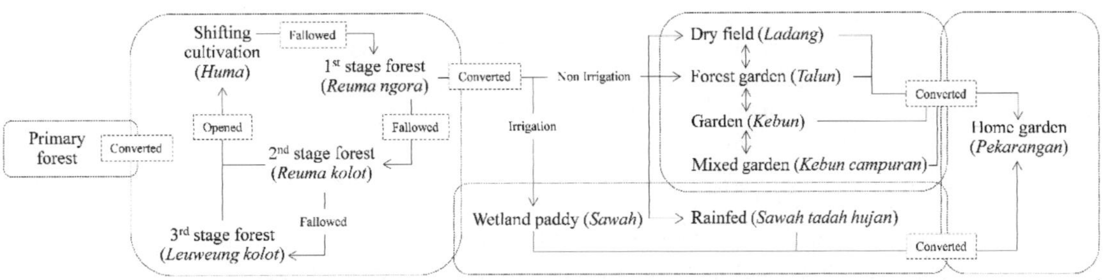

Shifting cultivation system	Forest garden, mixed garden, and paddy field system	Home garden system
The early stage of agroecosystem's formation is depend on people's perception to fulfil their basic needs (food, fuel, material for building). Thus, primary forest changed into secondary forest to be used for shifting cultivation system (*huma*).	The second stage is developed based on the trend of people's needs and the availability of water resources. In addition, people's preference of natural resources (vegetation) is also influences the formation. Mixed garden is more diverse than other dry land farming types due to people's preference to plant various plant species.	The last stage is developed based on the intensification of arable land in the small area due to more allocation for building construction.

Figure 2. Development of agroecosystem in *SundaParahiyangan* landscape

IOP Conf. Series: Earth and Environmental Science **91** (2017) 012016 doi:10.1088/1755-1315/91/1/012016

3.2. Agricultural society in SundaParahtyangan landscape

Socio-economic system of agriculture was expressed in the productive and reproductive activities. These activities constitute agricultural resource extraction activities to be used as a source of life. Productive activities carried out on farm, while reproductive activities conducted outside the land for agricultural resource utilization more feasible (off farm). Dominant land use as agricultural areas have not been able to provide optimal results despite a core source of income farm families. High production costs, lack of support for agricultural science and technology, as well as the lack of effective government's role in supporting the agricultural production have affected the low agriculture's role in improving farmer's life. Moreover, this study found that the ownership of arable land is relatively narrow with an average of 0.3 ha/farmer. This indicates the lack of power for farmers to support their basic needs, while the availability of land is important. A significant development strategy need to be implemented to empower the ability and capability of farmers as the key person in utilizing and managing agricultural landscape and its resources such as by implementing land reconsolidation or intensifying land [3][4][17].

Traditional ecological knowledge (TEK) related to agriculture in the spiritual-cultural system was expressed by the term of the local customs, activities, and cultural objects of agriculture [6][7]. The concept of *kabuyutan* is one of the TEKsthat implementedin the management of agroecosystems for generations. *Kabuyutan*was believed to be a model of managing agricultural landscapes to provide benefits in the sustainable condition. *Kabuyutan* concerned to protect the main elements in the landscape to protect a whole landscape. For example, the existence of spring water is an important element in the landscape that should be protected to ensure the availability of water resources. In term of vital element, the protection should be considered as a whole part of area, both its core, buffer, as well as its development zone. Therefore, the concept of *kabuyutan* is proposed to consider the elements of landscape as an integrated system that link each other.

The term of*kabuyutan* is not widely known by local people, but the existence is remaining in their knowledge. People reported that to implement this concept, three development steps should be taken into consideration (setting places, times, and actions). To set the place, the knowledge about the landscape should be mainly focussed on ecological aspect such as the ecological boundary (watershed) is prefer to be used rather than administrative boundary. In case of Ciomas Village, village area located in the area of Cidarma Watershed that connected to Citanduy River as a main watershed. Then, the main spring water in the landscape should be prioritized to protect both its existence and surrounding environment. Considering the location of this spring, the prohibited area (*leuweunglarangan*) and conserved area (*leuweungtutupan*) would be designated. These two types of protected area are distributed in every land types that considered to be protected such as springs, riverbanks, lakes, valleys, hills, steep slope, and even very gentle slope that located in settlement area (Figure 3).

The existence of *kabuyutan* was shown by the existence of some protected areas that culturally perceived as sacred places. These sacred places varied in size and have protected for generations due to the customary rules (*taboo*). People perceived that applying taboo is the appropriate cultural strategy to protect something, especially in order to prevent a destruction of place.They reported that *kabuyutan* (sacred places also named as *kabuyutan* by local people) provides many benefits such as water resources, especially for *kabuyutan* where located close to settlement or agricultural land area. Accordingly, implementation of this conceptin broader scale is suggested to ensure the hydrological system that will influence the whole ecosystem. Further, the potential of the community who will be enhanced to be more productive communitywith full of fairness and wisdomrather than exploitative consumers. Fairness means utilization based on carrying capacity of landscape, while wisdom means utilization by considering the use for future generations.After designating the place for protection function both prohibited (core zone) and conserved (buffer zone) as well as for limited production (development zone), a development time as well as some actions to be implemented for improvement as well as for recovery were set. People reported that setting of time and actions depend on the condition of landscape especially the degree of damages that occurred within the landscape. More damages need more time to recovery and more detail actions to be conducted [17][19].

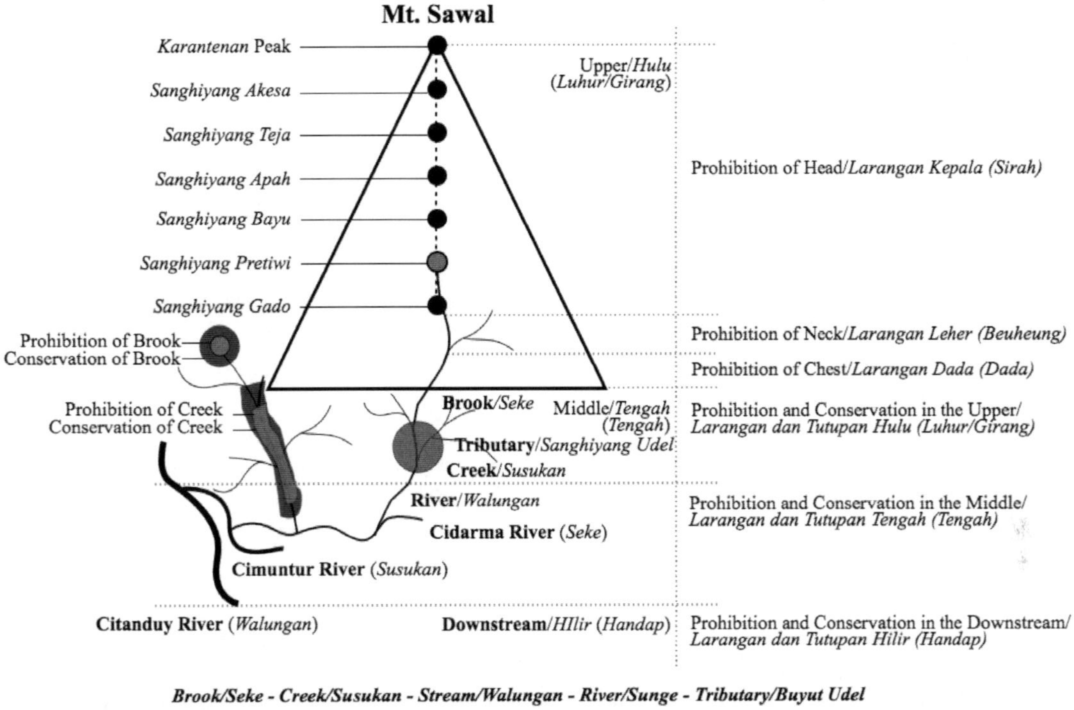

Figure 3. The concept of *kabuyutan* and its implementation in a landscape

3.3. Landscape sustainability analysis

Landscape Characteristic Assessment (LCA) method was used to identify the characteristic of *SundaParahiyangan* landscape. This method focused on exploration of main factors that influence the development of a landscape. Results indicated that *SundaParahiyangan* landscape is more likely a mountainous agricultural landscape due to three physical environment aspects (elevation, soil, and climate) that assumed have great influence to develop its characteristic. Ecologically, the existence of Mount Sawal providessufficient amount of water resources, fertile soil, as well as various type of plant and animal species. Culturally, natural resources have been used for long time period wisely in accordance with local people's TEK. They proposed a traditional concept of *kabuyutan* that focused on how managing landscape and its resources wisely.

According to the analysis of the LCAcriteria, Ciomas Village has a potential become a sustainable landscape to support agroecosystems. The condition of the abundant agricultural resources provides an opportunity for farmers to use agricultural resources for present andfuture needs. In addition, the availability of agricultural resources to provide a decent quality of farmer's life with the availability of jobs for productive and reproductive activities, as well as the availability of food with high quality, adequate nutrition, accessible, and affordable price [1][16]. Optimal conditions can be achieved by maintaining the quality and quantity of water and land resources as the main elements of the *SundaParahiyangan* agricultural landscape.

Further, based on the analysis of the sustainability of the Community Sustainability Assessment (CSA) method, *SundaParahiyangan*society indicates the level of infallible progress toward sustainability with a total value of 1184 (ecological aspect of 300, social aspect of 405, and spiritual aspect of 479). However, the improvement efforts in some aspects with value less than 50 is necessary to achieve the optimal sustainability aspects such as enhancement of local people's sense of belonging to their place, and provision of infrastructure (land) for supporting agricultural activity. According to USDA's sustainability criteria, result showed that the condition in Ciomas Village both landscape and their people has a potential towards sustainable. The landscape can provide sufficient amount of food, fuel, fiber, and building material, while the society can improve their landscape through their valuable TEK. Furthermore, those benefits can preserve the existence of agricultural economic that will improve the welfare of agricultural society in particular [17].

The concept of managing *SundaParahiyangan* agricultural landscapes applied by combining both traditional and modern model. TEK is the main capital that enhanced by modern science in order to respond to the ecological landscape dynamics [6][8][11]. The integration can be applied through the implementation of an integrated concept of *kabuyutan* with agroforestry farming systems in each agroecosystem by optimizing intake and efficiency of the input from the outside (low-external-input and sustainable agriculture/LEISA) (Figure 4) [20]. In addition, external influences such as the initiation of the rural development programs from government or private organization need to be applied by an active involvement of local people. Empowerment is proposed to strengthen cultural-spiritual society that can increase their productivity and work ethic, further welfare of agricultural society will be realized.

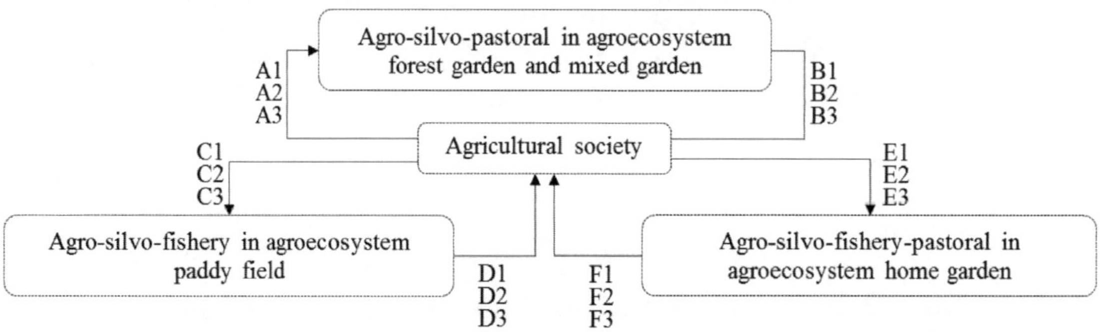

1. Sustainability of socio-ecological function: planting conservational plant species for water and land conservation as well as for biodiversity (A1, C1, E1); providing environmental services such as gen bank, fresh air and water, and fertile soil (B1, D1, F1)
2. Sustainability of socio-economic function: planting annual plant combined with perennial plant and middle husbandry (A2); planting local species of paddy combined with perennial plant and small fishery (C2); planting plant with various uses and function combined with small husbandry and fishery (E2); providing resources for basic people's need as well as economic benefits (B2, C2, E2)
3. Sustainability of spiritual-cultural function: planting local plant species based on TEK (A3, C3, E3); providing a place for appreciating their local spiritual-culture.

Figure 4. Concept of sustainable agricultural landscape management

4. Conclusion

This studyrevealed that the characteristicof *Sunda Parahiyangan* landscape is a mountainous agricultural landscape with water as the major landscape elements.This characteristic has been actualized in the various form of Sundanese agroecosystems,mainly forest garden (*kebun-talun*), mixed garden (kebun campuran), paddy field (*sawah*), and home garden (*pekarangan*) as a manifestation of local people's traditional ecological knowledge (TEK). The concept of *kabuyutan* as one of TEKs that focussed on revitalizingthe hydrological system is strongly suggested to be implemented in a broader scale. This concept is important to protect particular areas by designating as the sacred places that culturally perceived to be preseved (through application of a custumory rule, named as *pamali*). We concluded that the availability of agricultural land with fully supported by the existence of kabuyutan is essential to realize the sustainable agricultural landscape. We proposed an integrated farming system through optimally implementation of agroforestry combined with the LEISA system, as well asfully supportfrom government or private organization through attractive and participative rural development programs to achievea sustainable physical agricultural landscape with the strong society in term of socio-economic and cultural-spiritual aspect.

Acknowledgements
This research was supported by The Indonesia Endowment Fund for Education (LPDP).

References

[1] Food and Agriculture Organization 2010*The State of Food Insecurity in the World-Addressing food insecurity in protracted crises*(Rome: Food and Agriculture Organization of the United Nations) pp 54
[2] Statistics Indonesian 2011*Perkembangan beberapa indikator utama sosial-ekonomi Indonesia*
[3] Christanty L, Abdoellah O S, Marten G and Iskandar J 1986 Traditional agroforestry in West Java: the *pekarangan* (homegarden) and *kebun-talun* (perennial/annualrotation) cropping systems*Traditional Agriculture in Southeast Asia: A Human Ecology Perspective* ed Marten G G(Boulder and London:Westview Press) pp 132-156
[4] Abdoellah O S, Hadikusumah H Y, Takeuchi K, Okubo S and Perikesit 2006 Commercialization of homegardens in an Indonesian village: vegetation compotition and functional changes *Agroforest. Syst.* 68
[5] Billeter R 2008 Indicator for biodiversity in agricultural landscape: a pan-European study *Appl. Ecol.***45** pp 141-150
[6] Becker C D and Ghimire K 2003 Synergy Between Traditional Ecological Knowledge and Conservation Science Supports Forest Preservation in Ecuador*Conserv. Ecol.***8** pp 1-2
[7] Huntington H P 2000 Using traditional ecological knowledge in science: methods and applications. *Appl. Ecol.* **10** pp 1270-1274
[8] Mulyoutami E, Rismawan R and Joshi L 2009 Local knowledge and management of *simpukng* (forest garden) among the Dayak people in East Kalimantan, Indonesia*Forest Ecol. Man.* **257** p 255
[9] Rigg J 1862*A Dictionary of The Sunda Language of Java*(Batavia: Lange&Co.)
[10] Miskinis P 2011 Mathematical modeling of mountain height distribution on the Earth's surface. *Geologija***1** pp 21-26
[11] Walker, D. H., Sinclair, F. L., Joshi, L., and Ambrose, B. 1997. Prospect for The Use of Corporate Knowledge Bases in The Generation, Management, and Communication of Knowledge at A Front-Line Agriculture Research Center.*Agri. Syst.* 54: pp 291-312.
[12] Muleler J G, Assanou I H, Guimbo I D and Almedom A M 2009 Evaluating Rapid Participatory Rural Appraisal as an Assessment of Ethnoecological Knowledge and Local Biodiversity Patterns. *Conserv. Biology***1** pp 140–150
[13] Swanwcik C 2002*Landscape Character Assessment: Guidance for England and Scotland* (UK: The Countryside Agency and Schottist Natural Heritage)
[14] Global Ecovillage Nework 2008*Community Sustainability Assessment* (CSA)(USA: Ecovillage Network of the Americas) p 41
[15] United States Department of Agriculture 2007*Sustainable agriculture: definitions and terms*(United States of America: National Agricultural Library)
[16] Food and Agriculture Organization 2007SARD and Mountain Regions*Sustainable Agriculture and Rural Development (SARD)*(Policy Brief 19)(Rome: Food and Agriculture Organization of the United Nations) pp 1-4
[17] Turpin N and Oueslati W 2008 Editorial: Landscape in the Dynamic of Local Economies*Land. Research***3** pp 259-262
[18] Soemarwoto O and G R Conway 1992 The Javanese homegarden *J. Farming Syst. Research-Extension* **2** pp 95-118
[19] Finotto F 2011Landscape Assessment: The Ecological Profile*Landscape Indicators: Assessing and Monitoring Landscape Quality* ed Castella C and Peano A(Dordrecht: Springer Science Business Media B. V.)
[20] Reijntjes C, Haverkort B and Bayer A W 1992*Farming for Future: An Introduction to Low-External-Input and Sustainable Agriculture*(Netherland: ILEIA)

IOP Conf. Series: Earth and Environmental Science **91** (2017) 012017 doi:10.1088/1755-1315/91/1/012017

Analysis of Green Space Characteristic Effect to the Comfort Microclimate in the Simple Flats in Jakarta

Nenah Suminah[1], Bambang Sulistyantara[2], Tati Budiarti[2]

[1] Student Architecture Landscape Department, Bogor Agricultural University PO BOX 168, Bogor 16001, Indonesia
[2] Lecturer Architecture Landscape Department, Bogor Agricultural University PO BOX 168, Bogor 16001, Indonesia

* Corresponding author: ena_suminah@yahoo.co.id, Telp: +62 813 8127 0712

Abstract. Existence of green spaces on simple flats (Rusunawa), especially trees, greatly affect the ambient temperature outside and inside the building. Density of tree canopy can modify air temperature under trees or buildings by reducing solar radiation on building facade. The aims of this study is to analize the characteristics of green space in Rusunawa and its influence on the microclimate. The experiment was conducted at four Rusunawa building located in Jakarta, i.e. Rusunawa Jatirawasari, Tambora, Pulogebang, and Marunda Cluster A. Species and number of trees were identified, while the ground cover was identified by its species and coverage. Physical characteristics of trees were measured by tree thrunk, tree height, and width crown. Measurement of temperature, relative humidity (RH), and wind velocity was conducted at seven points in each Rusunawa to determine their effects. The results showed the presence of trees can modify the temperature of building outside similar with in the building. The decrease of temperature was influenced by green coverage area, green coverage index, and environmental condition around Simple flats. In this study, the best microclimate was modification by the ability to reduce largest temperature differences. Further more, green space in Rusunawa Marunda Cluster A was considered as the most effective one in reducing temperature.

Keywords: *simple flats, green space, temperature, tree*

1. Introduction

Jakarta as the capital city was very dynamic, infrastructure development continue to be in line with population growth. Due to limited land, development tends to be done by utilizing natural lands with ecological functions. The tendency for change the natural land as a potential green space in the city into smaller plots will continue if not controlled properly and will have an impact on the balance ecosystem of the city, and consequently the quality of the urban environment will decrease (flooding during the rainy season, UHI in the dry season, lack availability of clean water and quality of urban air) [1]. The development of Simple flats (Rusunawa) is one of Jakarta municipality's program in overcoming both high level of urbanization and land limitation issues. Simple flats that dominated by massive multi-storey buildings and pavement areas needs to be balanced by the development of green spaces.

Green space is one of the important indicators regarding quality of life to establish a sustainable landscape development in urban areas. Construction of the green space facility has begun now to meet the psychological needs and comfort for the residents. The presence of green spaces as part of the environment, not only contribute positively to the environment and aesthetic quality, but also a gathering place for residents to socialize and recreation. Green space in the outdoor is able to control and improve thermal comfort and air temperatures. For areas in the room, shade trees can reduce the radiation from the sun on the facade of the building, improve thermal comfort and saves energy spent to maintain a healthy environment [2]. Based on the above, it is necessary to analyze the influence of the temperature characteristics of green spaces in simple flats.

The objective of this study is:

1. analyze the characteristics of green space in Jakarta's Simple flats;
2. analyze the effect of green space on the comfort climate.

2. Methods

Observations in this study made in four simple flats (Rusunawa) in Jakarta. The fourth simple flats are Jatirawasari in Central Jakarta, Tambora in West Jakarta, Pulogebang in East Jakarta, and Marunda Cluster A in North Jakarta (Figure 1 (a)).

2.1 Characteristics of the green space

Green space characteristics obtained the large area of Simple flats, Basic Building Coefficient, and the physical characteristics of plants. Physical characteristics of trees were measured by tree thrunk, tree height, and width crown. Green coverage area obtained by multiplying the number of trees with an average surface area of tree canopy. Green coverage index calculated by divided green coverage area by its large area. The data obtained from each Simple flats were analyzed and presented in table.

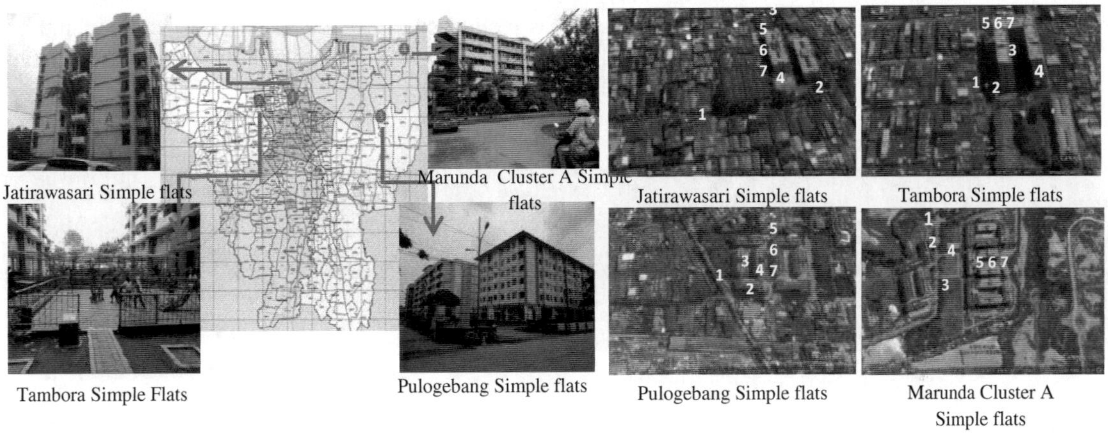

Figure 1. Site plan of experiment (a); seven observation points of site measurements of temperature, RH, and wind velocity in fourth Simple flats (1. control/outdoor; 2. below tree; 3. ground cover; 4. pavement; 5. 1st floor building; 6. 3st floor building; 7. 5st floor building)

2.2 Influence of the green space characterictic to the microclimate

Green space elements microclimate done by measuring temperature, relative humidity (RH) and wind speed three times in the morning, afternoon and evening at seven observation points: 1. outside area of Simple flats as a control; 2. under tree; 3. ground cover; 4. Pavement; 5.1st floor of the building; 6. 3rd floor of the building; 7. 5th floor of the building. The temperature on the outside area of Simple flats served as the control measurement to obtain relative temperature, i.e. the difference temperature between the control and six other observation points. Data measurement of temperature, RH and wind speed is done twice, in March and April 2016. Seven observation points of the data measurements can be seen in Figure 1 (b).

Anova and Duncan's Multiple Range Test (DMRT) did to see the difference between observation location variance. DMRT conducted to determine the best type based on its rank, the test is also used to see their influence among the treatments tested. Equipment used in the measurement of physical characteristics is gauge or meter roll to measure tree thrunk diametry and width crown, Abney level to measure the height of the plant. Measurement of climatological data is done by termohygrodigital to record the temperature and RH automatically. Anemometer used to measure wind speed. Climatological data measurement is done by placing the equipment at breast height, ± 1.1 m.

3. Results and Discussion

Simple flats in Jakarta generally already has a green space for playground, socialization, and climate amelioration. Types of trees present at front area around the fence on Tambora, Pulogebang and Marunda Cluster A Simple flats dominated by Sengon (*Albizia chinensis* (Osbeck) Merr.), whereas in Rusunawa Jatirawasari types of trees planted in the area of the fence varies from Angsana (*Pterocarpus indicus* Willd), starfruit (*Averrhoa carambola* L.), Dutch teak tree (*Guazuma ulmifolia* Lamk.), jambu air (*Syzygium aqueum* (Burm.f.) Alston), mango (*Mangifera indica* L.) and cherries (*Muntingia calabura* L.). Front area of Marunda Cluster A is also used by Farmers Group as agricultural land, planted with mango trees, eggplant and peppers, as well as the Guest House for hydroponic. Playing area and socializing in the fourth Simple flats generally located in the area between the blocks. In Pulogebang, they had Integrated Child Friendly Public located in the middle area of Simple flats. Children's playground in Jatirawasari located in the rear area of Simple flats, they also had socializing area at inter block building. Areas between blocks on Marunda Cluster A generally in the form field used for sports, playing areas/socialization. Marunda Cluster A Simple flats do not having playing ground area specifically for children. Playing Activity or socializing in Tambora is generally done in the field as well as a park area in the middle of the tower. The green space on the fourth Simple flats can be seen in Figure 2. Green space characteristics in four Rusunawa can be seen in Tabel 1.

Jatirawasari Simple flats Tambora Simple flats

Pulogebang Simple flats Marunda Cluster A Simple flats

Figure 2. Green space available in Simple flats

Comparison of the average tree thrunk, tree height and width crown or tree canopy of the four Rusunawa showed in Figure 3. Marunda Cluster A Simple flats still have many space to be planted in with trees that can be utilized to maintain the balanced ecosystem as a result of the construction of the buildings and other massive pavements such as roads and parking lots. Those flats has large area and small KDB value. Tambora has more number of trees than Pulogebang despite having a smaller area, it is because the trees in Tambora Simple flats especially the tower flats majority measured as small trees. The placement of the trees has not considered and less attention to it developments that will occur in a few years later. Although it has less number of trees than Tambora Simple flats, Pulogebang has a bigger an average tree thrunk, tree height and width crown than Rusunawa Tambora. Rusunawa Jatirawasari has the smallest number of trees because it does have the smallest area and is located in the middle of densely settlement, eventhough have trees in mature enough to have an average size of tree thrunk, tree height and width crown or tree canopies.

Table 1. Green space characteristics in fourth Simple flats

Characteristic	Jatirawasari	Tambora	Pulogebang	Marunda Cluster A
Large area	5.875 m^2	19.917 m^2	43.395 m^2	69.815 m^2
Basic Building Coefficient	50 %	31,49 %	20,94 %	20,9 %
Total number of tree	59	422	337	1156
Green coverage area	1102,96 m^2	785,64 m^2	1536,50 m^2	5402,65 m^2
Green coverage index	0,19	0,04	0,04	0,08
Dominant tree species	dutch teak tree (*Guazuma ulmifolia* Lamk.)	Orange trees (*Citrus* sp.)	Mango (*Mangifera indica* L.)	Mango (*Mangifera indica* L.)
Total area of ground cover	82,88 m^2	65,25 m^2	246,03 m^2	20 m^2
Total area of other plant	18 m^2 (productive plant)	32,85 m^2 (TOGA plant)	341,6 m^2 (productive plant)	1.440 m^2 (productive plant)

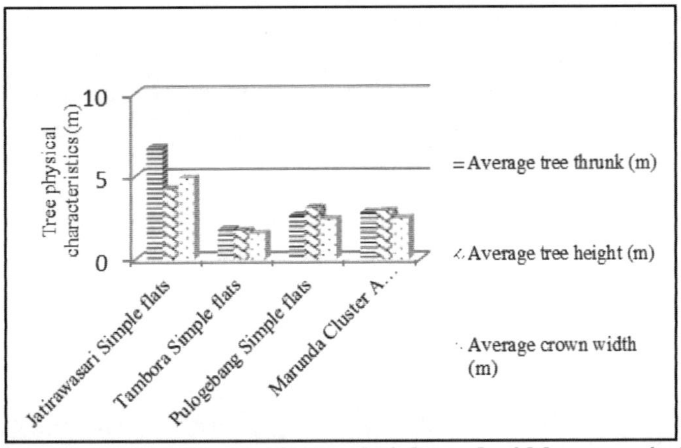

Figure 3. Comparison of average tree trunk, tree height and width crown in four locations of Simple flats

Observations of temperature, RH and wind speed on the field in twice (on March and April 2016 for each Rusunawa) shows the average temperature in Jatirawasari Simple flats 33,21°C, RH 63.48 %, and wind speed of 0.82 m/s. The average temperature in Tambora Simple flats 33,25°C, 65.14% RH, and wind speed of 0.94 m/s. Pulogebang Simple flats has an average temperature 32,66°C, 64.48% RH, and wind speed of 0.76 m/s. The average temperature in Marunda Cluster A Simple flats 34°C, RH 67.21%, and wind speed of 0.7 m/s. Figure 4 (a), (b), and (c) shows the results of measurements of temperature, RH, and the wind speed at each measurement point in four Simple flats. Comparison of the differences temperature at four Simple flats in six measurement locations to control shown in Figure 4(d). Measurement of the temperature in the outdoor space of Simple flats, trees provide the greatest temperature reduction effect to the control, with an average reducing temperature is 4.2°C. Measurement of the temperature in ground cover area provide an average reducing temperature of

IOP Conf. Series: Earth and Environmental Science **91** (2017) 012017 doi:10.1088/1755-1315/91/1/012017

2.39°C, while measurements on the pavement gives the average reducing temperature of 1.09°C. Measurement of the temperature in the area of the building gives the average reducing temperature by 4-5°C. Trees shades can improve the energy balance of human heat when doing outdoor activities, as well as improve the conditions of space in the building with reduce solar radiation on the facade of the building [3]. Shading by trees – provided by single trees, clusters of trees or urban forests – may decrease air temperature and reduce considerably the solar radiation income of the ground and other surfaces in the shade [2,4]. Shading interventions greatly reduce vulnerability to heat stree, and changed thermal sensation ratings from danger to vulnerable [5]. Apartment in massive building below tree canopy correctly manages the building's envelope, it makes indoor temperatures become cooler [6]. Trees contribute to their environment by supplying the needed oxygen, absorbing CO2, improving air quality, enhancing the microclimate, conserving water, preserving soil and filtering air pollution as well as increasing the quality of life by hosting natural elements and wildlife habitats into urban settings [3]. Urban morphology, materials, urban forests and the local microclimate must be in balance with the built environment, which promotes sustainability of the city [10].

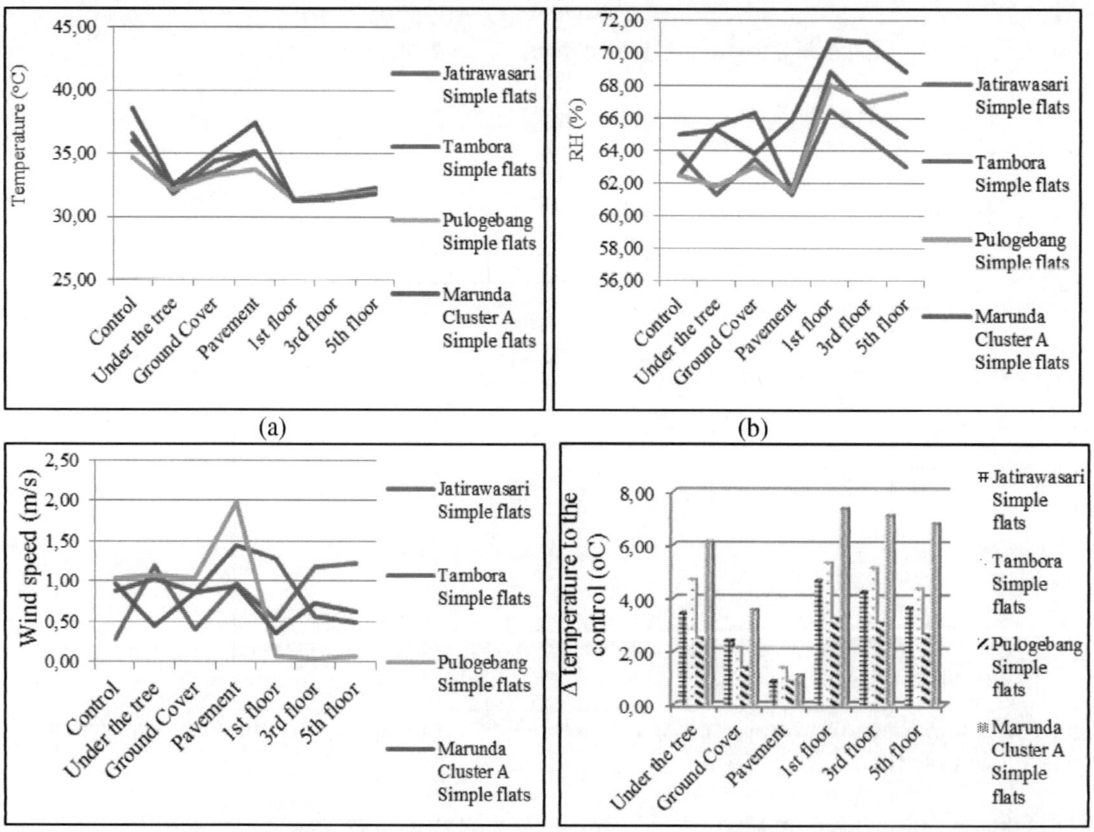

Figure 4. The average temperature (a), RH (b), and wind speed (c) at seven observation points in four locations of Simple flats; and the difference temperature in the six-point observations againts control at four locations of Simple flats (d)

In tropical climates, vegetation has significant effects to influence the microclimate. Vegetation cover can influence urban microclimates directly by shading surfaces and channeling wind, and indirectly by evapotranspiration of water [7]. Vegetation in built-up area especially in urban centres has positif impact and significantly important to help mitigate the phenomena of UHI [8]. Trees, grass and shrubs will reduce air temperatures near the house and provide evaporative cooling. The type and species of trees used is very important. They should be selected according to the amount of shade they provide as well as their aesthetic appearance. Shrubs and other low growing foliage provide shading during the morning and late afternoon when the sun is low in the sky. The evapotranspiration process can reduce surrounding air temperature [7]. In hot and humid climates, high-branching and wide canopy trees and low ground covers should be used to promote shade and encourage wind [9].

Results of statistical analysis showed that the point of observation location and group of Simple flats significantly affect the difference in temperature. Based on DMRT test of influence point of observation location to the temperature differences (Table 2), the location within the building, e.g 1st floor, 3rd floor, and 5th floor of the building, and location under the trees has the greatest influence on the temperature differences, the four locations are located within the same subset, meaning that the temperature under the tree is not significantly different from the temperature inside the building. Ground cover have a considerable influence on the temperature difference, followed by pavement. Duncan test of Simple flats group's influence to temperature difference (Table 3) shows Marunda Cluster A Simple flats provide the greatest temperature difference, followed by Tambora and Jatirawasari in the same subset, and the smallest one was Pulogebang. Based on Table 1, Marunda Cluster A has the highest number of trees, broadest green coverage, and 2nd rank green coverage index. The environment around Marunda Cluster A still has enough green space and big trees on the green street. Others environmental Simple flats dominated by dense settlements, the surrounding area is dominated by buildings and pavement. Other environmental factors that can affect the results of this research are wind direction and the direction of the measurement position, unfortunately these two things did not measure in this research. Direction and wind speed affect the range effect of vegetation on air cooling. Wind direction and its speed are very important to be considered in the placement of a tree or forest vegetation in the city. It is also determined the effectiveness of vegetation in controlling the temperature [10].

Table 2. Influence of treatment observation location to the differences temperature by DMRT test

Observation location	**Subset**		
	1	**2**	**3**
Pavement	1,0875		
Ground cover		2,3875	
Under tree			4,1975
5th floor of the building			4,3775
3rd floor of the building			4,8900
1st floor of the building			5,1500

Table 3. Influence of Simple flats to the differences temperature by DMRT test

Simple Flats	**Subset**		
	1	**2**	**3**
Pulogebang	2,3000		
Jatirawasari		3,2317	
Tambora		3,8500	
Marunda			5,3450

Green coverage area, green coverage index, and environmental conditions around Simple flats affects the average temperature difference in Simple flats. When the condition of the green space tightly, it can reducing the air temperature highly. The condition described by the density of vegetation and green coverage by tree canopy. Dense vegetation and its wider canopy cover, can reduce air temperature [11]. Selection of the tree species in the green space should also be considered in order to maximize its influence in reducing the temperature. Plants with dense canopy is more effective in reducing urban air temperatures. The size and shape of the tree crown can improve thermal comfort in a microclimate, as well as the size and shape of the leaves, tree thrunk, and the permeability of the crown [2].

4. Conclusions

Marunda Cluster A Simple flats has the largest number of trees, followed by Tambora, Pulogebang, and Jatirawasari. Marunda Cluster A Simple flats has the largest green coverage area, followed by Jatirawasari, Pulogebang, and Tambora. Pulogebang and Tambora Simple flats has the lowest green coverage index, followed by Marunda Cluster A, and Jatirawasari as the highest one.

Trees provide the greatest temperature reduction effect to the control, with an average reducing temperature is 4.2°C. Green space in Marunda Cluster A Simple flats is considered the most effective one in decreasing air temperature because it has the best green coverage area, 2nd green coverage index, and has better environmental condition than other Simple flats.

5. References

[1] Joga N and Ismaun I 2011 *RTH 30 %! Resolusi (Kota) Hijau* (Jakarta: Gramedia Pustaka Utama) p 280

[2] de Abreu-Harbich LV, Labaki LC, Matzarakis A 2015 Effect of tree planting design and tree species on human thermal comfort in the tropics *J. Landscape and Urban Planning* 138: 99-109 http://dx.doi.org/10.1016/j.landurbplan.2015.02.008

[3] Nasir RA, Sabarinah SA, Azni ZA, Norhati I 2015 Adapting human comfort in an urban area: the role of tree shades towards urban regeneration *Procedia Social and Behavioral Science* 170: 369-380, doi:10.1016/j.sbspro.2015.01.047

[4] Takács Á, Kiss M, Hof A, Tanács E, Gulyás A, and Kántor N 2016 Microclimate modification by urban shade trees – an integrated approach to aid ecosystem service based decision-making. *Procedia Environmental Sciences* 32 (2016) 97 – 109 *International Conference – Environment at a Crossroads: SMART approaches for a sustainable future* doi:10.1016/j.proenv.2016.03.015

[5] Brown RD, Vanos J, Kenny N, and Lenzholzer S 2015 Designing urban parks that ameliorate the effects of climate change *Lanscape and Urban Planning* 138 (2015): 118-131 http://dx.doi.org/10.1016/j.landurbplan.2015.02.006

[6] Balter J, Ganem C, and Discoli C 2016 On high-rise residential buildings in an oasis city: Thermal and energy assessment of different envelope materiality above and below tree canopy *Energy and Buildings* 113 (2016): 61-73 http://dx.doi.org/10.1016/j.enbuild.2015.11.011

[7] Misni, A. 2013. Modifying the outdoor temperature around single-family residences: The influence of landscaping. Procedia Social and Behavioral Sciences 105: 664 – 673, doi:10.1016/j.sbspro.2013.11.069

[8] Buyadi SNA, Mohd W M N W and Misni A 2013 Green spaces growth impact on the urban microclimate *Procedia - Social and Behavioral Sciences* 105 (2013) 547 – 557 *AicE-Bs2013London Asia Pacific International Conference on Environment-Behaviour Studies* University of Westminster, London UK 4-6 Sept 2013 "From Research to Practice" doi:10.1016/j.sbspro.2013.11.058

[9] Buyadi S N A, Mohamad W M N W, and Misni A 2013 Impact of land use changes on the surface temperature distribution of area surrounding National Botanic Gardens, Shah Alam. *Journal of ASIAN Behavioural Studies: Sustainability Science and Management* 4(11) doi:10.1016/j.sbspro.2013.07.225

[10] Tauhid 2008 Kajian Jarak Jangkau Efek Vegetasi Pohon Terhadap Suhu Udara Pada Siang Hari di Perkotaan Tesis (Semarang : Universitas Diponegoro) p 140

[11] Masruroh H 2013 Hubungan ruang terbuka hijau (RTH) dengan suhu dan kelembapan dalam kajian iklim mikro di kota malangp p 11 *Online Journal UM* http://www.jurnal-online.um.ac.id/data/artikel/artikelAAB854200FE4D0214566D2C428CC8D5D.pdf

IOP Conf. Series: Earth and Environmental Science **91** (2017) 012018 doi:10.1088/1755-1315/91/1/012018

Identification Of Minangkabau Landscape Characters

Asrina M[1], Gunawan A[2] and Munandar Aris[2]

[1]Graduate Students of Landscape Architecture, Department of Landscape Architecture, Faculty of Agriculture, Bogor Agricultural University (IPB)
[2]Department of Landscape Architecture, Faculty of Agriculture, Bogor Agricultural University (IPB)

Email: amerisaasrina@gmail.com

Abstract. Minangkabau is one of cultures in indonesia which occupies landscape intact. Landscape of Minangkabau have a very close relationship with the culture of the people. Uniqueness of Minangkabau culture and landscape forming an inseparable characterunity. The landscape is necessarily identified to know the inherent landscape characters. The objective of this study was to identify the character of the Minangkabau landscape characterizes its uniqueness. The study was conducted by using descriptive method comprised literature review and field observation. Observed the landscape characters comprised two main features, they were major and minor features. Indetification of the features was conducted in two original areas (darek) of the Minangkabau traditional society. The research results showed that major features or natural features of the landscape were predominantly landform, landcover, and hidrology. All luhak (districts) of Minangkabau showed similar main features such as hill, canyon, lake, valley, and forest. The existence of natural features such as hills, canyon and valleys characterizes the nature of minangkabau landscape. Minor features formed by Minangkabau cultural society were agricultural land and settlement. Rumah gadang (big house) is one of famous minor features characterizes the Minangkabau culture. In addition, several historical artefacts of building and others structure may strengthen uniqueness of the Minangkabau landscape character, such as The royal palace, inscription, and tunnels.

1. Introduction

Minangkabau is one of the oldest cultures in Indonesia. Minangkabau is the largest matrilineal society in the world has a strong culture [7]. Minangkabau landscape affected by the culture of society, similarly otherwise seen from natural and non-natural features [9] in the landscape. Minangkabau society see the land not only as the shelter but also as a symbol of social status. Communal land management system at this tribe, resulting in the tradition of land use can still maintained until now. This makes the Minang has an identity that is different from other areas in Indonesia [6]. Land-use regulations in Minangkabau is based on the adage "*nan rato kaparumahan, tabu tumbuah dinan lereng, kok manggu kapakuburan, nan bancah ditanami sawah*" which means that the flat land used for housing, sloping land overgrown plants, dry land used for burials and wetlands used for rice paddies [2] [6]. From this adage can be seen that the land use pattern is done by dividing the group of functions based on the condition of the land. This proverb is used on land use patterns in each nagari located in darek (the core area of Minangkabau). The uniqueness of the Minangkabau landscape in terms of the values physical nature associated with culture and history of the people causing the landscape has a strong character that indicates Minangkabau culture.

A landscape has a strong character can be caused by the force of natural features (major features) and minor features (minor feature) interacting. The landscape character shaped by landscape elements such as landforms, landcover, hidrology, settlements, and artifacts of historical / cultural [3] [5] [9]. Spatial association of these elements form a very strong character. With the value of the uniqueness Minangkabau landscape's be interesting to study.

This study was conducted to identify and analyze the Minangkabau landscape character by looking at the major features and minor features result from the people and nature interaction .

2. Research Methods

2.1. Study area

The study were made in the province of West Sumatra where the majority (approximately 95%) the population of West Sumatra is Minangkabau ethnic. Geertz (1981) explained that The central area minangkabau culture is a valley surrounding Mount Merapi, near the town of Bukittinggi. Original areas of the Minangkabau culture called darek territory (land) which this region has a high level of originality, is different from the seacoast region which is a region division with levels of acculturation from outside is very high. Darek region is divided into three luhak now called District, including Tanah Datar, 50 koto, and Agam (Fig 1). Three luhak are agricultural areas with high enough population density and teachings tradition held tight enough. the Site chosen because of the diversity of the unique landscape, semi-natural, cultural features and shape changes that occurred in the last few years. The topography in site moderately change between 0 and 2,891 meters above sea level with Mount Merapi as the highest point.

2.2. Method of the study

The method used in this research is descriptive qualitative of the landscape character approach proposed by Starke and Simonds [9]. The study was conducted through two main stages, by using descriptive method comprised literature review and field observasion.

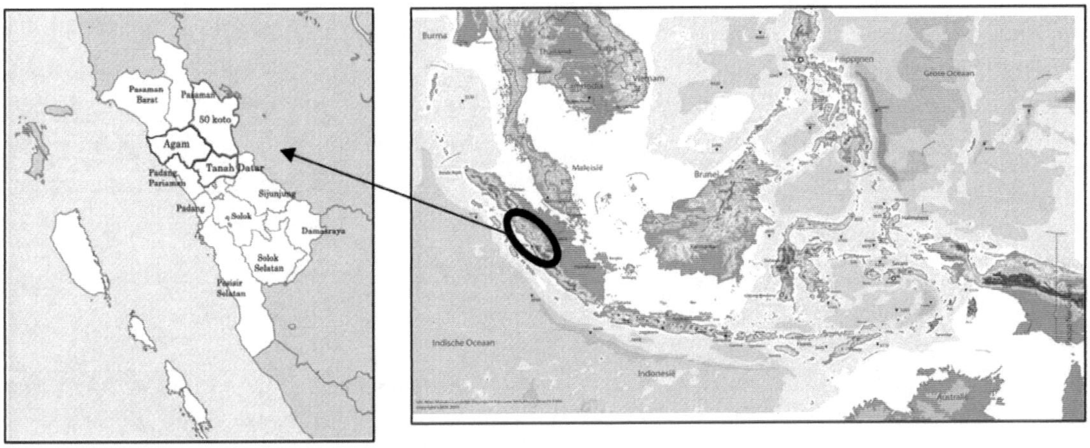

Figure 1. Location of research area

The study of literature is done by collecting a variety of information Minangkabau landscape character through a variety of literature and official documents of the local government, in this case the Regional Development Planning Agency (Bappeda). Field studies conducted with a visit to the Minangkabau landscape. Photographing various landscape character made during the field observation. The result of literature review and field observasion an analysis with Landscape Character approach Starke and Simonds [9], the character is formed through the features of nature and features of non-natural (artificial), or a combination of both. In detail, these features can be seen in Table 1 below.

Table 1. Nature and Man-Made Features

No	Features	Sub-Features	Elements
1	Nature Feature	Topography	mountain, hill, valley, canyon
		Land cover	forest,
		Hidrologi (water body)	lake, river, sea
2	Man-made Feature	Agriculture	paddy fields, farm, plantation
		Traditional Building	home, settlement
		Artifacts	monuments, caves, tunnel

3. Result and Discussion

Minangkabau is one Indonesia culture is located in West Sumatra Province. Minangkabau world is divided into two regions are Pasisia territory (coastal) or seacoast and darek territory (land), which consists of three luhak Tanah Datar, Agam, and 50 Koto. Minangkabau Landscape has a complete feature in the third luhak (districts) as formulated by Starke and Simonds (2013), includes natural features and man-made features. Overall the features, constructing landscape character that provides cultural identity of Minangkabau. The results of the study literature and field observations are presented in Tables 2 and 3.

3.1. Major Feature

Minangkabau landscape is dominated by the major features. Compilation of the results study literature and field observation is shown in Table 2. The main features of the Minangkabau

IOP Conf. Series: Earth and Environmental Science **91** (2017) 012018 doi:10.1088/1755-1315/91/1/012018

landscape based on topography, land cover and hydrology. These features are based on topography includes mountains, hills, valleys and canyons. Mountain features has in the three (districts), includes Mount Marapi, Mount Singgalang, Mount Sago, Mount Bungsu, dan Mount Sanggul. Mountain Features were the most prominent elements as they giving the very strong landscape character identity [9], especially the Minangkabau landscape character.

Table 2. Major Features Three Luhak (district) Minangkabau

Major Feature		Luhak (Districts)			Informations
		Tanah Datar	Agam	50 Koto	
Topography	Mountain	✓	✓	✓	Mount Marapi, Mount Singgalang, Mount Sago (Tanah Datar, Agam) Mount Sago, Mount Bungsu, Mount Sanggul (50 koto)
	Hill	✓	✓	✓	Aua Hill, Batu patah Hill (Tanah Datar), Bulek Hill (50 koto)
	Valley	✓	✓	✓	Anai Valley (Tanah Datar), Permai valley (Agam), Harau Valley (50 koto)
	Canyon		✓		Sihanok Canyon (Agam)
Land Cover	Forest	✓	✓	✓	Tropical Forest
Hydrology	Lake	✓	✓		Singkarak Lake (Tanah Datar), Maninjau Lake(Agam)
	River	✓	✓	✓	Batang ombilin, Batang anai,Batang sinamar, Batang selo, Batang sumpur (Tanah Datar), Batang antokan, Batang katutulan, Batang agam(Agam), Batang sinama, Batang lampasi (50 Koto)
	Sea		✓		Tiku (Agam)

a. Tanah Datar District

Luhak Tanah Datar that is now called Tanah Datar district known as "LUHAK Nan Tuo" is one area that is located in the middle of West Sumatra Province. This district is located in a basin of mountains and hills that circle with a height ranging from 100-2891 meters above sea level.The position of tanah datar district is located between 3 pieces of the mountain (mount Merapi, Mount Singgalang dan Munt Sago) and two hills (Aua hill and Batu Patah Hill) so that the topography varies from flat, undulating, hilly to mountainous. Mountains, hills and flat landform in the middle of the natural features that characterize the landscape Luhak Tanah Datar [8].

The natural landscape elements is a strong supporting characters in tanah datar district are the water element in the form of lakes, rivers, and waterfalls. The famous lake in this luhak is Singkarak Lake with an area of Approximately 1,000 hectares (Fig 2a). The famous river in this area is Anai river where there is a very beautiful valley with waterfalls, is Anai valley. Lakes, rivers, and waterfalls displaying a very high aesthetic quality through the support element in the form of forest vegetation is still covered in full by the standing trees.

IOP Conf. Series: Earth and Environmental Science **91** (2017) 012018 doi:10.1088/1755-1315/91/1/012018

Tanah Datar district landscape character does not reflect the flatness of the landscape (flat landform = level), because overall very varied landform. However, in the middle of this luhak there is a relatively flat area that is currently occupied as residential areas or urban. Flat region surrounded by hills and mountains and is equipped with a water element in the form of lakes and rivers characterize the landscape displays the typical visual landscape [5].

b. Agam District

Luhak Agam character very varied landscape that stretches from the coast to the mountains (0-2877 meters above sea level). The elements of the landscape which are natural features include mountains, hills, valleys, canyons, lakes, rivers, forests, and beaches. Mount Singgalang and Mount Merapi is a mountain elements in luhak that are also contained in luhak Tanah Datar, because the location of both the mountain located on the border luhak Agam and luhak Tanah Datar. Maninjau lake is a landscape element of water is very famous, and the lake area is 9,950 hectares. The lake is located at an altitude of about 500 meters and is surrounded by hills. A very dense forest located in the hills surrounding the Maninjau lake and in the two mountains.

In this luhak also include valleys, is the permai valley. The valley is not as big and famous as Anai valley on luhak Tanah Datar. Natural landscape elements feature very famous both within and oversea is Sianok Canyon. The canyon has a similar character to the Grand Canyon in the United States, therefore, known as the Grand Canyon of Minangkabau (Figure 2b). Several rivers strengthen this luhak character, the Antokan river, Katutulan River, and the Agam River. Natural elements that do not exist in other luhak is the beach. Up to 43 km long beach and overlooking the Indian Ocean. Landscape character luhak Agam very strong is Mount Merapi, Maninjau lake, Sianok Canyon, and the Tiku beach. Most of these natural features already become a Minangkabau world landmark.

c. 50 koto District

Luhak 50 koto is the Easternmost district in the province of West Sumatra, the which is the main gateway belt of landline. the topography 50 koto district of varies between flat, undulating, hilly and mountainous (120-2271 meters above sea level). the strong Natural features in this luhak is topography (landform). Mount Sago, Mount Bungsu and Mount Sanggul is a naturally occurring element that visually looks very strong. Geological structure, this area belongs to the Payakumbuh basin area so there are many valleys and rivers. One element of the landscape in the form of a famous valley in lareh 50 koto is Harau Valley (Fig 2c). This valley covers an area of 300 hectares. Status area of the valley is a nature reserve. Good soil on the floor of valley lay the fields, into an exotic panorama region. The cliffs of the valley almost upright shapes and colorful moss due to rain and running water from the top of the cliff. The Wall of Harau Valley long stretches, interspersed with a few waterfalls. Elements of the river that is known at this lareh is Sinama river and Lampasi river. Features of land cover in the form of the element was dominant forest cover the hills and mountains row. The elements that are not owned by luhak are canyon, lakes, beaches.

(a) Singkarak Lake (a) Sianok Canyon (b) Harau Valley

Figure 2. Main Features Minangkabau Landscape Characters

3.2. Minor Feature

Besides major features, minor features in minagkabau landscape also display a strong character. Minor features is a form of land use by humans, which a landscape is modified for human needs as well as to enhance the beauty of an area (Starke and Simonds 2013). Some minor feature that shows the character of Minangkabau landscape are agriculture (paddy field), settlement (traditional house), and artifacts in the form of buildings, tunnels, and inscriptions (Table 3).

Table 3. Minor Features Three Luhak (district) Minangkabau

Minor Features		Luhak (district)			Informations
		Tanah Datar	Agam	50 Koto	
1. Agriculture	Paddy field	✓	✓	✓	Tanah Datar, Agam, 50 Koto
2. Settlement	traditional house	✓	✓	✓	Rumah Gadang
3. Artefact	buildings	✓	✓		Jam Gadang, Pagaruyung palace
	tunnels		✓		Japanese Hole
	inscriptions	✓	✓		Adityawarman Inscription, Menhir

Paddy fields location in the three luhak always close to housing, because the most of Minangkabau people are farmers. The reasons paddy fields location close to the settlement because of need to control the form of a permanent water supply and management fields. Land use patterns in Minangkabau governed by Nagari [6]. Nagari is composed of a central housing, agricultural land and forest areas are located on the outside. This makes the paddy fields landscape character in Minangkabau is quite unique because it follows the arrangement pattern of Nagari. The complex Minangkabau settlement shaped is Nagari. Nagari is a collection of several villages, while the village is a compilation of some of Hamlet. Minangkabau Settlements composed by several elements among which the Rumah Gadang (big house). Rumah gadang is a Minangkabau traditional home. This house the residence of Minang community are located in nagari.

Physically, Rumah Gadang has a roof gonjong like a horn of buffalo tapering upward. But of the three lareh there are different characters. Differences of rumah gadang seen in the form of arrangement,in tanah datar district has a wings on the left and right a building, while in the agam and 50 koto district a smaller because it does not wear wings on either end of the building, or do not have a raised floor on both ends of the building (Figure 3).

Luhak Tanah Datar Luhak Agam Luhak 50 Koto

Figure 3. Minor features : Minangkabau traditional house

Minangkabau region a long time ago is a kingdom centered in pagaruyung. So many historical relics left behind at that time, one of them are Pagaruyung Palace, tombs of kings and some inscriptions relics of the king Adityawarman. There are other historical relics in the Japanese colonial period that also characterizes the Minangkabau region, Jam gadang and Japanese holes is located in luhak Agam. Jam gadang looming in the Agam district become an icon for the Minangkabau region beside that it has the unique shape with roof gonjong in the top hours.

Minangkabau natural landscape features form the character of a very "distinct" from the viewpoint of class variety of ratings Visual Resource Management. This is because the landform features a very diverse and their elements of vegetation and water [1], such as the Moun Marapi, Anai Valley and Harau valley, Sianok Canyon, as well as Singkarak lake and Maninjau lake. In addition, some elements of the man-made features to support very strongly the formation of Minangkabau landscape character as the rumah gadang and Jam gadang (gadang hours), even these elements become the icon of the Minangkabau culture. Other elements both natural features and man-made features is unique, but not strongly shape the character of the landscape.

4. Conclusion

Minangkabau landscape character is shown by the existences of natureand man-made features strongly interacting and making an iconof the landscape. Dominant the naturefeatures of the landscape consist of landforms and hydrology with the elements such as Mt. Marapi, Lembah Anai, Harau Valley, Canyon Sianok, Singkarak lake, and the lake Maninjau. The man-made features that make the icon and character of the cultural landscape of Minangkabau are the Jam Gadang, Rumah Gadang, and Pagaruyung Palace.

References

[1] [USDA] 1973. *National Forest Landscape Management: Volume 2, Chapter 1, The Visual Management System* USDA Forest Service, USAAtik, M., Is¸ ıklı, R.C., Ortaçes¸ me, V., Yıldırım, E., 2015. Definition of landscape character areas and types in Side region, Antalya-Turkey with regard to land use planning. Land Use Policy 44, 90e100.

[2] Basri H 2003 *Kitab Undang-Undang Hukum Adat Minangkabau* KAUHAD

[3] Brown G and Brabyn L 2012 An analysis of the relationships between multiple values and physical landscapes at a regional scale using public participation GIS and landscape character classification *Landsc. Urban Plan.* 107 (3), pp 317-331.

[4] Geertz H 1981 *Aneka Budaya dan Komunitas di Indonesia* (Depok: Yayasan Ilmu-Ilmu Sosial & FIS-UI)

[5] Jellema A, Stobbelaar D J, Groot J C J, Rossing W A H 2009 Landscape character assessment using region growing techniques in geographical information systems. *J. Environ. Manag* 90, pp 161-174.

[6] Kosmaryandi N 2005 Kajian Penggunaan Lahan Tradisional Minangkabau Berdasarkan Kondisi Tanahnya *Media Konservasi Vol. X, No. 2:* pp 77 – 81

[7] Muhamed K 1978 *Migration and the Matrillineal System of Negeri Sembilan, Malaysia* [Unpublished Ph.D Thesis] (United States: University of Pittsburgh)

[8] Nugraha H, Wacano D, Dipayana AG, Cahyadi A, Mutaqin BW, Larasati A 2015 Geomorphometric characteristics of landslides in the Tinalah Watershed, Menoreh Mountains, Yogyakarta, Indonesia *Procedia Environmental Sciences 28* pp 578 – 586

[9] Starke BW and JO Simonds 2013 *Landscape Architecture* (New York: Mc Graw-Hill Book Co., Inc)

IOP Conf. Series: Earth and Environmental Science **91** (2017) 012019 doi:10.1088/1755-1315/91/1/012019

Local Knowledge About The Structure, Function And Conversion Of Landscape In The Karangwangi Village, Cianjur, West Java, Indonesia

Fatiya Ulfa Dwi Amelia[1, 2] and Johan Iskandar[3, 4]

[1] Post Graduate and Environmental Study, Universitas Padjadjaran, West Java, Indonesia
[2]Jalan Sekeloa Selatan I Bandung 40134, West Java, Indonesia.
[3]Depertment of Biology, Faculty of Mathematics and Natural Science and Postgraduate of Environmental Study (PSMIL & DIL) and Institute of Ecology PPSDAL, Universitas Padjadjaran, Indonesia
[4] Jl. Raya Bandung-Sumedang Km 21, Jatinangor, Sumedang 45363, West Java, Indonesia.

Email : fatiyaamelia@gmail.com

Abstract. Karangwangi people is one of indigenous people in West Java who has local knowledge about their nature thoroughly. They have local tradition about landscape ecosystem arrangement that based on *sakral* (sacred) norm. With this rule, local people will always try to preserve the sustainability of their natural environment. However, modernization, increasing population, decreasing forest, and increasing market economic penetration, causing this rule and structure of landscape in Karangwangi village has changed. Land conversion in Karangwangi was occur because of settlement and land investment by people outside the village. These behavior changes in tradition and landscape (structure, function and conversion) in Karangwangi may impact on their daily activities, and so do the changes in daily activities can change their behavior in tradition and landscape. This research was undertaken in the Village of Karangwangi, Sub-district of Cidaun, District of Cianjur, Province of West Java, Indonesia. This paper aims to identify how indigenous people in Karangwangi understand kinds of landscape and another conversion that was happen as a result of management. The method used in this paper is qualitative with ethno ecological approach. The resulted of the study show that local people in Karangwangi Village understand how chronological of landscape structure, function and conversion.

1. Introduction

Indonesia is one of country that has various indigenous people with various special characteristics. Every indigenous people has different way of social life, economic, culture and all of it is depend on the ecosystem where they live (Waluyo 2008). This dependence creates an interaction system between social and biophysics (Marten 2001). Naturally, human always take the benefit out of nature for granted, especially things for construction materials, water, food, medicine, and handicraft that they are sourced from nature (Sastrapranaja *et al.* 1979). This causes the social life, culture, economic, technology and politic to grow and develop in every ethnic that eventually influenced their perspective to manage the environment. Such knowledge discover the value of culture ideas in social system (Suparlan 1980; Ramdhan *et al.* 2015). Interaction between indigenous people and their environment is part of ethno ecology.

Karangwangi people is one of indigenous people in West Java who has local knowledge about their nature thoroughly. They are using Bahasa Sunda as local language in their daily activities and has many local traditions like such traditional ceremony for planting and harvesting of paddy (Iskandar et al. 2016). One of their local tradition in Karangwangi is landscape ecosystem arrangement that based on *sakral* (sacred) norm. With this rule, local people will always try to preserve the sustainability of their natural environment. However, modernization, increasing population, decreasing forest, and increasing market economic penetration, causing this rule and structure of landscape in Karangwangi village has changed. Land conversion in Karangwangi was occur because of settlement and land investment by people outside the village. These behavior changes in tradition and landscape (structure,

IOP Conf. Series: Earth and Environmental Science **91** (2017) 012019 doi:10.1088/1755-1315/91/1/012019

function and conversion) in Karangwangi may impact on their daily activities, and so do the changes in daily activities can change their behavior in tradition and landscape. This paper aims to identify how indigenous people in Karangwangi understand kinds of landscape and another conversion that was happen as a result of management by using ethno ecology approach.

2. Materials and Methods

This research was undertaken in the Village of Karangwangi (*desa*), Sub-district (*kecamatan*) of Cidaun, District (*kabupaten*) of Cianjur, Province (*provinsi*) of West Java, Indonesia. Geographically, this village is located at approximately latitude 7°25'- 7°30' S and longitude 107°23'-107°25' E. Administratively, Karangwangi Village was expanded by Cidamar Village which area about 2.300,17 hectares. Topographically, Karangwangi Village consist of moderate to very steep slope hills, and at altitude between 200-275 m above sea level (asl). This village is passed by two rivers, they are Cikawung River and Cisela River. The Karangwangi Village is west bordering to Cidamar Village, north to Cimargang Village, east to Garut District and south by the Indonesian Ocean (Figure 1). In 2015, there are 5.587 people inhabited in this village with 1.817 households (Iskandar and Iskandar 2016).

The method used in this paper is qualitative with ethno ecological approach. The writer want to understand the local knowledge about various type of landscape such as structure, function and its changes within people in Karangwangi Village until now. The collecting data technique in the field was done by conducting in-depth interview with local people and literature study. The data was analyzed in descriptive.

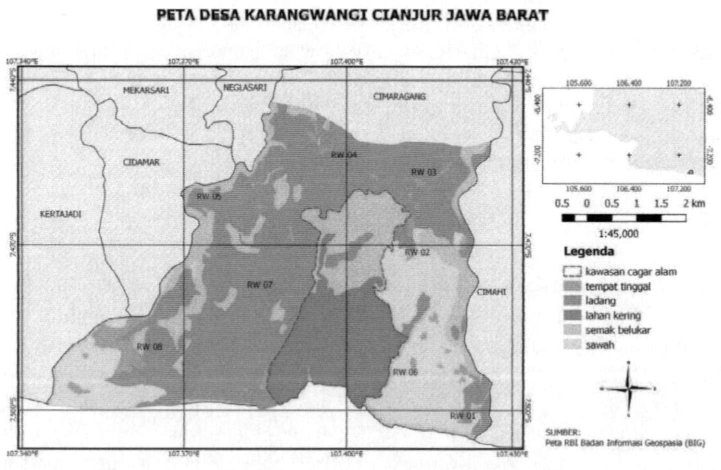

Figure 1. Research Location, Karangwang Village, Cianjur, West Java, Indonesia
Source : Iskandar and Iskandar 2016

3. Results and Discussions

Modernization, population growth, decreasing forest, and increasing market economic penetration, were phenomenon that occur in Karangwangi, which impact in the changes of structure and landscape function from time to time. Based on in-depth interview which was conducted together with local people in Karangwangi, landscape dynamic was known starts from forest (*leuweung*), swidden field (*huma*), garden (*kebon*), rice field (*sawah*), and homegarden (*pekarangan*).

3.1 Structural and functional shifting from forest landscape (leuweung) into swidden field (huma)

In the ancient time, West Java area was dominated by forest. The people is rely on it to fulfill their necessity. One of the methods to harness the forest is with farming. Local people called the forest with *leuweung*. In 1950s and 1960s, Karangwangi people was usually plant paddy in the forest by nomad. It is because the forest condition was still huge and widespread and the population was still low. This method was called *ngahuma*. Each household, was permitted to open the forest as *huma* every year within 0, 5 – 1 ha areas.

The secondary forest which was called *jami* is a *huma* which had just left behind or fallowed less than one year and still had paddy residual straw on it. *Jami* which was fallowed more than one year will become secondary forest and be called as *reuma*. The secondary forest which fulfill by bushes was called *rungkun*, *ruyuk* or *dungus* and if the field has 'old enough', which was characterized by underbrush, can be used as *huma* once again.

Usually *huma* was planted with variety of local paddy (*Oryza sativa* L), such as *pare sintung* and *pare jampang*. The paddy which was harvested from *huma* was not allowed to be sold into the market and must be stored at paddy barn (*leuit*) in bundles (*gedengan pare*). Besides paddy, there were poly culture system in *huma* such as *suuk* (*Arachis hypogaea* L), *sampeu* (*Manihot esculenta* Crantz), *cengek* (*Capsicum frutescens* L), *wijen* (*Sesamum indicum* L), *hiris* (*Cajanus cajan* (L) Huth), *turubuk* (*Saccharum edule* Hassk), and *hajeli* (*Coix lacryma-jobi* L) plants. These kind of plants depend on the farmers' diligence, the more diligent the farmer, the more diverse the *huma* would become. Unlike the paddy, these plants may allowed to be sold into the market.

However, population growth and modernization in Karangwangi Village are causing decreasing forest area. Eventually it also the reason of the decreasing of *huma* system in Karangwangi. Nowadays, *huma* system can only be found in special places such as *tegalan* field or *kebon kai*. Paddy barns was no longer used by people, because farmers are not harvesting their paddy in a bunch anymore, they are collecting the paddy in sacks. Paddy seed which is planted in the field are superior seeds with 100 days age. The planting time usually occur on October or December, and harvested on February and March. After paddy was planted, *huma* field usually was planted again with soya bean (*Glycine max* (L) Merill), *kacang susuk*, etc. Nowadays, rice field system is more advance in Karangwangi Village than *huma* system.

3.2 Structural and functional shifting from swidden field landscape (huma) into rice field (sawah)

The presence of rice field (*sawah*) system was developed a little late for Karangwangi people compared to swidden field (*huma*) system which was firstly applied. The change of *huma* system to *sawah* system applied by repaired and build an irrigation system for their farm. In the beginning, Karangwangi people who switch to *sawah* system was few in number. Based on information from interviewee, sawah system even seldom found in the high income farmer. It was because expense to pay the laborer when it needed to made new field with *sawah* system is too expensive. There is no significant difference between management of *sawah* system and *huma* system because the farmer used the same organic farming system by using various kinds of local paddy and fertilizers from livestock manure and organic waste. The distant of planting and harvesting was still the same, once a year. And the distant between harvesting and planting was use to aquaculture and cultivating *genjer* (*Limnocharis flava* (L.) Buchenau) or just fallow it.

The significant change was happen to Karangwangi farmer in early 1970s when the green revolution and the five farming programs (*panca usaha tani*) were introduced. The impact of these programs was 'similarity' in farming industry. The *panca usaha tani* program are involved in the rice intensification such as : (1) use of modern varieties paddy, such as IR, PB, and other genetically engineered seeds from laboratory; (2) use of chemical fertilizer, like urea, TSP, etc.; (3) improvement in land preparation (using pesticide); (4) improvement in irrigation; and (5) improvement in crop maintenance. Another significant change of this program can be seen in the method of farmer in Karangwangi Village. Usually they were using local paddy seeds for their rice filed, then it was changed into modern varieties seed. Moreover, after green revolution, pesticide often found in the farming process by Karangwangi people more often. The planting cycles was changed as well, where originally planting was conducted once a year, now it has become multiple time in a year, depend on *sawah* location and water supply. Originally in *rendeng* season (planting season) was occur on

November or December and harvesting season occur within the next three months. The second planting cycle was done on March or April and harvesting on August. This planting will continue only if the *sawah* location is near to the water sources. Meanwhile for sawah which located far from water source and was not obtain enough irrigation, farmers will use the field for planting *palawija* like maize or fallow it until the next season

3.3 Structural and functional shifting from swidden field (huma) landscape into garden (kebon)

In 1990s, varieties of local paddy was rarely planted in *huma* because of the long harvesting time which was around 4-5 months. So that, local people was changed with new varieties of paddy which harvesting after 100 days planted. For increasing the income, at the same time people was planted beans and chili. The *huma* was left behind will be open field which it is called *tegal* or *tegalan*. If rainy season come, *tegal* will planted with various seasonal crops which it is called *kebon*.

In 2000s, Department of Forestry was issued a policy such reforestation in former field of huma. Some of timber plants were introduced by Department of Forestry such as albizia/*jengjen* (*Paraserianthes falcataria* (L) I Nielsen), *jabon* (*Anthocephalos candaba*), and mahogany (*Switenia mahagoni* (L) Jacq). Then, it made people often to plant albizia in tegalan which was usually called with *kebon albasiah/jengjeng*. Similarly in the field which was planted with bamboo (awi) it was called *kebon awi*. And then, *kebon* kai which was planted by sugar palm (*Arenga pinnata* (Wumb) Merr), coconut (*Cocos nucifera* L), peuteuy (*Parkia speciosa* Hassk) and mangos (*Mangifera indica* L). In the past, the function of field area with various timber plants and seasonal plant, fruits which were used for full fill the long term necessity it was called *talun*. But, this name was not familiar for young generation and they rather know *kebon kai, kebon awi, kebon albasiah*.

Nowadays, majority status of *tegalan* or *kebon* in Karangwangi have turned into land owned due to implementation of fallow certification by Karangwangi Village policy in 2013. But, there are still some fallow field that has not been certified which are far from settlements, it called *taneuh bodas* or *taneuh pangangonan*. Based on history, that field was buffalo grazing land (*pangangonan*), while in 1985/1986 buffalo was not be cared anymore by local people because there are plow machine as the substitute. Now, when rainy season in November, these fallow land more planted with seasonal crops like *suuk* (*Arachis hypogaea* L), maize (*Zea mays* L), *bonteng* (*Cucumis sativus* L), *kacang panjang* (*Vigna cylindrica* (L) Skeel), *waluh* (*Cucurbita moschata* (Duch) Poir), and sesame (*Sesamum orientale* L). Whereas for second times crops (*morekat*), these *kebon* were planted by *kacang hejo* (*Vigna radiata*), *suuk* and sesame. The fertilizer was used for *kebon* is inorganic fertilizer are urea, NPK (nitrogen, phosphor, kalium) and pesticide to eradicate pests. Furthermore, *kebon kai* was planted albizia timber in monoculture and *tumpangsari* with mahogany, or mixed with seasonal crops like cardamom (*Elettaria cardmommum* Maton) and beans. For indigenous people in Karangwangi Village, albizia has economic value which can be sold and used for fulfilling daily needed such as construction material, *kosen*, table material, chair, etc.

3.4 Structural and functional shifting from garden (kebon) landscape into home garden (buruan)

Karangwangis' villager refer the land across their house which they are using to plant *palawija* as *buruan*. Inside *buruan*, we can find various plants such as *peuteuy selong* (*Leucaena leucocephala*), *peuteuy* (*Parkia speciosa* Hassk), guava (*Psidium guajava* L), sapodilla (*Manilkara achras* (Mill) Fosberg), mangos (*Manggifera indica* L), citrus (*Citrus aurantifolia* (Christin and Panz), *jeruk purut* (*Citrus hysterix* Dc), pineapple (*Ananas comosus* (L) Merr), coconut (*Cocos nucifera* L), cassava (*Manihot esculenta* Crantz), *surawung* (*Ocimum basilicum* L), cardamon (*Elettaria cardmommum* Maton), galangal (*Languas galanga*), ginger (*Zingiber officinale* Roxb), *cengek* (*Capsicum frutescens* L), turmeric (*Curcuma domestica* Val), dan lemongrass (*Cymbopogon citratus*). *Buruan* also known for its high in plants variation (cultivar). For example at least there are 12 variations of banana that can be found in *buruan* such as *cau kapas, cau kosta, cau nangka, cau raja bulu, cau saba/jimluk/jibeuh, cau ambon, cau angleng, cau galek, cau raja cere, cau beureum/gember, cau kapok* and *cau muli*. Not only bananas, the coconut also has its own variations including *kalapa puyuh, kalapa gading, kalapa beureum/merah,* dan *kalapa hejo*. Home garden system in Karangwangi is not just about plants, it's also includes variety of livestock such as chicken and lamb. Chickens that lives in home garden also varies in Karangwangi, we can find several type of chicken such as *ayam lisung, ayam tukung, ayam kate* dan *ayam aduan*. Karangwangis' villager used to domesticate those animals inside cages near their houses. However, sometimes chickens and lamb are set free by its owner to find its own food.

Karangwangis' local people is never using pesticide for plants on their home garden. They choose animal waste as their source for fertilizer for their plants. This is because home garden system doesn't require much attention as intensive as *sawah* system. Moreover, it is because home garden system is barely attack by pests. Karangwangis' villager gets a huge benefits from their home garden, such as from fruits, spices, herbal medicine, plants for traditional ceremonies, household materials, and firewood that they are planted. Fruits that can be found in home garden system are bananas and sapodilla. Clove, *cikur*, ginger, citrus, *surawung,* galangal and lemon grass are spices that usually can be found in home garden system. We also can find several herbal medicine from this place, for example kibeling (*Sericocalyx crispus* (L) Bremek), ginger (*Zingiber officinale* Roscoe), galangal (*Alpina galanga* (L) Willd), panglay (*Zingiber cassumunar* Roxb) and citrus (*Citrus aurantifolia* (Christin and Panz). Some plants from home garden are also used for ritual or traditional ceremony such as roses (*Rosa hybrida* Hort), betel (*Piper betel* L), areca nut (*Areca catechu* L), coconut (*Cocos nucifera* L) and banana (*Musa paradisiaca* L).

There is no specific time for harvesting those fruits or plants in home garden-system. It is because all of those plants have different flowering and different season to grow its flower and fruits. Some plants such as *lalab,* spices and some fruits such as papaya can be found every time whole year.

3.5 Discussions

Culture and landscape are always closely related, it is because every changes in landscape is happened because there is changes in culture of mankind. That mankind has a significant role in shaping pattern of nature and landscape process (Dewi 2009). Landscape changes also occur in Karangwangi Village, some of the reasons are modernization, population increase, rapid economic growth, and also the forest which become lesser every day. Karangwangis' Villager are able to identify and differentiate types of landscape. The ability to differentiate types of landscape is based on several indicators which is based on function, characteristic and also ownership. Moreover Karangwangis' Villager are also capable to explain the chronological order of landscape shifting from forest landscape into *huma* (partial), *huma* landscape into rice field landscape, *huma* landscape into garden landscape and garden landscape into home garden landscape (Figure 2).

Landscape shifting from forest into *huma* had happened since ancient time up until 18th century, especially in West Java. People in South Banten and South Priangan area chose to become farmer/*peladang* (*ngahuma*) for living (Terra 1958; Iskandar 2009). Such shifting also occurred in Karangwangis' Village because of its people found benefits from the forest. By having farm in the forest they got natural nutrient from the forest to grow their crops. *Huma* system is considered as a better system for ecology because it integrated into general structure of the forest. If it can't adapt so it will return (Geertz 1963). We can see it also happened in transformation from *huma* into *jami*, and becomes *reuma* and transform into *rungkun/ruyuk/dungus* and eventually that's all transform into *huma* again.

According to Pranowo (1985) there are two factors which affecting the amount of time to process *huma*: type of the forest and population. Primary and secondary forest have different amount of time to be processed into *huma*. Primary forest takes 2-4 years before it can be used as *huma*. Meanwhile, secondary forest takes less than 2 years. This process is caused by different amount contents of natural fertilizer in side those land forest. Primary forest contains more natural fertilizer than the one found in secondary fertilizer. Population density also affects the amount of time to process the land. The more population an area has, the more time it takes to process the land, and vice versa. It happens because the opportunity to open a new forest is becoming small. Eventually it affects the amount of *huma* in Karangwangi's Village and later transformed into rice field.

If we compared rice field with *huma*, rice field produces more rice because its' high frequency in planting and harvesting throughout the year (up to 3 times a year). Rice field is well known to be stable and durable, it can produce the amount of rice that never decrease in terms of harvesting (Gertz 1963). Besides the transformation from landscape into rice field/*sawah*, there is also cultural transformation/ shifts in Karangwangi. People in Karangwangi usually held a ritual every planting season in order to get the blessing. There are some items that people use for offering (*sesuguhan*) in that ritual such as green coconut (*Cocos nucifera* L), basil sleaf (*Ocimum bacilicum* L), and tobacco (*Nicotiana tabacum* L). The *huma* ceremony usually in *pupuhan*, a *sakral* place (sacred) which is bordered with *hanjuang* (*Cordyline fruricosa* (L) A.Chev.), a place for paddy planting (*ngaseuk*). The day and direction for *ngaseuk* start from *pupuhunan* direction that usually determined by calculation of *naptu tempat* and *naptu hari* which was considered bring lucky.

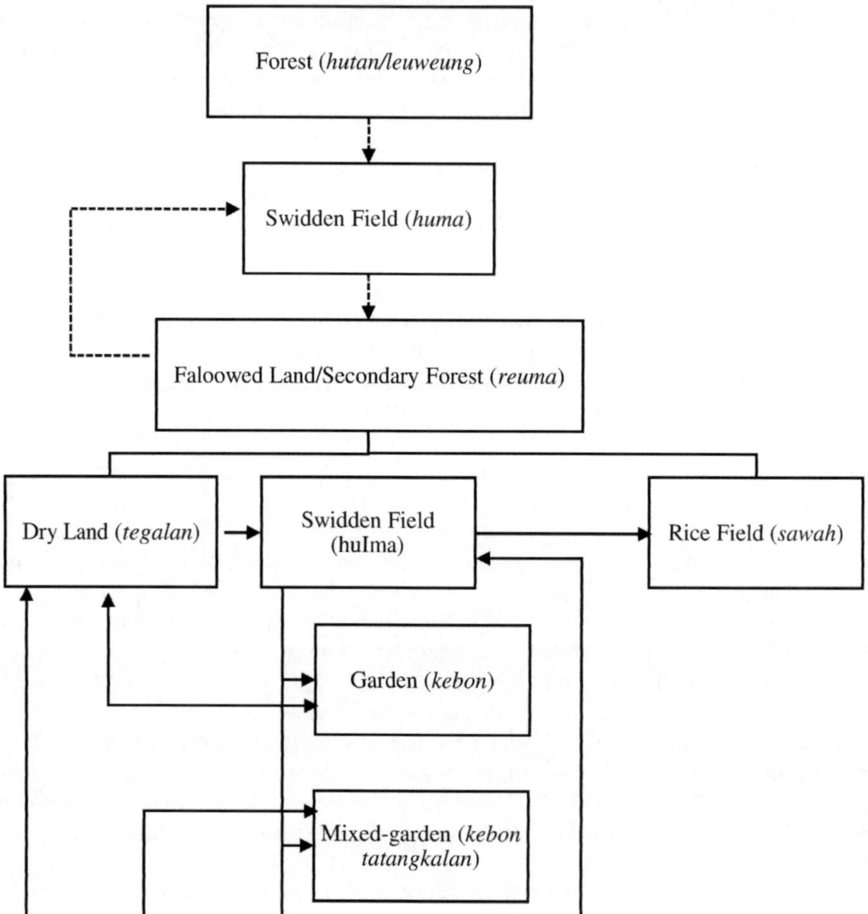

Figure 2. The Structure, Functions and Conversion Landcape in Karangwangi Village, Cianjur, West Java, Indonesia (was modified from Iskandar et al. 2016)

The entry of modernization such intensive agriculture which is supported by the government through the Green Revolution and *panca usaha tani* has worsened soil quality in Karangwangi Village. It is caused monoculture, pesticide and fertilizers which was increasingly used by local people. Now, it is quite dilemma for farmers become dependence on pesticides for pests that continuously come.

Besides being into *sawah*, *huma* also turned into gardens. This was occurred because the garden system adapted form of increasing population and the market economy that is growing rapidly. Consequently, the local people make the selection and introduction of new crops which more profitable economically and perform new inputs such as inorganic fertilizers and pesticides to suit market demand (Iskandar 2009).

The more dense the population, the higher the demand for settlement. This also happened in the Karangwangi Village with new settlements which reduces the area of the garden. To meet its needs, the community has its home garden are planted with a wide variety of crops such as fruits, spices and plants ceremonies and the livestock. This home garden system is not managed intensively and are rarely attacked by pests because of its poly culture. Local people of Karangwangi used to harvest itself the result of home garden for satisfy personal or family needs. The existence of this home garden is very good for the ecology because of it's still natural without inorganic fertilizers or pesticide and can maintain the diversity of the local plants.

The results of this study provide valuable information that helps us to understand the local knowledge of the landscape of the rural population, especially the local or indigenous people of the Karangwangi Village. Preferably, this local knowledge can be integrated together with western science to improve the economic (economically viable), friendly to the environment (ecologically sound) and socially equitable for rural people (socially justly).

4. Acknowledgments

This study is one of topics of the Academic Leadership Grant (ALG) program of Prof. Johan Iskandar, funded by DIPA Universitas Padjadjaran (UNPAD), Indonesia fiscal year 2015. Therefore, on this occasion we would like to thank rector of UNPAD, Prof. Dr. med. Tri Hanggono Achmad, who has provided ALG program as implementation to achieve World Class University. In addition, we also would like to thank the village head of Karangwangi Village and his staff, along with the informants of Karangwangi who have kindly helped us to provide information.

5. References

[1] Geertz C 1963 *Agriculture involution: The Process of Ecological Change in Indonesia* ed S Supomo (Indonesia: Bhratara Karya Aksara) chapter 2 pp 16-39

[2]Iskandar J 2009 *Ekologi manusia dan pembangunan berkelanjutan* (Bandung: Universitas Padjadjaran) chapter 8 pp 148-153

[3]Iskandar J, Iskandar B S and Pertasmita R 2016 *Response to environmental and socio-economic changes in the Karangwangi traditional agroforestry system South Cianjur West Java* vol **17** 1 (Indonesia: Biodiversitas) pp 332-341

[4]Iskandar J and Iskandar B S 2016 *Conf. Etnoekologi dan pengelolaan agroekosistem oleh penduduk desa Karangwangi Kecamatan Cidaun Cianjur Selatan Jawa Barat* (Makalah)

[5]Marten G G 2001 *Human ecology: Basic Concept for Sustainable Development* (London: Earthscan Publication Ltd)

[6]Dewi R 2009 *Perencanaan lanskap untuk pengembangan wisata alam di sempadan Sungai Kemiri Kecamatan Margadana-Kota Tegal* (Thesis)

[7]Pranowo A P D S 1985 *Manusia dan hutan : Proses Perubahan Ekologi di Lereng Gunung Merapi* (Indonesia : Gadjah Mada University Press) chapter 3 pp 44-46

[8]Ramdhan B, Chikmawati T and Waluyo E B 2015 *Perspektif kultural pengelolaan lingkungan pada masyarakat pada masyarakat adat Cikondang Kabupaten Bandung Jawa Barat* vol **1** 1 (Indonesia: Sumberdaya Hayati) pp 7-14

[9]Sastrapadja et al 1979 *Tanaman pekarangan* LIPI Bogor

[10]Waluyo E B 2008 *Rev. Research ethnobotany in Indonesia and the future perspective* vol **9** (Indonesia: Biodiversitas) pp 59-63

Landscape Management of Public Open Space in Bogor Heritage City

F D Pusparini[1], Nurhayati[1] and H S Arifin[1]

[1] Department of Landscape Architecture, Faculty of Agriculture, Bogor Agricultural University, Kampus IPB Darmaga, Bogor 16680, Indonesia

E-mail: flourentina@gmail.com

Abstract. Public open space landscape plays important role in Bogor Heritage City. Although these spaces can carry various kind of public activities and enhance environment quality, they are fragile to disturbance and changing due to city development. Therefore, as an effort to maintain public open space landscape in 8 zones within Heritage City of Bogor, management of public open space landscape is necessary. Syntesys of public open space landscape management had held by using Analytical Hierarchy Process (AHP). The AHP result showed public open space landscape in Palace and Botanical Garden of Bogor zone (20%) is in high priority to be maintained. The other zone are Situ Gede (16,9%), Plan Karsten (12,3%), European Settlement (12%), Suryakencana (11,2%), Batu Tulis (10,2%), Empang (9.2%) and West Development (8,2%). Ecological function (34,8%) becomes the most important function to be maintained (34.8%) then social culture (31.2%), economic (17%), and aesthetics (17%).

1. Introduction

The history of Bogor began from a resting place of Governor-General of the Dutch East Indies called Buitenzorg in the former capital of the Pajajaran Kingdom in 17th Century. In 1817 Stamford Raffles built a botanical gardens in Buitenzorg as a major research center for agriculture and horticulture. After Dutch reoccupied Buitenzorg from British, they developed Buitenzorg as modern city in 19th Century. The city landscape was designed to highlight the beauty of the rivers and views of Mount Salak. The remain of Buitenzorg which can still be seen today are city spatial pattern, historical buildings and some public open spaces such as streets, parks, square and waterfront.

In 2012 the Directorate General of Spatial Planning Ministry of Public Works and Indonesian Heritage Preservation Council (BPPI) initiated Planning and Preservation of Heritage Cities Program (P3KP). Bogor was chosen as one of 10 pilot project of Heritage Cities in Indonesia. The term of Heritage City used to define a city that puts heritage conservation activities as a key strategy into sustainable cities. Heritage can be natural heritage, cultural heritage (physical and non-physical) and landscape heritage. Through this program spatial plan which is owned by the city is expected to boost the quality and quantity of Heritage Region. Since the establishment of Heritage City in 2012, Bogor has conducted some actions due to Heritage City Program. However, the maintaining of public open space in Heritage Region of Bogor gets less attention than building structures.

Public open space is an open public land in the form of parks, sports fields and other open spaces that can be accessed and used by people without limitation of space, time, and costs (Permen PU No 12 Year 2009). Forms of public open space consists of green open space (RTH) and non-green open space (RTNH). Public open space has ecological, economic, social, cultural, aesthetic and emergency function.

Public open space is an important element in urban environment that contributes positively to the quality of life in the city [1]. Beside enhancing the identity and shape of the city, public open space is also a social infrastructure, improving economic conditions, aesthetic, ecological protection, and containers of cultural activity [2]. Public open space is fragile to disruptions and changes that occur in the city. The aim of this study is to propose a landscape management strategy of public open space in Heritage Region of Bogor.

2. Method

2.1. Study Area

The study was conducted in 8 zones in Heritage City of Bogor, West Java, Indonesia (Figure 1). Six zones have been established by Planning and Preservation of Heritage Cities Program (P3KP) in 2013, they are : 1) Palace and Botanical Garden of Bogor; 2) Arabic Settlement of Empang; 3) Chinatown of Suryakencana; 4) European settlement; 5) West Development; and 6) Karsten Plan. Two more zones are proposed zone which are strategic area of the environment and culture that defined in the RTRW Bogor 2011 - 2031, they are (1) Batu Tulis and (2) Situ Gede. Situ Gede had been established as natural heritage based on PERWALI BOGOR (City Mayor Regulation) No 17 Year 2015. The study was conducted from December 2015 until May 2016.

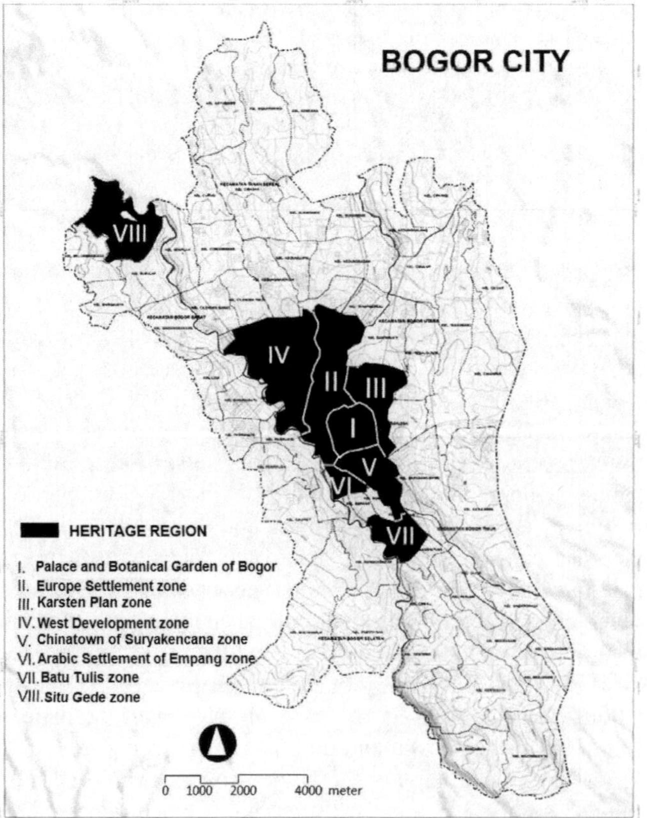

Figure 1. Location of 8 zones in Heritage City of Bogor

2.2 Syntesys of Landscape Management strategy of Public Open Space

Strategic management of public open space in Heritage Region conducted with an approach to the various stakeholders and relevant experts. The methods were interviews and Analytical Hierarchy Process (AHP) by using Expert Choice software. The AHP questionnaire aims to determine the priority function of public open space landscape. The choices of the function are: 1) ecology, 2) socio-cultural, 3) economic, and 4) aesthetics. Each function is subdivided into several alternatives. Specialists who were asked to participate in assessing the AHP are: 1) the experts from Regional Development Planning Departement of Bogor (BAPPEDA), 2) City Park Department of Bogor (DKP), and 3) Cultural Community of Bogor.

3. Result and discussion

From 8 zones of study location, the European settlement zone has more diverse forms of public open space than the other zones (Table 1). Pedestrian path, waterfront path, parks and waterfront parks are forms of public open space which is almost exist in each zone. Bike path, plaza, playground and sport field are still rare to be built. Linear shaped of public open space i.e pedestrian paths, bike path, waterfront paths and plazas that exist today are mostly situated in historical value street that has existed since the Dutch colonial era. Some parks, waterfront parks and sports field are renovated parks from the old ones that also has existed since Dutch colonial era.

Table 1. Public Open Space Matrix in Heritage Region of Bogor

	Zone							
Form of Public Open Space	**Palace and Botanical Garden of Bogor**	**European Settlement**	**Karsten Plan**	**West Development**	**Suryakencana**	**Empang**	**Batu Tulis**	**Situ Gede**
Pedestrian path	exist, historical value	exist, historical value	exist, historical value	exist, historical value	exist, historical value	exist, historical value	exist, historical value	
Bike path		exist, historical value						
Waterfront path		exist, historical value	exist, historical value			exist, historical value	exist, historical value	
Plaza	exist, historical value	exist, historical value						
Park		exist, historical value	exist, historical value			exist, historical value		
Playground			exist, historical value					
Waterfront Park		exist, historical value	exist, historical value				exist, historical value	exist
Sport field		exist, historical value	exist, historical value	exist				

[cross-hatch] exist [vertical lines] historical value

After the implementation of the Heritage Cities Action in 2015, addition of several public open spaces has been carried out. Some public open spaces which are the improvement from the previous one are : Tugu Kujang plaza (Palace and Bogor Botanical Gardens zone), Taman Ekspresi (Karsten Plan zone), Taman Air Mancur (Europe Settlement zone), and Garden and Suryakencana Gate (Suryakencana zone) (Table 2). The Parks and plazas are utilized by the citizen to carry out various social and cultural activities now. Giving new function (adaptive re-use) to new public open space has positive impact to rivive the old heritage region [3]. Taman Air Mancur and Taman Kencana are actively used as social and cultural spaces. Alun – alun, a historical square park in Empang zone is in bad condition because of lack management.

Table 2. List of Public Open Space Objects in Heritage Region of Bogor

Form of Public Open Space	Zone	Name	Leght/ Square	Management authority
Pedestrian path	Palace and Botanical Garden of Bogor	Juanda street	3.329,18 m	Bina Marga Departement
		O. I. Dinata street	1.478,80 m	Bina Marga Departement
		Jalak Harupat street	1.970,65 m	Bina Marga Departement
		Pajajaran street	663,56 m	Bina Marga Departement
	European Settlement	Kapten Muslihat street	883,50 m	Bina Marga Departement
		Dewi Sartika street	1.719,18 m	Bina Marga Departement
		Pengadilan Steet	974,94 m	Bina Marga Departement
		Sawojajar street	1.060,35 m	Bina Marga Departement
		Jend.Sudirman street	2.645,10 m	Bina Marga Departement
		Ahmad Yani street	1.793,05 m	Bina Marga Departement
	Karsten Plan	Sempur street	675,66 m	Bina Marga Departement
		Pajajaran street	2.360,00 m	Bina Marga Departement
	West Development	Jembatan Merah street	1.550,73 m	Bina Marga Departement
		Merdeka street	2.841,24 m	Bina Marga Departement
		Semeru street	5.029,26 m	Bina Marga Departement
	Suryakencana	Suryakencana street	1.966,02 m	Bina Marga Departement
		Siliwangi street	3.120,00 m	Bina Marga Departement
	Empang	R. Saleh street	756,04 m	Bina Marga Departement
		Empang street	329,44 m	Bina Marga Departement
		Pahlawan street	1.619,86 m	Bina Marga Departement
	Batu Tulis	Batu Tulis street	1.382,86 m	Bina Marga Departement
		Cipaku street	650,82 m	Bina Marga Departement
		S. Danasasmita street	404,92 m	Bina Marga Departement
Bike path	European Settlement	Kapten Muslihat street	415,68 m	Bina Marga Departement
		Juanda street	133,71 m	Bina Marga Departement
Waterfront path	Europe Settlement	Paledang street	385,62 m	Bina Marga Departement
		J. Sudirman street	304,06 m	Bina Marga Departement
	Karsten Plan	Sempur Kidul street	326,74 m	Bina Marga Departement
	Empang	Pulo Empang street	382,01 m	Bina Marga Departement
	Batu Tulis	Cipaku street	650,82 m	Bina Marga Departement
Plaza	Palace and Botanical Garden of Bogor	Tugu Kujang plaza	602,94 m	Bina Marga Departement
	Europe Settlement	Nyi Raja Permas	251,70 m	Bina Marga Departement
Park	European Settlement	Taman Air Mancur	3.036 m^2	City Park Departement
		Kapten Muslihat	7.852 m^2	City Park Departement
	Karsten Plan	Taman Kencana	4.795 m^2	City Park Departement
		Taman Ekspresi	1.098 m^2	City Park Departement
	Empang	Alun – alun Empang	2.747 m^2	Local community
Playground	Karsten Plan	Sempur	1.398 m^2	City Park Departement
Waterfront Park	European Settlement	Taman Peranginan	1.699 m^2	City Park Departement
		Taman Cipakancilan	1.194 m^2	City Park Departement
	Karsten Plan	Sempur	1.202 m^2	City Park Departement
	Batu Tulis	Taman Cipaku	5.600 m^2	City Park Departement
Sport field	Europe Settlement	Pajajaran	10.554 m^2	Youth and Sports Dept.
	Karsten Plan	Sempur	38.831 m^2	City Park Departement

A landscape management strategy is proposed to maintain these public open space by using Analytical Hierarchy Process (AHP). Based on the AHP combined results, public open space landscape of Palace and Botanical Garden of Bogor zone (20%) is the first priority to be maintained (Figure 2). Another zones are Situ Gede (16.9%), Karsten Plan (12.3%), European settlement (12%), Suryakencana (11.2%), Batu Tulis (10.2%), Empang (9.2%), and West Development (8.2%).

Figure 2. Priority zone in Heritage Region of Bogor City

For priority function of public open space, ecological function is the first priority function which should be prioritized in the Heritage Region of Bogor with a value of 34.8% (Figure 3). The next priority after the ecological function are the social culture (31.2%), economy (17%), and aesthetics (17%).

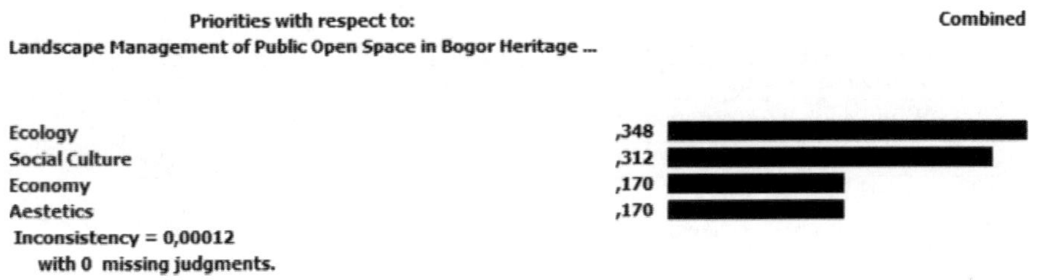

Figure 3. AHP result of priority function of public open space management

Based on the AHP Result, Palace and Botanical Gardens of Bogor is the first priority in the management of public open space landscape. Landscape Palace and Botanical Gardens is a milestone development of the city. It also the green open space (RTH) which dominates in the city. Therefore, to protect the historical landscape, public open space located in the outer ring of the zone must be preserved and enhanced.

Ecological function is in high priority of the management of public open space landscape. Public open space landscape of Bogor is dominated by the presence of green open space (RTH) in the form of street greenery and parks. They support the ecological functions of the city. Water infiltration, oxygen supply, protection of natural forms, temperature/climate comfort, and biodiversity conservation are several alternatives to improve the quality of ecological function in Heritage Region (Table 3). Public open space on the edge of the river serves as a water catchment, forms a protective nature and biodiversity conservation. Parks and street greenery also serves as a water catchment and create a comfortable temperature/climate in Heritage Region. These public open spaces need to be maintained and enhanced in order to make a better quality of ecological function of Bogor Heritage City in the future.

Social Culture function of public open space is the second priority function in Heritage Region. This function provide a wide range of social and cultural activities in public open space. Evacuation, socialization of citizens, cultural activities, education and research, recreation, and sports are a few alternative of social culture function in the Heritage Region (Table 3). Some parks that have been renovated now actively used to carry out various social and cultural activities.

Economic and aesthetic function of public open space in Heritage Region take the third priority to be managed (Table 3). After the establishment of the Bogor Heritage Cities, tourism as an important contributor sector to the heritage city's economy can be increased by improving the quality of the public open space around the historical and cultural sites. Pedestrian path, plazas and parks need to be improved in order to give comfort to the travellers. Public open space at Heritage Region also needs to be managed to empower informal economy industries related to preserve the cultural heritage products such as traditional food and crafts. Management of public open space around natural site and historical/cultural site will also enhance the beauty of the urban environment and architectural of the Heritage City. Creativity of citizen is needed to enhance the environment beauty of Heritage Region. Adding the element of public space that represents the uniqueness of each zone can be done to highlight the aesthetic in each zone. To highlight the traditional character in the Empang zone could be done by restoring Alun – alun square park into new public open space. The existence of the Alun - alun as a public open space will make the visual character of the Muslim region can be easily recognized [4].

Table 3. Function alternatives of AHP result

Function		**%**
Ecological		*34,80*
alternatives	Water infiltration	31,70
	Oxygen supply	24,30
	Protection of natural forms	20,60
	Temperature/climate comfort	12,40
	Biodiversity conservation	11,10
Social culture		*31,20*
alternatives	Evacuation	23,30
	Socialization of citizens	22,60
	Cultural activities	18,30
	Education and research	13,90
	Recreation	12,30
	Sports	9,60
economic		*17,00*
alternatives	Tourism	47,60
	Empower informal economy industries	27,60
	Urban agriculture	24,80
aestetics		*17,00*
alternatives	The beauty of urban environment	37,90
	The beauty of urban architecture	33,10
	Creativity of citizen	29,00

4. Conclusion and recommendation

This study concludes public open space plays important roles to Bogor Heritage City. Palace and Botanical Gardens of Bogor is the first priority zone in management of public open space landscape. Landscape Palace and Botanical Gardens is a milestone development of the city. It also the green open space (RTH) which dominates in the city. Therefore, to protect the historical landscape, public open space located in the outer ring of the zone must be maintained and enhanced. Ecological function is in high priority of the management of public open space landscape. The presence of green open space (RTH) in the form of street greenery and parks support the ecological functions. These public open spaces need to be enhanced in order to make a better quality of ecological function in Bogor Heritage City. Tourism as an important contributor sector to the heritage city's economy can be increased by improving the quality of the public open space around the historical and cultural sites. Management of public open space around natural site and historical/cultural site will also enhance the beauty of the urban environment and architectural of the Heritage City. Adding the element of public space that represents the uniqueness of each zone can be done to highlight the aesthetic in each zone.

Acknowledgements
I would like to express my gratitude to AHP questionaire respondents from Regional Planning Departement of Bogor, City Park Department of Bogor, and Cultural Community of Bogor.

References
[1] Nasution AD, Zahrah W. 2012. Public Open Space Privatization and Quality of Life, Case Study Merdeka Square Medan. *Procedia - Social and Behavioral Sciences* 36 Page 466 – 475
[2] Lee YL, Saida I, Kubotab A. 2013. The Roles of Cultural Spaces in Malaysia's Historic Towns: The case of Kuala Dungun and Taiping. *Procedia - Social and Behavioral Sciences* 85 (2013). Page 602 – 625.
[3] Cho M, Shin S. 2014. Conservation or economization? Industrial heritage conservation in Incheon, Korea. *Habitat International* 41 (2014). hlm 69-76
[4] Between Colonial, Moslem, and Post-Independence Era, Which Layer of Urban Patterns Should Be Conserved. Di dalam *Procedia - Social and Behavioral Sciences*68 (2012). hlm.775 – 789

Study on Ecological Design Concept of Buton Sultanate Cityscape Based on Local Culture

A Mansyur[1], A Gunawan[2] and A Munandar[2]

[1]Local Planning & Development Board, Bau-Bau City, Province of Southeast Celebes
[2]Lecture, Department of Landscape Architecture, IPB

E-mail: mansyur.butur@gmailcom

Abstract. Buton Sultanate Cityscape was constituted of man-made landscape constructed in the era of Buton Sultanate in 1322. It is one of the Indonesian heritage networks proposed to be the world heritage city. The Sultanate cityscape should have the concept of traditional city and refer to the ecological principles. This research was conducted to analyze elements and spatial patterns of Sultanate cityscape based on the ecological principles (eco-design). Descriptive method was utilized in the research by conducting in-depth interviews with the local custom figures and experts of the local culture, literature reviews, and field observations. The main elements of Buton Sultanate Cityscape consisted of palaces, city square, mosque, cemeteries, and settlements, while the supporting elements located outside the city border include mountains, valleys, rivers, and forests. City square is located in the city center surrounded by the palace, cemetery, and mosque. The main pattern of city circulation pattern has formed a simple figure of human body. Ecological principles can be examined from the housing layout paralleled to the road, direction of most city gates facing the east and forests, and the city wall pattern which is closely related to the religious matter.

1. Introduction

The Buton Sultanate is a manmade city of the Buton Sultanate era in 1322. In a range of periods, this Sultanate city has been established as one of the networks of Indonesian Heritage Cities supported by the Ministry of Public Works of Indonesia. Its establishment was predicted to occur, for Buton Sultanate city offers wealthy historical artifacts and cultural life, some of which can be studied further; including city forts, sultanate palaces, houses on stilts, mosques, cemeteries, and etc. [12].

Heritage city along with its landscapes is an essential unification in city planning [5] [10], for it has considered nature in its planning since past times, which is currently known as the concept of ecological design [2] [3]. Cultural and ecological values in a traditional landscape are strongly attached to the community life [9]. A number of researches related to the heritage cities have been conducted, particularly regarding the sultanate cities [1] [6], however, a study on the ecological design concept on cityscape by referring to the cultural information has not been performed.

Subsequently, this study aims to: (1) identify the elements of the landscape and its layout, (2) analyze the spatial pattern to form the sultanate cityscape, and analyze its landscape characteristics in reference to the local culture as the indicators of current ecodesign concept.

2. Research Methodology

2.1 Time and Place of Research
The research took place in Buton Sultanate cityscape in Baubau City, Southeast Sulawesi, which is commonly known as Keraton Fort with an area of 4.0 ha located at the coordinates of 5.21 LU – 5.30 LS and 122.30 - 122.45 BT (Figure 1). Thus, the research was conducted from May to September 2015.

2.2 Research Method
This research was conducted on the basis of Buton culture and history, by using the content analysis to explore the cityscape elements and its siting, and cityscape ecological concept in reference to the local wisdoms. By adopting descriptive method, the research was conducted in three steps i.e. (1) reviewing literatures, (2) interviewing the local custom figures and experts of the local culture, and (3) observing the citycape.

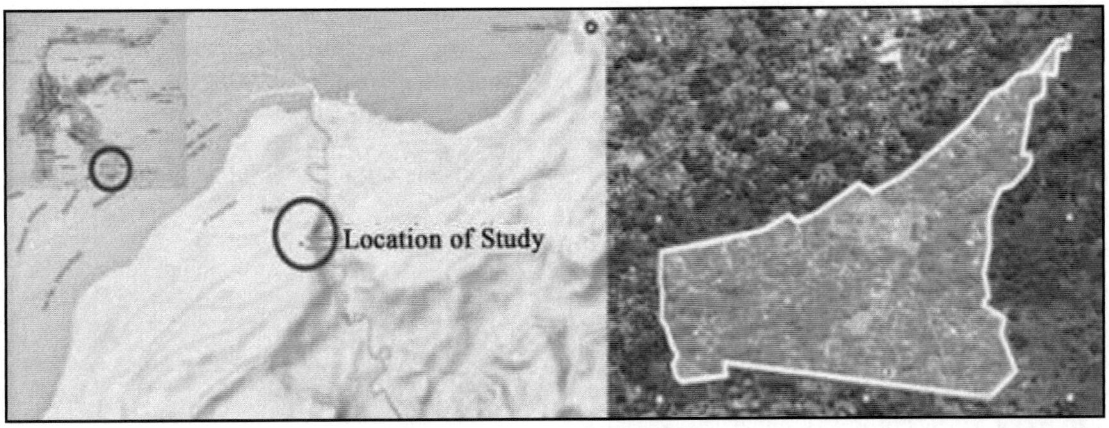

Figure 1. Location of Study

2.2.1 Literature Review
Desk study was initially conducted to achieve the research objectives i.e. elements which formed Buton sultanate cityscape at its triumph period and location. In addition, the research also examined the pattern of city space with its respective functions in reference to the cultural information. The cultural information was collected from customary manuscripts, history and culture books, and journals related to Buton customs.

2.2.2 Interview
The interviews with 23 key informants consisting of the leaders of indigenous organizations, religious leaders, cultural figures, historians, craftsmen, artists, manuscript guards, hereditary officials of sultanate era, and local governments were conducted. Furthermore, questions leading to confirm the results obtained in the previous stage, literature studies, whether it is appropriate or there is any difference and addition were then administered to them. In-depth interviews were also conducted to reveal the geomancy meaning and cultural functions of landscape elements.

2.2.3 Field Observation
Field observation was performed in reference to the results of the research in the previous two stages i.e. literature studies and interviews. It was primarily conducted to verify the elements of the cityscape and its layout based on cultural information, and to examine cityscape space and characteristics at present time. Further, it was also performed by utilizing photo images downloaded from the google map.

2.3 Cityscape Analysis
Content dan descriptive analyses were used in this research. The former was conducted at the initial and second stages of the research while the latter was conducted at the third stage. The contents of the document were analyzed, and what kinds of elements and spaces contained in the cityscape of Buton sultanate during its glory were also analyzed. In addition to the presence of elements and spaces, their functions and philosophies were also examined further. In the third stage of study, the pattern of the cityscape was analyzed by adopting Lynch's approach in imagable city [7], covering the circulation paths, regional borders (edges), district parts, city activity center (nodes), and landmarks. The analysis was performed on the ancient maps, images, and field observations. The eco-design concept on the cityscape was observed in reference to the effect of element layouts and spatial patterns on the city environment quality [11].

3. Result And Discussion

In an overview, Baubau city has a convenient climate condition with city temperatures ranging from 19.8 to 37.4°C, wind speeds of 2.0 to 4.1 knots, and annual rainfalls ranging from 1,500 to 2,500 mm (Central Bureau of Statistics, 2015). Buton sultanate city is located in the southern part of Baubau at an altitude of 104 m above sea level and has undulating landform. Moreover, the sultanate city is surrounded by cliffs and valleys (northern and eastern sides), hills and forests (eastern and southern sides), ditches and settlements (southern and western sides), and cemeteries (western side). These characteristics of cityscape are primarily referred to the significant portion of cultural philosophy.

3.1 Elements of the Cityscape
Spatial cityscape elements and patterns were compiled in accordance with the on-going development of Buton culture at that particular time. The elements of sultanate cityscape cover the city borders (edges), settlement (districs), city square (node), roads (paths), mosques and palaces (landmarks) (Table 1). It corresponds to what Lynch called as imagibility of the city. See this following Figure 2.

Table 1. Elements of Buton Sultanate Cityscape

No.	Cityscape Elements	Lynch's approach	Existence of the elements (on the basis of)		
			Literature	Interview	Observation
1.	Fortress	Edges	●	●	●
2.	Gates/Baluara		●	●	●
3.	Settlement	Districts	●	●	●
4.	Cemetery		●	●	●
5.	City square (alun-alun)	Node	●	●	●
6.	Green Corridor	Paths	●	●	●
7.	Street		●	●	●
8.	Mosque	Landmarks	●	●	●
9.	Palace		●	●	●

Figure 2. Elements of Buton Sultanate Cityscape

IOP Conf. Series: Earth and Environmental Science **91** (2017) 012021 doi:10.1088/1755-1315/91/1/012021

Out of the nine elements, fortress is the cityscape element which strongly indicates the uniqueness of the Buton Sultanate City. It is a city bordering element made of Stone Mountain (karst) on which bastions and gates stand. Thus, its construction was conducted during the era of Sultan Buton the fifth, Sultan Gafarul Wadudu (1632-1645). The fortress was built with an adjustment to the land contour and the edge of the hills reaching up to 2,740 m, and with varying dimensions; i.e. the height of 1 to 4.5 m and width (thickness) of 0.5-2 m (Makassar Cultural Heritage Preservation Center, 2013).

Furthermore, the fortress is the sultanate city border which was constructed to provide protection for the community from pirate encounters, the Dutch colony [12]. It mainly consists of 16 bastions and 12 gates. All of the 12 gates indicate the 12 entrances (holes) to a human body (interview with La Suluhu). This means that if there is something entering one's body illegally, the body will react to neutralize, destroy, or release the substance, and similar events occurred in the respective gates. In reference to the religious character of Buton Muslim community, they significantly uphold the prophet Muhammad SAW by carving his name on the fortress wall pattern and city shape. At the northend of the east wall, the Arabic letters of 'mim', 'ha', and 'mim' were written, while the city shape as a whole formed the letter 'dal', so that when they were combined, they formed the word of 'Muhammad'. The area which formed the first letter of "mim" is perceived as a 'sacred place' because it functions as the cemeteries of the sultan and his family (interview with Mujazi).

In addition to the cemeteries of the sultan and his family mentioned above, there was also another sultan cemetery located in the city center, at the border of the city square. The city square was an open space in the city center where the sultan gathered his people. It is the unifying element of the city as proposed by Lynch [7] to be referred to as the 'node'. Hence, the city elements in the forms of pathways included the roads and green lanes. What referred to as green lanes were the lines of the coconut trees alongside the fortress, which were practically planted as the requirement for the newlyweds obliged to plant three coconut trees. Meanwhile, elements which served as 'landmarks' in this sultanate city were the Sultanate Palace and Mosque. Both buildings are perceived to be highly significant compared to other city elements. During the glory of sultanate, the house roof belonging to the common people could not be higher than that of the sultan's, for it was believed that if its height exceeded the sultan's house roof, it would not give any blessings (results from the interviews).

3.2 City Land Uses

The spatial pattern of Buton sultanate city formed the Arabic letter of 'dal' which, according to the result of interviews with local custom leaders, symbolized the unity within the human physical body. The existing land uses in the city at the sultanate era covered the city square, settlements, sultanate palace, and plantation (Figure 3). It is discovered that the settlements highly dominated the land uses in the city; including the settlements of the Prime Ministers (Bonto Siolimbona), Home Affair Minister (Bonto Yinunca), and Liaison Ministers (Bonto Lencina Kanjawari). Subsequently, Bonto Siolimbona region, located in the northern part of the city, became a shield for the sultanate area. Thus, its area location was surrounded by each settlement of the ministers, precisely in the city center, close to the city square and plantation. The city square in Buton sultanate city was similar to that of commonly found in Java, although the spatial use of its surrounding was not exactly the same. It is presumed that there was an influence of Majapahit kingdom in its glory period. Furthermore, a plantation is one of the land uses serving as a green area of the sultanate city, with breadfruit as its main commodity.

IOP Conf. Series: Earth and Environmental Science **91** (2017) 012021 doi:10.1088/1755-1315/91/1/012021

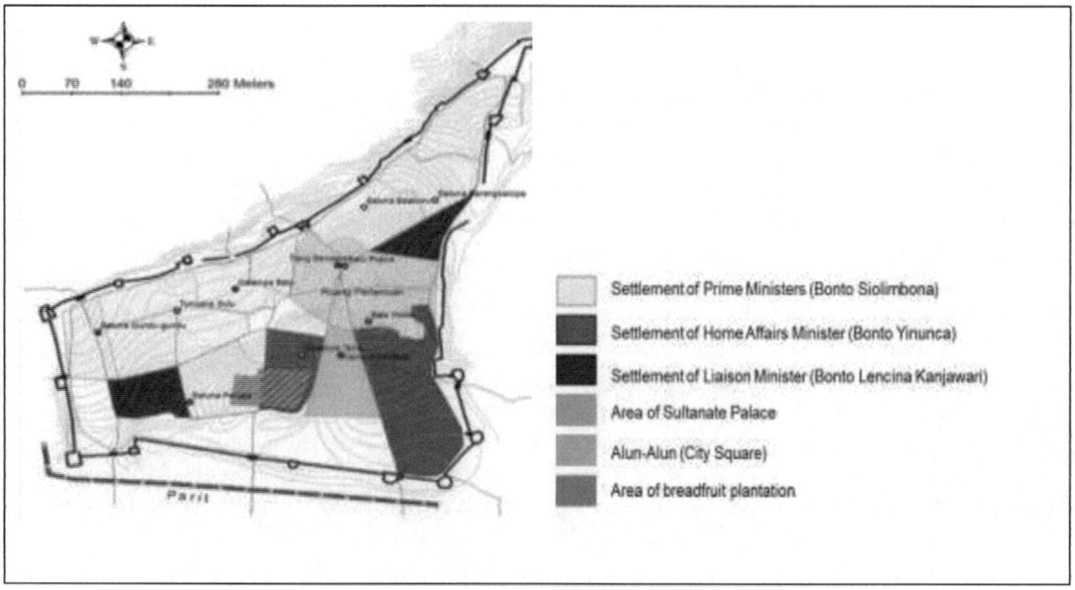

Figure 3. The Cityscape Land Uses

3.3 City Pattern

The city pattern could be formed by lanes or river pathways [8]. Lanes in Buton sultanate city indicated a unique pattern i.e. the pattern of human skeleton. The geomancy of Buton Sultanate cityscape referred to the philosophy of human body and spirit. In addition, the geomancy of fortress wall elements referred to the philosophy of human body identical with Arabic letter philosophy. The respective pattern was generated from a combination of Arabic letters described earlier i.e. "mim", "ha", "mim", and "dal" (Figure 4). These letters were written in the traditional manuscript, and the pattern of human body was attached to the main circulation pattern of Buton sultanate city at that period. The head part was the palace area and connected by road to the city center, or city square. In the body part, lanes were identified moving across the east and the west, as parts of the arms. Lastly, the leg part in the pattern resembled lanes moving to the north side.

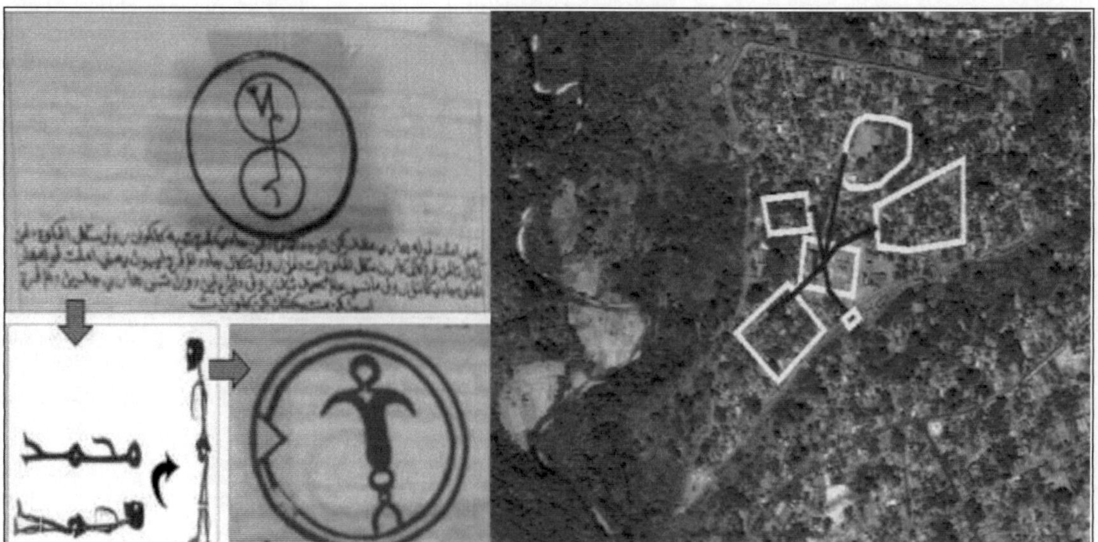

Figure 4. The City Pattern

3.4 Ecodesign Concept

Buton sultanate city was constructed in reference to cultural and religious approach adopted by the Sultan and his people. Such approach is consistent with the eco-design concept which is currently echoed and becomes a trend in city planning and design. It primarily covers the concept-based approach to the issues, principles, and process of eco-design [4] [11]. Thus, the pattern of Buton sultanate city forming the Arabic letter of "dal" enabled the city at the north side to border diagonally from the northeast to the southwest with the longest side of the fortress. The longest side faces the sea and settlements outside the Sultanate palace. In reference to the palace position on high ground, the palace was able to observe and secure situation from enemy encounters at the coast and examine the condition of the people outside the palace. This notion is significantly consistent with the concept of eco-design proposed by VanderRyn and Cowan [11] stating that 'solutions grow from place'. At the eastside of city wall, there is a spread of forest and hills. This condition is perceived to be beneficial to provide fresh air surrounding the palace. Such design concept combines the power of nature with the city needs, which is later referred to as the eco-design concept of 'design with nature' [11].

Figure 5. Open Space Between The Houses

The housing layout towards the street at the sultanate era can be examined in Figure 5. Three houses were facing one another with a green open space in the center. Two houses were bordered with roads while facing the open space. This was a common pattern in the neighborhood of the palace. The open space could serve as grounds and community spaces. Although the road had not produced harmful pollution at the sultanate era, if the pattern was applied to the present conditions, it is capable in helping to avoid or reduce smoke and dust pollution in the road, which is considered to be an important eco-design issue as argued by VanderRyn and Cowan [11].

Natural elements, e.g. mountains or hills, are the symbols of high status. Subsequently, the Sultanate Palace was situated on top of a hill. In addition to its symbolism, in reference to its function, the position was suitable in monitoring the whole city from the palace, and the music sound of traditional procession performed in the palace could be heard clearly by the community within and around the keratin fortress. The palace was oriented and designed eastward, for it captured the sunlight from the morning to noon symbolizing notions of youth and glory. These matters were efforts to design an ecological city at the sultanate era which can be adopted as a city design reference at the present time. Ecodesign formulation proposed by VanderRyn and Cowan [11] was also in accordance with the concept of Buton city at the sultanate era.

4. Conclusions

The Buton sultanate local culture was significantly concerned with the ecological environment of the cityscape. This is reflected from the city-forming elements and city space and circulation patterns. The main elements forming the sultanate cityscape identified in reference to the local culture include (1) the city fortress covering the fortress walls, gates, and bastions; (2) settlements and cemeteries; (3) city square; (4) roads and green lanes; and (5) palaces and mosques.

The pattern of sultanate cityscape is considered to be unique with a significant influence of religious belief held by the sultan and his people i.e. Islam. Thus, the geomancy form of Buton sultanate city shapes the Arabic letter of "dal", while its main circulation pattern resembles a human body derived from the Arabic letters as well.

The eco-design concept in the Buton sultanate was indicated on the city layout in an environment supporting the ecological system of the city, settlement patterns, and location of the palace which supports the city monitoring and information system

References

[1] Awat R and Nur M 2012 *Menemukan Kearifan Lingkungan dan Pola Pemukiman Keraton Buton.* Darmawan Y danMukmin F, editor. (BauBau (ID) RESPECT)
[2] Baran M, Yildirim M, and Yilmaz A 2011 Evaluation of Ecological Design Stategies in Traditional Houses in Diyarbakir, Turkey *Elsavier: Journal of Cleaner Production* **19**: 609-619.
[3] Barnett and Beastley 2015 *Ecodesign for Cities and Suburbs* (Washington, DC. Island Press)
[4] Beck T 2013 *Principles of Ecological Landscape Design* (Washington, DC. Island Press)
[5] Braaksma et al 2015 Production of Local Heritage: A Case Study in The Netherlands Routledge *Landscape Research* **41**: 64-78
[6] Ernawi I S 2009 *Kearifan Lokal dalam Perspektif Penataan Ruang* Wikantiyoso R danTutuko P, editor. (Malang: GrupKonservasiArsitekturdan Kota.)
[7] Lynch K 1960 Image of the city.
[8] Marshall 2005 *Streets and Patterns* (New York. Spon Press)
[9] Spirn AW 1998 *The Language of Landscape* (United State of America: Thomson-Shore,Inc)
[10] Starke and Simonds 2013 *Landscape Architecture. A Manual of Enivronmental Planning and Design. Fifth Edition* (New York: McGraw Hill Education)
[11] Vander R and Cowan 1996 *Ecological Design* (Washington, DC: Island Press)
[12] Zahari A M 1977 *Sejarah dan Adat Fiy Darul Butuni (Buton).* Edisi I-III. (Jakarta: DepartemenPendidikandanKebudayaanRepublik Indonesia)

IOP Conf. Series: Earth and Environmental Science **91** (2017) 012022 doi:10.1088/1755-1315/91/1/012022

Taneyan Lanjhang, Study of Home Garden Design Based Local Culture of Madura

R T Maningtyas[1] and A Gunawan[1]

[1]Department of Landscape Architecture, Faculty of Agriculture, Bogor Agricultural University

Email : rosyi.damayanti@yahoo.com

Abstract. This research aims to study the arrangement of landscape elements in Madura home garden and the underlying philosophy of the existence of these elements to formulate a concept of Madura home garden in accordance with the culture. Data about traditional culture, the character of the community, and the arrangement of the landscape around the home garden obtained through library research, field observation, and interviews of certain resource persons through purposive sampling techniques. The results showed that the Madura developed a pattern of home garden arrangement called *taneyan lanjhang*. Each taneyan lanjhang at least consist of several elements, *namely langghar, roma, dapor, kandhang*, *taneyan*, and organic fences. The elements are placed in a certain position in the direction of east-west and north-south by the Madura concept of life *bappa-babbhu-guru-rato* (father-mother-teacher-queen). The concept proposed residential garden is a garden house that is functional and aesthetic. This concept *taneyan lanjhang* split into five space, which is a public space, private space, semi-public spaces, room service, and a buffer space. The concept of circulation in taneyan lanjhang made according to the axis pattern that directs entrance to the *langghar* (prayer room) as a focal point.

1. Introduction

Madura is one of Indonesian ethnic that has strong character. The character appears in the daily nature through interaction with others, attitude of life, including the territorial arrangement [4]. In addition, the Madura population is quite large and spread throughout Indonesia. Madura Island is an island with a population density is quite high. besides the Madura also scattered in various regions in Indonesia, especially in Java [1].

The Madura have a unique culture. Madura community is embracing the bilateral patrilineal kinship. This kinship is taking into account the male lineage similar to women. They are known as a tenacious and high self-esteem, but also religious and faithful [4]. They live in groups with their family and field. They worked as farmers, traders, craftsmen, or fishermen for their livelihood. The uniqueness of Madura is how they honor their parents. They have motto : *bappa, babbhu, guru, rato* that's mean order the person to be honored was father, mother, teacher, and leader [5]. This concept implemented in every part of life and culture, including in the landscape.

The recently research on Madura focused on aspects of economic, anthropology or traditional architecture [1] [2] [3] [4] [5] [6] [7], while the spatial aspects of home garden have not heeded. Therefore, the study ought to examine the home garden design Madura and mapping the elements that exist to produce a traditional home garden concept of Madura by considering aesthetic values of the garden.

2. Purpose

This research has several objectives:

1. reviewing the elements of the Madura home garden;
2. studied symbolism and its layout elements spatially in the home garden Madura
3. drafting the design concept of traditional Madura home garden that suits the character of Madura

3. Outcome and Benefit

This research is expected to provide information and an overview of the Madura traditional home garden design so that it can become a benchmark for garden design based on the cultural environment of Madura. Thus, this study is expected to further strengthen cultural values and raised the traditional values of Madura as a part of national culture.

4. Methodes

4.1 Location and Time of Study

Research conducted on the island of Madura to visit four districts, namely Bangkalan, Sampang, Pamekasan, and Sumenep. The study was conducted in May-September, 2010.

4.2 Method

The method used is descriptive method through field observation, literature review, and interviews. Field observation done by conducting site visits at some locations still maintain the local culture in structuring the landscape of residential houses, namely Sub Blega in Bangkalan; Subdistrict Sampang Sampang; District of Pamekasan, District Tlanakan, Pamekasan; Subdistrict Lanteng and Sumenep. The literature review done by searching articles or handbook about Madura culture. Interviews were conduct by purposive sampling techniques, namely the cultural experts Madura.

5. Result and Discussion

5.1 *Key Elements on Madura Settlement*

The traditional settlement of Madura is formed by groups of houses. Each group bounded by hedges surrounding the complex of buildings. Every group consists of several independent families, but has relation to one another. In general, Madura culture recognizes two kinds of family settlement patterns, kampong *mejhi* and *taneyan lanjhang*. The major difference between the two patterns lies in the existence of their house building. *Kampung mejh*i usually have only one house and located far from another group (isolated), while *taneyan lanjhang* consists of several houses were built from west to east.

Taneyan lanjhang consists of several elements, arranged in a specific pattern. This pattern is fixed long ago and passed on from generation to generation. The main elements in *taneyan lanjhang* are:

a. *Roma*

Roma is a term for a family house in taneyan. Such homes are divided into two, namely the main house and the house girls. The big house called the *roma tongghu*. Usually the main house was built in the west on the north site with facing south. While the houses of daughters built to the east of the main house with a lined pattern. Roma consists of one main door, one bedroom, and terrace. The bedroom is

IOP Conf. Series: Earth and Environmental Science **91** (2017) 012022 doi:10.1088/1755-1315/91/1/012022

a single room that serves as bedroom for women and children, while the terrace serves as a reception room for female guests. The layout of *roma* in the *taneyan* (site) can see in Figure 1.

b. Langghar

Langghar is a building that functioned as a place of daily prayer for the residents and a marker for an independent *taneyan*. In addition *langghar* also serves as a center of activity in *taneyan* and a place to receive guests. This function causes the *langghar* also acts as a gazebo for the complex *taneyan*. In general *langghar* built at the west end *taneyan* face to face with the entrance. The *langghar* buildings tend to be higher than the surrounding buildings. This shows a very important role *langghar* for *taneyan*. The activities undertaken in langghar is a place to rest after work, deliberation, chatting, receiving guests, and worship. The layout of langghar in the taneyan (site) can see in Figure 2.

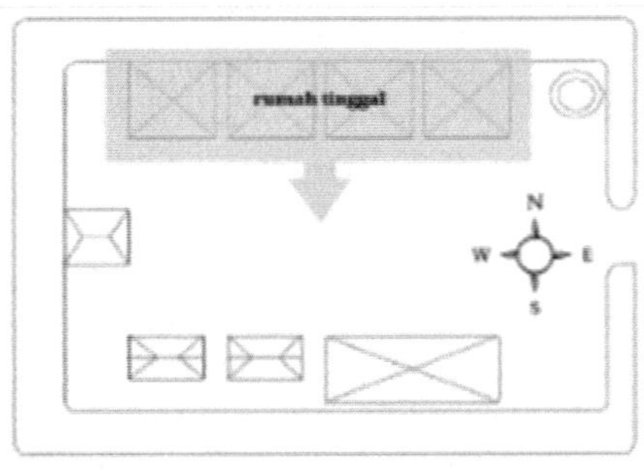

Figure 1. Layout of roma in the *taneyan* (site)

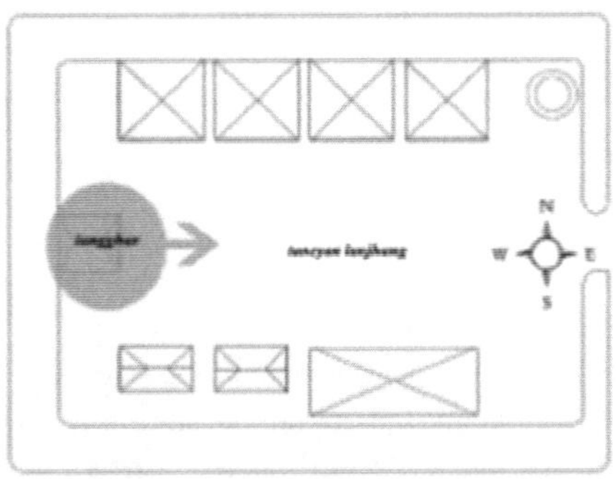

Figure 2. Layout of *langghar* in the *taneyan* (site)

c. *Dapor* and *kandang*

dapor is the term for the owners kitchen. The kitchen is built to deal with their respective homes on the south side *taneyan*. *Kandang* is term for the cage. It built on the south side *taneyan* facing the north. Often kandang also built beside to the kitchen. The existence of the cage itself is now adjusting to the livelihoods of the owners taneyan.

d. Taneyan

Taneyan is a courtyard surrounded by buildings, empty expanse of land. *Taneyan* usually used for drying agricultural products. In recent development, *taneyan* begin decorated with various types of potted plants arranged along the side *taneyan*, but still leaving space left blank elongated middle section. The other function is to perform traditional rituals or family celebration.

e. Organic fences

Organic fence is a row of trees or shrubs that grew close. The organic fence around taneyan deliberately planted and allowed to grow large and tight. It planted as boundary area as well as a protective *taneyan* from harm animals or enemies from outside. The vegetation used as a organic fence is is bamboo (*Bambusa bambos*), tree horse (*Lannea coromandelica*), Sarikaya (*Annona squamosa*), or hibiscus plants (*Hibiscus rosa-sinensis*).

5.2 Symbolism of Madura Home Garden Elements and Layout

Madura home garden laid out according to the concept b*appa, babbhu, guru, rato* which is the concept of life/motto of the Madura. The *roma tongghu* is a symbol of parents (*bappa-babbhu* / father and mother) should be respected so it is placed in the north-west of *taneyan*. The west is the symbol of the sacred space for the people of Madura, while the east is a symbol of profane space. Daughter's house built

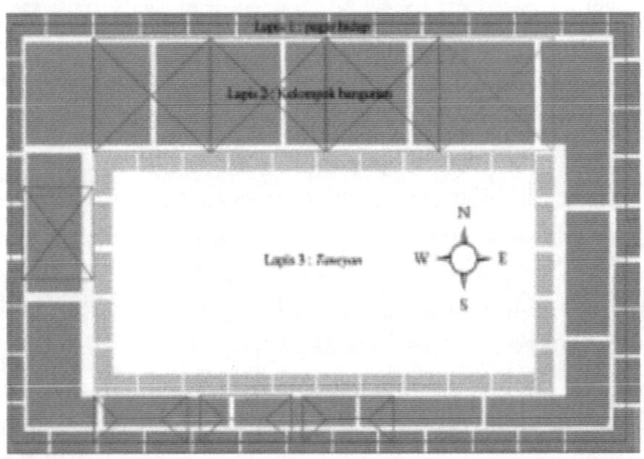

Figure 3. The form of protection area in *taneyan lanjhang*

successively from west to east beside *roma tongghu*, to show deference to older siblings. *Roma tongghu* and daughter's house is a space that is private and only accessible by the resident.

Langghar is a symbol of respect to teachers and leaders (*guru-rato*). *Langghar* area is a semi-public activity centre and has a high value that is placed on the west end of *taneyan*. In the *langghar* area on-going process of socialization and transfer of cultural values of the parents, religious scholars, or the village head to the residents and society. *Langghar* built opposite the entrance taneyan as the oversight function of taneyan owner to the entire area within taneyan.

Taneyan and organic fence is a symbol of defense *taneyan lanjhang* of external danger. *Taneyan* and organic fence serves protection to the resident in taneyan of the danger of wild animals or enemies come. Taneyan is a public area that can be accessed by anyone. The form of protection area can be seen in Figure 3.

5.3 Design concept

The basic concept is applied in the design of *taneyan lanjhang* is a garden that is not only functional but also aesthetically according to environmental conditions of Madura culture. Functional means the garden is able to provide physical benefits for their owners, while the aesthetic means the garden is able to provide visual beauty but it also contains cultural values Madura.

The concept of space is recommended for Madura home gardens are public spaces and semi-public space is between private space and room service. The fourth room is surrounded by a buffer space. Public and semi-public space is a place to socialize with each other so it is a very important space for the settlement of collective existence as *taneyan lanjhang*. The Figure 4 shows the spatial concept of *taneyan lanjhang*.

The concept of vegetation in *taneyan lanjhang* divided into several functions, architectural function, production function, controls the microclimate, and enhancing aesthetics. The four functions are arranged according to the concept of the yard, where the type of vegetation that have either form of crop production

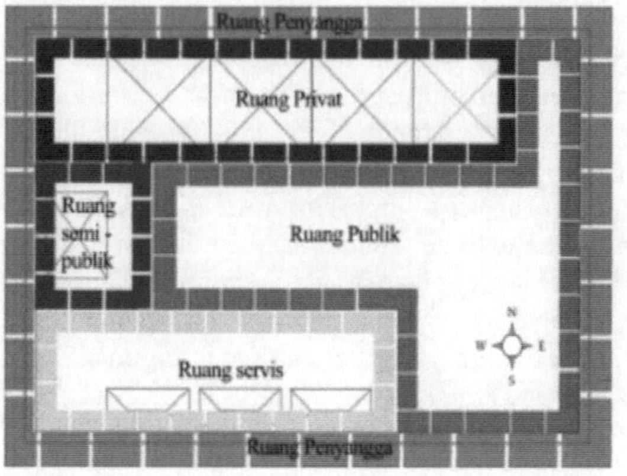

Figure 4. Spatial concept of *taneyan lanjhang*

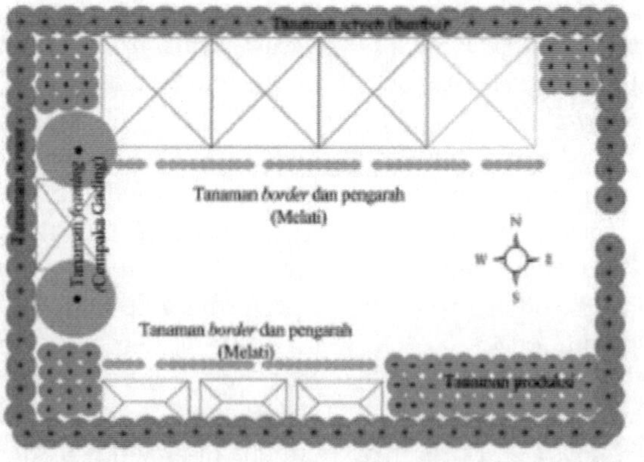

Figure 5. Vegetation concept of *taneyan lanjhang*

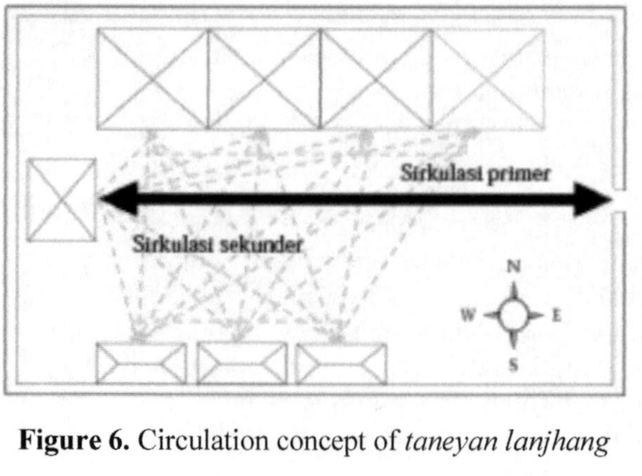

Figure 6. Circulation concept of *taneyan lanjhang*

or herbaceous perennials. The concept of vegetation can be seen in Figure 5.

The concept of circulation serves the movement of the *taneyan lanjhang*. Circulation is divided into primary and secondary circulation circulation. Primary circulation formed by axis that connects the entrance to the semi-public spaces through public space. Secondary circulation is imaginary links between elements. The concept of circulation can be seen in Figure 6.

Based on the basic concept of the proposed development of the concept space, the concept of vegetation, and the concept of circulation which has been described, the design of Madura home garden should refer to the following concepts:

1. The public spaces shaped axis that connects the entrance to the room semi-public
2. Axis formed an open space so that the view extends and not blocked
3. The public and semi-public space is between private space and room service to accommodate the services of daily needs without disturbing Private activity
4. Vegetation is applied to the buffer area and sideways in private, semi-public, and service area
5. Additional elements can be put on public space but not disturbing the axis.

The design concept of taneyan lanjhang as Madura home garden can be seen in Figure 7.

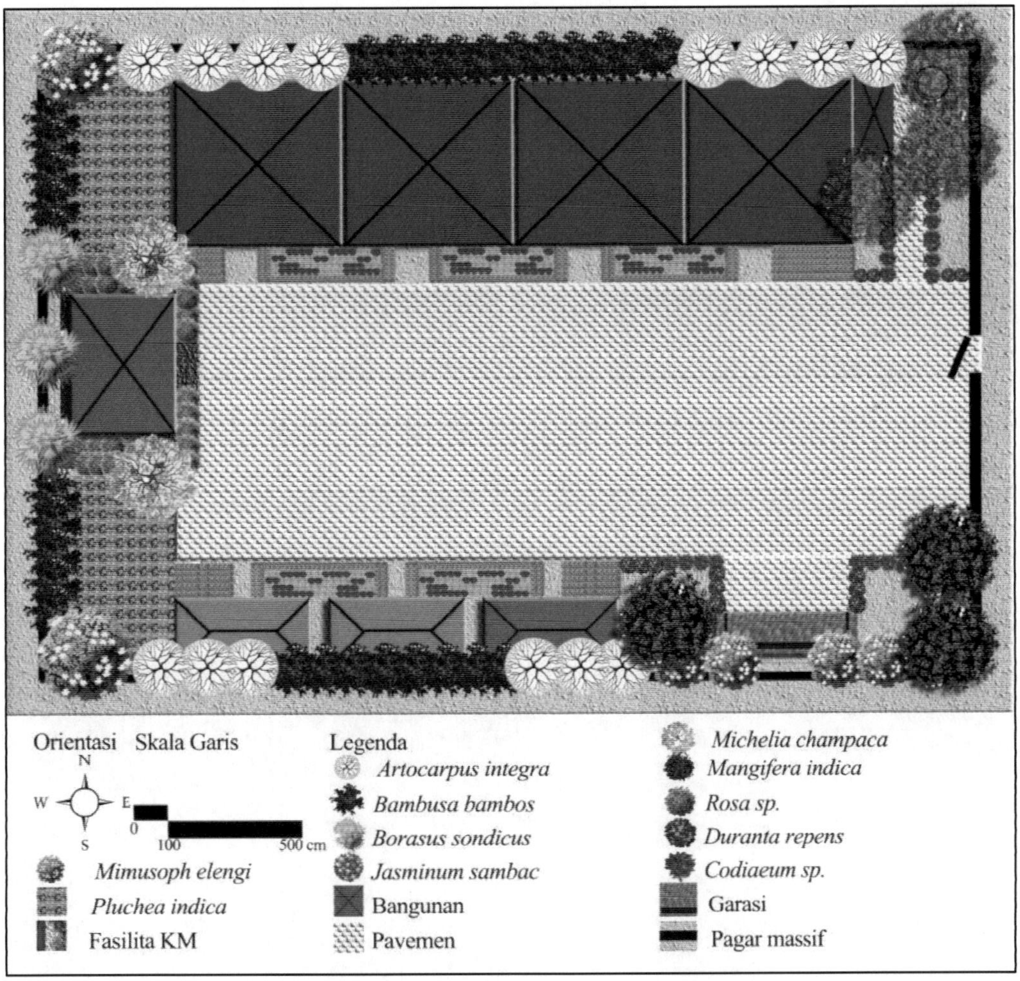

Figure 7. Design concept of *taneyan lanjhang*

6. Conclusion

Madura home garden formed by groupof houses, called *taneyan lanjhang*. *Taneyan lanjhang* consist of the basic elements of *roma*, *langghar, dapor, kandang, taneyan,* and *organic fences*. All of the elements are laid out according to the rules of the west-east and Madura life concept, *bappa-babbhu-guru-rato*. Elements that have the highest value is usually located in the western part taneyan.
Based on the type of activity and function space, proposed a concept of the park is a public space in the form of axis that connects the entrance with a semi-public space. Axis formed an open space that extends and unobstructed view. Green system formed by perennials, ground cover and shrubs. Besides additional elements may be added in the public space for not disturb the axis.

References

[1] De Jonge H 1989 *Madura dalam Empat Zaman : Pedagang, Perkembangan Ekonomi, dan Islam, Suatu Studi Antropologi Ekonomi*. Jakarta:PT Gramedia
[2] Hastijanti R 2005 *Pengaruh Ritual Carok terhadap Pemukiman Tradisional Madura*. Surabaya : Dimensi Teknik Arsitektur Volume 33 No. 1, Juli
[3] Jordaan RE 1979 *Tentang Rumah Tradisional Madura*[makalah seminar].Departemen Pendidikan dan Kebudayaan R.I. : Jakarta
[4] Rifai MA 2007 *Manusia Madura, Pembawaan, Perilaku, Etos Kerja, Penampilan, dan Pandangan Hidupnya Seperti dicitrakan Peribahasanya.* Yogyakarta : Pilar Media.
[5] Sadik S 1996 *Madura dalam Sebuah Potre*t. Surabaya : CV Kurnia
[6] Tulistyantoro L 2005 *Makna Ruang pada Tanean lanjang di Madura*. Surabaya : Dimensi Interior Vol.3 No.2 Desember
[7] Wiyata L 2002 *Carok : Konflik Kekerasan dan Harga Diri Orang Madura*. Yogyakarta : LKIS

Community Movement in Applying Mosquito Net on House Ventilations: An Initial Support for Green Architecture to Decrease Dengue Disease in Bandung Indonesia

F R Rinawan[1], I P P Dewi[2], G Z Haifa[3], K D Suharno[3], K Oktavinus[3] and P S Lyn[3]

[1]Department of Public Health, Faculty of Medicine, Universitas Padjadjaran, Indonesia
[2]Primary healthcare center, Puskesmas Ujungberung Indah, Bandung, Indonesia
[3]Medical profession program,Faculty of Medicine, Universitas Padjadjaran, Indonesia

E-mail: f.rinawan@unpad.ac.id

Abstract. Green architecture still has risk to dengue disease when trees cover house roofs' gutter. This study was aimed to continue a geographical information system (GIS) and remote sensing (RS) study on roofs factor association with dengue disease by initiating community movement in applyingmosquito net on house ventilations to cut the disease transmission and mosquito breeding sites inside house. Our methods was an operational research in which improvement of interventions, policies and regulations towards dengue disease prevention is our intended endpoint. Several steps were conducted such as: (1) research problems formulation from GIS-RS analysis from previous phase research in Bandung city, (2) informal and formal approach to community leaders and primary healthcare centre (Puskesmas), (3) Video education and focus group discussion (FGD), (4) initial application of mosquito nets on house in communities; and (5) advocacy to Mayor of Bandung city (was on progress).Our study resulted several supports: one of sub-city leaders (Camat) in the city, village leaders (Lurah), and sub-village leaders (Ketua RW) of 5 villages (kelurahan), one kelurahan which mainly comprised formal settlements needed more efforts, which was experts on dengue disease from university to directly explain the mosquito nets application to its community. Informal leaders in all kelurahan's community suggested only mothers movement was not enough, thus, youths in community was mentioned to help the community movement on the mosquito nets application.

1. Introduction

Dengue disease is a female *Aedes* species mosquito-borne disease caused by dengue virus. It transmits the virus from human to human by mosquito biting[1]. The disease may cause dangerous conditions to human when bleeding and shock occur [2, 3]. *Aedes* mosquitoes can disperse house to house to bite human because human is their primary mode. They seek human because of their thermal stimuli, carbon dioxide and acid odor stimuli [4]. The dispersal depends on the distance between houses. Crowded population such as in urban may lessen the mosquito dispersal. Rural communities where distance between houses is farther can also modify mosquito dispersal into longer distance [5]. The peak of dengue disease infections can be found when the dispersal is less than 100 meters [6, 7]. The nature cause of biting is that they need to continue their breeding phase from eggs, larvae, pupae and adult[8, 9].

IOP Conf. Series: Earth and Environmental Science **91** (2017) 012023 doi:10.1088/1755-1315/91/1/012023

Houschold is a major factor of *Aedes* species mosquito breeding sites in stagnant water [10]. Blocked roof gutter, for example by falling leaves from trees is one of causes of stagnant water [9, 11].Mosquitoes that emerge from breeding sites outside house can enter house through any ventilations for biting and finding inside-house breeding sites purposes. This conditions caused by house ventilations design is not covered by mosquito net [12]. However, application of mosquito net is effective for houses with regular function such as family activities. It is not effective to house with shop activities because door and ventilations are usually open. Additionally, house with poor hygiene behavior such as shown by poor garbage disposal is at-high-risk of dengue[13]. A study by Rinawan, et al about geographical information system (GIS) and remote sensing (RS) was applied to identify risks of houses as the breeding site of *Aedes* mosquito depicting results that Bandung city has higher risk to dengue disease. The breeding site of the mosquito at a house can be a risk ofdisease escalating to surrounding neighbourhood [6].

Green architecture of houses should also can prevent dengue disease by mosquito net on ventilations, or reduce the risk of mosquito biting by mosquito trap or distractordesign on house. Green architecture goals is not only good for environment, but also for human's health [14]. In its initiation, people movement that consists of youths and adult is crucial. It can be supported by scientist and or practitioner who can help lessen the gap between people in society and government [15]. This study was aimed to continue GIS-RS study on roofs factor association with dengue disease by initiating people movement to use mosquito net on house ventilations to cut the disease transmission and mosquito breeding sites inside house.

2. Method

Our methods was an operational research in which improvement of interventions, policies and regulations towards dengue disease prevention is our intended endpoint. Several steps were conducted such as: (1) research problems formulation from GIS-RS analysis from previous phase research in Bandung city, (2) informal and formal approach to community leaders and primary healthcare centre (Puskesmas), (3) Video education and focus group discussion (FGD), (4) initial application of mosquito nets on house in communities; and (5) advocacy to Mayor of Bandung city (was on progress).

2.1. Study Area

We started in one of subs-city (Kecamatan), Ujungberung of Bandung city where inhabited by 71,254 people in 612.79 hectares or approximately 116 inhabitants per hectare, which is less than the total density of Bandung city (2,455,517 inhabitantsin 16,731 hectares total area), 146 inhabitants per hectare. The city is the capital of West Java province in Indonesia, the third crowded city after Surabaya and Jakarta.

Figure 1. Bandung city map with Kelurahan administration border detail. The blue line highlight depicts the border of Kecamatan Ujungberung

IOP Conf. Series: Earth and Environmental Science **91** (2017) 012023 doi:10.1088/1755-1315/91/1/012023

Figure 1 depicts the city map with village (Kelurahan) administration border detail. Blue highlight is 5 Kelurahans in KecamatanUjungberung. Figure 2 illustrates the Kecamatan of Ujungberung using Red, Green and Blue color composite of Worldview 2 satellite data. Every Kelurahan has at least seven sub-Kelurahan (RW) depending on the size of the area. The northern part, Kelurahan Pasirwangi, is the highest contour reaching 925 meters above sea level, and the southern part, Kelurahan Cigending and some parts of Pasanggrahan is 687.5 meters above sea level.

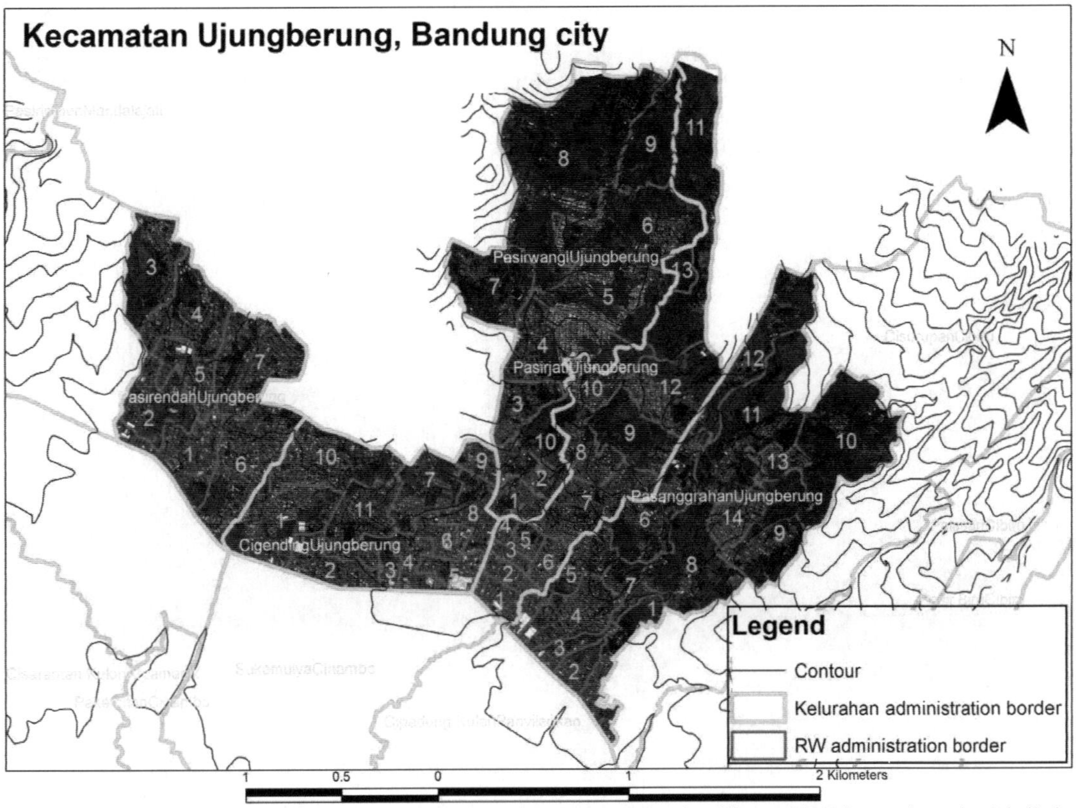

Figure 2. Administration border of villages (Kelurahan) and sub-villages (RW) numbered with light blue number on a Red, Green, and Blue color composite Worldview 2 satellite data, and contour data

2.2. Data and Analysis Procedure

Our data on the first phase: research problems formulation from GIS-RS analysis using dengue data from Bandung city health office. We also used Worldview 2 satellite data from Bandung city development planning agency to identify houses. Clearance of these data were already issued in the previous work by the government. On the first phase, we found that Kecamatan Ujungberung was at-high-risk of dengue disease infection chain. This problems was analysed using previous publication by Rinawan et al [6].

The second phase which was actually concurrent preparation of interventions, were informal and formal approaches to community leaders and primary healthcare center (Puskesmas). This was conducted by voluntarily expressing our intention to head of Puskesmas and of Kecamatan (Camat) in order to initiate community movement to prevent dengue disease in the Kecamatan.

The third phase data includes participation of mothers' community from each of Kelurahan in Kecamatan(67 persons) in health education using video that was made based on Health Beliefs Model (HBM) [16]regarding dengue disease symptoms, risk, severity, mosquito screen use on house ventilations. We evaluated the video by pretesting and post-testing using 15 knowledge 'true' and 'false' questions based on HBM.The true-and-false questions were: 1. Dengue disease is caused by virus, 2. dengue disease is transmitted by physical contact, 3. One of dengue disease's symptoms is under skin bleeding, 4. First symptoms is fever, may be continued with under skin bleeding and or stomach ache, 5. If no management after dengue disease, every patient will be cured spontaneously, 6. Dengue disease can be fatal, 7. *Aedes* mosquito is more attracted to dark place, 8. Hematemesis is one of dengue disease complications, 9. *Aedes* larvae is found in dirty water, 10. Method of identifying larvae is only using bare-eyes, 11. Water receptacles cleaning is done every 1-month period, 12. Prevention is only done when dengue disease is found, 13. Mosquito net use is part of prevention steps, 14. Mosquito net is effective if used for small ventilation only, 15. Monthly report on *Aedes* larva to Puskesmas is one of important movement in dengue disease prevention.We rated the scores and compared them using Wilcoxon signed rank test (a non-parametric test) because the data was not in normal distribution (Kolmogorof-Smirnov $p< 0.05$). These 15 questions were initially made with local content, and their construct to identify them as a set of 'knowledge' factor questions have not yet been measured.

After the video education, we continued with FGD to assess what was the problem in practicing community movement in dengue disease prevention. We did not record their voice when conducting FGD, only wrote it. Figure 3 shows the seating position when performing video education and FGD.

Figure 3. The seating position when video education and FGD

The fourth phase was initial application of mosquito nets on house in communities by discussing which house has better understanding and cooperation; andfifth, advocacy to Mayor of Bandung city was still on progress.

3. Results

Our study continuation resulted several supports: one of Camat of Kecamatan Ujungberung, village leaders (Lurah), and sub-village leaders (Ketua RW). Of 5 villages (kelurahan), one kelurahan which mainly comprised formal settlements needed more efforts, which was experts on dengue disease from university to directly explain the mosquito nets application to its community.

Table 2 shows participation of mothers community in pre and posttest, which was high enough 64 of 67 persons in the test. The 3 were late. 55 persons (86% of 64 persons) can answer all questions of pre and posttest.

Table 1. Participation of mothers community

Participation	N
Invitations	67
Participate in test	64
Full answers on pre and post-test	55

Table 2 shows ranks of posttest minus pretest using Wilcoxon Signed Ranks Test. It resulted significance ($p = 0.003$) with higher positive ranks of 32 persons and the rest negative and ties ranks.

Table 2. Ranks of posttest minus pretest using Wilcoxon Signed Rank Test.

Ranks	N
Negative ranks	12
Positive ranks	32
Ties	11
Total	55

From the FGD, informal leaders in all kelurahan's community suggested only cadres' movement was not enough, thus, youthscommunity such as youth organization in community (KarangTaruna) was mentioned to help the community movement on the mosquito nets application. The community wanted to initiate the mosquito net application even though they thought it would be not effective because only cover small ventilation using mosquito net from kinds of "tile" clothes. Figure 4 shows the small ventilations where they initiated community movement on the use of "tile" clothes as mosquito net.

Figure 4. The small ventilations on the upper part of window

The process of informal and formal advocacy was also done to the head of Bandung city health office, resulting informal manner on approaching the mayor to decide meeting time because formal manner may require much longer time to meet the Mayor.

4. Discussions

Combination of informal and formal networking to community leaders and government authorities can support initiation of a program [17-19]. In our result, mother community knowledge can be changed into better understanding using video. The role of researcher from university, professionals, and primary health care centre can be a middle networking catalyst besides the role of youth with creativity in educating mothers while advocating government authorities[15]. The middle networking can engender bridging ideas from community to government including by performing health advocacy that green architecture of a house can also be a healthy-oriented architecture [14], for example using mosquito net on house ventilations and other tools design to decrease mosquito bite and infections[12].

Only mosquito net application is not enough in green architecture because not every house characteristics is mosquito-proof [13]. However, to start with small steps is better to stimulate community movement than no start at all, or waiting too long to start a step. More importantly, the community and primary health care centre have collaborated to perform a program. Top-down program from government may not be effective without movement response. Henceforth, government should support stimulation of community to initiate a small movement then a bottom-up ideas may arise to lessen the gap between top-down decision and bottom-up ideas [15, 20].

5. Conclusions

We conclude that this work on operational research still needs to come up with further collaborations with creative ideas from creative youth, researcher, and professionals. Therefore can support more a health-oriented green architecture program for community. It may be easier to educate mothers community to understand dengue disease, but it is still questionable on how to educate youths in community including to collaborate them with mothers community.

Acknowledgments

This study was supported by PuskesmasUjungberung Indah, KecamatanUjungberung Bandung city, and Faculty of Medicine, UniversitasPadjadjaran Indonesia.

References

[1] Getis A, Morrison AC, Gray K, Scott TW 2003 Characteristics of the spatial pattern of the dengue vector, Aedes aegypti, in Iquitos, Peru *American Journal of Tropical Medicine and Hygiene* 2003;69(5):494-505.

[2] Zakaria Z, Zainordin NA, Sim BLH, Zaid M, Haridan US, Aziz ATA, et al. 2014 An evaluation of the World Health Organization's 1997 and 2009 dengue classifications in hospitalized dengue patients in Malaysia *Journal of Infection in Developing Countries* 2014;8(7):869-75.

[3] Sahana KS, Sujatha R 2015 Clinical Profile of Dengue Among Children According to Revised WHO Classification: Analysis of a 2012 Outbreak from Southern India *Indian Journal of Pediatrics* 2015;82(2):109-13.

[4] Verdonschot PFM, Besse-Lototskaya AA 2014 Flight distance of mosquitoes (Culicidae): A metadata analysis to support the management of barrier zones around rewetted and newly constructed wetlands *Limnologica* 2014;45:69-79.

[5] Harrington LC, Scott TW, Lerdthusnee K, Coleman RC, Costero A, Clark GG, et al. 2005 Dispersal of the dengue vector Aedes aegypti within and between rural communities *American Journal of Tropical Medicine and Hygiene* 2005;72(2):209-20.

[6] Rinawan FR, Tateishi R, Raksanagara AS, Agustian D, Alsaaideh B, Natalia YA, et al. 2015 Pitch and Flat Roof Factors' Association with Spatiotemporal Patterns of Dengue Disease Analysed Using Pan-Sharpened Worldview 2 Imagery *Isprs International Journal of Geo-Information* 2015;4(4):2586-603.
[7] Tran A, Deparis X, Dussart P, Morvan J, Rabarison P, Remy F, et al. 2004 Dengue spatial and temporal patterns, French Guiana, 2001 *Emerging Infectious Diseases* 2004;10(4):615-21.
[8] Hugo LE, Jeffery JAL, Trewin BJ, Wockner LF, Nguyen Thi Y, Nguyen Hoang L, et al. Adult Survivorship of the Dengue Mosquito Aedes aegypti Varies Seasonally in Central Vietnam. Plos Neglected Tropical Diseases. 2014;8(2).
[9] Montgomery BL, Ritchie SA 2002 Roof gutters: A key container for Aedes aegypti and Ochlerotatus notoscriptus (Diptera : Culicidae) in Australia *American Journal of Tropical Medicine and Hygiene* 2002;67(3):244-6.
[10] Fuller DO, Troyo A, Calderon-Arguedas O, Beier JC 2010 Dengue vector (Aedes aegypti) larval habitats in an urban environment of Costa Rica analysed with ASTER and QuickBird imagery *International Journal of Remote Sensing* 2010;31(1):3-11.
[11] Li Y, Kamara F, Zhou G, Puthiyakunnon S, Li C, Liu Y, et al. 2014 Urbanization Increases Aedes albopictus Larval Habitats and Accelerates Mosquito Development and Survivorship *Plos Neglected Tropical Diseases* 2014;8(11).
[12] Manrique-Saide P, Coleman P, McCall PJ, Lenhart A, Vazquez-Prokopec G, Davies CR 2014 Multi-scale analysis of the associations among egg, larval and pupal surveys and the presence and abundance of adult female Aedes aegypti (Stegomyia aegypti) in the city of Merida, Mexico *Medical and Veterinary Entomology* 2014;28(3):264-72.
[13] Thammapalo S, Chongsuvivatwong V, Geater A, Dueravee M 2008 Environmental factors and incidence of dengue fever and dengue haemorrhagic fever in an urban area, Southern Thailand *Epidemiol Infect* 2008;136(1):135-43.
[14] Ragheb A, El-Shimy H, Ragheb G 2016 Green Architecture: A Concept of Sustainability *Procedia - Social and Behavioral Sciences* 2016;216:778-87.
[15] Cohendet P, Grandadam D, Simon L 2010 The Anatomy of the Creative City *Industry & Innovation* 2010;17(1):91-111.
[16] Glanz K, Rimer BK, Viswanath K 2008 Health Behavior and Health Education: Theory, Research, and Practice:(Wiley)
[17] Stoto M A 2008 Regionalization in Local Public Health Systems: Variation in Rationale, Implementation, and Impact on Public Health Preparedness *Public Health Reports* (1974-). 2008;123(4):441-9.
[18] Suwanbamrung C, Dumpan A, Thammapalo S, Sumrongtong R, Phedkeang P 2011 A model of community capacity building for sustainable dengue problem solution in Southern Thailand *Health* 2011;03(09):584-601.
[19] Harris L, Coles A-M, Dickson K 2000 Building Innovation Networks: Issues of Strategy and Expertise *Technology Analysis & Strategic Management* 2000;12(2):229-41.
[20] Heintze C, Velasco Garrido M, Kroeger A 2007 What do community-based dengue control programmes achieve? A systematic review of published evaluations *Trans R Soc Trop Med Hyg* 2007;101(4):317-25.

IOP Conf. Series: Earth and Environmental Science **91** (2017) 012024 doi:10.1088/1755-1315/91/1/012024

Study Of Lampungnese Traditional Home Garden Design

R A Pratiwi[1] and Gunawan[2]
[1] Graduate Student of Department of Landscape Architecture, Faculty of Agriculture, Bogor Agricultural University (IPB);
[2] Department of Landscape Architecture, Faculty of Agriculture, Bogor Agricultural University (IPB).

E-mail: adetiyapratiwi@gmailcom

Abstract. Lampung is one area in Indonesia which has a traditional culture that comes from two groups of descents, they are ulun Lampung Pepadun and ulun Lampung Saibatin. Lampungnese traditional culture has been well-known by Indonesian people for its traditional dances, traditional clothing, or traditional home architecture. However, Lampungnese traditional home garden recently may not yet been described. Information related to Lampungnese traditional home garden is still very limited and it does not yet represented the culture based design concept. This research was directed to identify the elements of the home garden and map it into design concept of the Lampungnese traditional home garden based on information of Lampungnese traditional culture. The study was conducted by using descriptive approach through literature review, interviews and cultural exploration, as well as field observation. The study was able to identify the elements forming the Lampungnese traditional home garden, namely *gakhang hadap*, *walai*, outdoor kitchenette, firewood place, outdoor kitchen, livestock barns, as well as plants. Space layout of the home garden comprises front yard (*tengahbah/terambah/beruan*), side yard (*kebik/kakebik*), and backyard (*kudan/juyu/kebon*). Each element of the garden is located in the right place of the space layout.

1. Introduction

The geographical conditions of Indonesia with a naturally formed landscapes form a variety of patterns and behaviours in the life of a community or ethnic group. The existence of any of these ethnic groups have distinctive cultural character portrayed in regional traditions, social activities, as well as the layout of the settlement that is used to support the communities' life. One of the ethnic groups in Indonesia which has a distinctive culture is the Lampungnese people. Historically, the word Lampung comes from *anjak lambung*, which means comes from above. It is intended that Lampungnese people were come from high areas of or mountainous regions. Lampungnese people are divided into two main groups, ulun Lampung Pepadun and ulun Lampung Saibatin. Within these two groups, there is still a division based on the inhabited territories by each group. This indigenous territories diversity gives a distinctive feature for the traditional settlement in each region. Traditional houses are structures that coherent and sustainable to the environtment and to the people [1]. Traditional architecture is a manifestation and physical representation of people's culture, which include houses and other built forms [2].

Vernacular architecture is built to meet the specific needs of its inhabitants to accommodate the values and way of life based on culture [3]. Lampungnese traditional house is seen as a form of adaptation to environment, where the geographical conditions of Lampung Province is one area that is prone to earthquakes [4]. This threatening environment make traditional houses in Indonesia are built to adapt the conditions [5]. However, lack of study about traditional home garden affects in the absence characteristic of the Indonesian garden in general [6]. Ignorance about the value of culture in the architecture and the traditional home gardens results in lower community efforts to preserve the cultural heritage traditions [7]. In order to obtain a real picture of traditional Lampungnese home garden, the study about it have to be conducted, so it can be preserved and can be applied by Lampungnese people. Objectives of this study were to identify elements, layout, and interpretation and to compose them into a traditional Lampungnese home garden design concept.

2. Methods

This study was conducted at several locations in the Lampung province, namely Wana Village and Sukadana Darat in East Lampung district, Pekon Kenali in West Lampung district and Negara Tulang Bawang in North Lampung district, started from the fourth week of January 2014 until the fourth week of March 2014. This study was conducted by using descriptive approach through literature review, interviews and cultural exploration, as well as field observation. Information obtained through historical and cultural exploration with a comparative approach to compare the past and future phenomenon [8].

The study was started with preparatory phase which includes problem definition and goal setting, followed by doing a preliminary study to collect secondary data about history and culture associated with the Lampungnese traditional home garden design. The next step in this study was data collection through the literature study, interviews with experts, as well as field observation. Studies conducted to explore written sources which are obtained from journals and books recommended by the experts. Interviews were conducted to obtain cultural and historical information about Lampung. This information was obtained from village chiefs in the traditional villages and historians, followed to their knowledge about culture and history of Lampung. This stage was conducted using in-depth interview directly and deeply to the informant related about the character and culture of the Lampungnese people, spatial layout of the settlement, forming elements, typical plant that is grown around the houses, as well as cultural activities that are still carried out by the Lampungnese people. The information was gathered by using open-ended question and interactively collaborating with the informants [9].

Field observation was conducted to determine the composition of architectural elements and home garden elements of the traditional Lampungnese home garden. At this stage, a checklist of the elements should be done based on the results of the literature study and interviews about cultural aspects. Field observation conducted in five areas, which are selected based on their potency and character where the people are still doing the cultural activities. Sample houses are selected based on their authenticity. Analysis is done by identifying the elements and layouts that should be exist based of the cultural information and compared it with the current conditions in the field. The final results of the study were synthesized in the form of a concept that can be used as a basis for designing a Lampungnese traditional home garden.

3. Results

3.1. Literature Review

Based on the literature study, it is known that spatial planning both interior and exterior of the traditional houses is influenced by kinship or values among family members and relatives. The literature study related to spatial elements and symbols on Lampungnese traditional home garden can be seen in Table 1. The placement of the constituent elements of the Lampungnese traditional home garden has meaning and cultural philosophy related to functions and beliefs as well as the habits of local communities.

IOP Conf. Series: Earth and Environmental Science **91** (2017) 012024 doi:10.1088/1755-1315/91/1/012024

Tabel 1. The spatial components comparison of Lampungnese traditional home garden based on literature study

Components	Literature Study
Boundaries	Area around the houses is usually left open, without fences because the strong kinship bonds [11] [12] [18].
Architecture	Each houses has *tepas* and *tadah embun* (porch) [12] ; at the end of the stairs there are *gakhang hadap* [12] [13] [14]; beside or behind kitchen there are *gakhang dapur/kudan* [12]; buffalo skulls and ornamental carving are the symbols of the house owner's pride [14] ; traditional houses are stilt houses as adaptation to the environment [14] ; house is divided into lower part of the house, centre part as the place of living, and higher part as the sacred place for god [15].
Space	Spatial arrangements inside the house are influenced by kinship [12].
Bah lamban	The lower part of the house or *bah lamban* was not initially used, it was just a form of adaptation [11]; Yields are kept in *bah lamban* [11] [16].
Kitchen and *gakhang*	Separate kitchen is connected to the house by *geragal/jembat,* and for some occasions, temporary kitchen can be built outside the house [15] [16].
Symbol	*lamban* symbolize harmony as a place to live for a family as well as harmony with nature and environtment [15]; a house also symbolize it's owner social status and pride [14].
Yard	Front yard or *tengahbah/terambah* is commonly used for drying crops and make *teratak* [12] [13]; Back yard is called *kudan/juyu* [14]; side yard is called *kebik/kakebik* [14] ; some *walai ramik* are built around houses, some of them are built outside the village [14] [15] [16].
Orientation	Oriented or close to the river [10] [11] [14] [16] ; linear, with houses are facing each other[12]; linear, close to the river because of the necessity of water [13].
Activities	Lifecycle ceremonies; maternity ceremony, birth and infant ceremony, childhood ceremony, adulthood ceremony [17].

3.2. Interview Results

The interview results related to spatial elements and symbols on Lampungnese traditional home garden can be seen in Table 2.

Table 2. The spatial components comparison of Lampungnese traditional home garden based on interview results

Components	Interview Results
Boundaries	Traditional house has no obvious boundaries with other houses, sometimes people use hedgerows[3] ; the plant that are usually used for *langen* or the hedgerows is *jarak kuto* (*Jatropha curcas*)[5]
Architecture	Traditional stilt houses is an adaptation form to the environment[1,2,7]
Space	Spatial arrangements inside the house are influenced by kinship[4]
Bah lamban	*Bah lamban* is functioned as a storage area[2,4,6,7]
Kitchen and *gakhang*	the kitchen is built in the left side of back house close to *gakhang*[6]
Symbols	house is divided into lower part of the house, centre part as the place of living, and higher part as the sacred place for god[1] ; different floor level represent different purity level[4] ; highest part of the house (roof) is used to keep the sacred things (*pamanohan*) [3]

Yard	*Beruan* or front yard, side yard, as well as back yard or *kebon* are left open for crops sun-drying purpose[4] ; all farming activities are done on the farm, not on the yard[5] ; front yard is left empty to show the luxuriousity of the house[8] ; The houses' owners only grow beneficial plants (food and medicinal plants) on their yards[3,4] ; crops are grown on the farm, far away from house [5,6]
Orientation	Houses at Pekon Kenali are used to be oriented to Gunung Pesagi (south)[2] ; other traditional houses are used to be oriented to the river because the necessity of water[5,7] ; oriented to the village chief's house[6] ; circulation around the house is usually straight[4], to the right side of the house[2]
Activities	traditional wedding ceremony[5,8]

1. Budi Supriyanto, S.SOS, M.Hum (Lampung Museum Staff)
2. Mad Sa'ari Glr. Batin Setia (village chief, *lamban pesagi* owner)
3. Habiburrakhman
4. Iskandar Zulkarnain
5. Hj.Uzunuhir, S.Pd Glr. Suttan Lepus (*Lamban gajah meghem* owner)
6. Fasykinar Bahari Glr. Dalom Putra Wiranegara (village chief at Negeri Olok gading)
7. Arsyad Glr. Suntan Ratu Putra (village chief at Negara Tulang Bawang)
8. Amasin Glr. Suntan Alif (village chief at Negara Tulang Bawang)

3.3. Field Observation

Field observation results showed the spatial pattern around the house and its components, and the constituent elements in the villages which were used as a reference (Table 3). Most of the traditional houses in the village which can be found today have clear boundaries. Although most houses have clear boundaries, the fences are generally not too high and not all of them have gates. Unlike other villages, in Pekon Kenali houses have no clear boundaries. The settlements are built linear to follow the roadway with the houses are built facing each other, and there was no clear boundary between the houses (Figure 1).

Table 3. The spatial components comparison of Lampungnese traditional home garden in all villages

Components	Villages			
	Pekon Kenali	Kampung Wana	Sukadana Darat	Negara Tulang Bawang
Boundaries	Concrete fences, metal fences, or wood fences	Inconsistent, some houses have hedgerows, concrete fences, some houses has no boundaries at all	Concrete fences, metal fences, or wood fences	Inconsistent, some houses have hedgerows, concrete fences, some houses has no boundaries at all
Architecture	*Lamban Pesagi*	Common traditional houses	*Lamban gajah mekhem*	Common traditional houses
Bah lamban	Exist and utilized	Exist and utilized	Exist and not utilized	Exist and utilized
Kitchen and *gakhang*	Exist, located in the back of the house, unite with house	Exist, located in the back of the house, unite with house	Exist, located in the back of the house, unite with house	Exist, located in the back of the house, unite with house
Symbol	*Cagak lamban*	Buffalo skulls, ornamental carvings	Buffalo skulls, ornamental carvings	Buffalo skulls, ornamental carvings
Yard	Front yard; exist, extensive ; Side yard: exist, extensive ; back yard: exist, narrow	Front yard; exist, narrow ; Side yard: exist or non exist ; back yard: exist or non exist	Front yard; non exist ; Side yard: exist, extensive ; back yard: exist or non exist	Front yard; exist, extensive ; Side yard: exist, extensive ; back yard: exist or non exist

IOP Conf. Series: Earth and Environmental Science **91** (2017) 012024 doi:10.1088/1755-1315/91/1/012024

Orientation	Oriented to the street ; circulation directed to the side of the house	Oriented to the street ; circulation straight toward the front stairs	Oriented to the street ; circulation straight toward the house (no stairs)	Oriented to the street ; circulation straight toward the front stairs

Figure 1 (left) linear settlement pattern; (middle) house with fence but no gate; (right) houses without any boundaries

Circulations around Lampung traditional houses are generally do not have a specific pattern. Although sometimes the traditional house has two staircases and two entrances, circulation still form a straight line directly toward the main door. At the end of the stairs, the first part of the house is called *tepas* (porch). Houses are generally divided into three main parts: the front, centre, and back part of the house. The front part of the house is starting from *tepas* to the first room after the entrance. The centre part is a second room after the entrance, as well as the back part is the kitchen and *gakhang* (Figure 2). *Gakhang* is a place to wash the dishes which its floor is made of bamboo so that the water is wasted down straight. The lower part of the house or *bah lamban* is usually functioned as storage or left empty and open. Space layout of the home garden comprises front yard (*tengahbah/terambah/beruan*), side yard (*kebik/kakebik*), and backyard (*kudan/juyu/kebon*).

Figure 2. (left) kitchen; (middle) *gakhang;* (right) *bah lamban*

Traditional houses are usually in rectangular shape stilt house with 2 meters high columns, where the shorter side is the front side of the house. After comparing the results of the literature study and interviews with field observation, it is found that the current condition shows a mismatch conditions that should be exist based on the culture. There have been many changes in houses and their surroundings, it is suspected that those changes happened because of easier acces to the village. House yards are usually left empty without any boundaries so that neighbour can pass through it. Yards can be categorized as a public space, with houses built close to each other. As a result of this proximity, the ownership boundaries seemed unreal. This kind of spatial layout is found in almost all traditional settlements. This happens because most of the villagers are family related. Based on the comparison of the literature study, interviews, and field observation, forming elements in Lampungnese traditional home garden are *gakhang hadap*, *walai*, outdoor kitchenette, firewood place, outdoor kitchen, livestock barns, as well as plants (Table 3). These components are exist to support and fulfill the needs of the community, both everyday needs and the needs of traditional ceremonies. The placement of these components does not always have a certain meaning, but can adapt to the needs and environmental conditions.

Table 4. Comparison of the constituent elements of Lampungnese traditional home garden

Elements	Literature Study	Interviews	Field Observation
Gakhang hadap	✓	✓	
Walai	✓	✓	
Outdoor kitchenette	✓		
Firewood place		✓	✓
Outdoor kitchenette	✓		
Livestock barns	✓	✓	✓
Plants		✓	

There were no *gakhang hadap* and *walai* found on field observation. Whereas *gakhang hadap* characterizes traditional customs and beliefs of *ulun* Lampung, house is a space with higher hierarchy with its surrounding environment so that when entering the house someone must be in a clean state by washing their feet on *gakhang hadap*. Meanwhile the absence of *walai* happens because people nowadays prefer to keep their rice in the house not in a separate building, and not all the people in the village have their own rice fields. Therefore, to strengthen the character of Lampungnese traditional home garden *walai* existence as a constituent element must be present. In particular purposes outdoor kitchenette is needed in Lampungnese traditional home. For this purpose the side yard near the kitchen in the house can be used. Outdoor kitchenettes which are usually used to make brown sugar and copra are usually built permanent and located behind the house closed to *walai* [16]. To support cooking activities in both indoor and outdoor kitchenettes, *ulun* Lampung are still using firewoods. Firewood in large quantities can't be kept in the house and require a special storage. From the results of field observation, a place to store firewood is usually placed in *bah lamban* right under the kitchen or in the yard beside the house near the kitchen.

A garden is a micro-scale landscape, so that the constituent elements can be likened to the landscape forming element, such as landforms, plant materials, structures, hard elements, and water elements [19]. Home garden is a form of outdoor space around the house, so that its utilization and beauty are determined by the owner [20]. There are several principles that should be applied constantly in designing the garden. These principles include the design framework (order), repetition, rhythm, unity, balance, proportion and scale, as well as emphasize [21]. The application of these principles in the home garden is to make the house harmony with the environment, to form a space by arranging elements in that area, to optimize the use of the site, as well as to build outdoor space that is useful and appropriate to the needs of home owners. Aside from making it aesthetically pleasing, the design must also be able to meet the needs of human beings and should be functional [22].

3.4. Design Concept

Space in a house can generally be divided into four, private space, public space, semi-public space, and service space [23]. Public space is an area that is accessible to everyone and usually located at the front of the house. While the service space is an area that supports the activities of the home owner, private space is a room that can only be accessed by the home owner. Semi-public space is an area of activity for homeowners, usually also accommodates any social activities. Side yard and *bah lamban* are usually used as a place to socialize in any ceremonies. In certain situations, *sebai* (women) group used to cook together around the courtyard next to the open kitchen or under *sudung* and *bah lamban*. In the side yard near the kitchen anyway to ease the activity is usually built a place to store firewood. Therefore, the side yard of the house as well as all *bah lamban* can be classified as semi-public space.

In contrast to the side yard, front yard can be classified as a public space because it is open and freely accessible by anyone. Meanwhile, elements such as *walai*, outdoor kitchenette, and livestock barns are usually located in the back yard. Those elements are elements that are built and used to support the live of house owner so that it can be classified as elements in service area. Based on these considerations, the classification of space in the house and Lampungnese traditional home garden is done. When compared with the concept of space by [23], it shows that Lampung traditional houses accommodate more space for public and semi-public use. The placement of each element is depend to the needs and habits of the people. Based on the spatial layout in terms of culture, front and side yard are left open or planted with grass. Plants that have a meaning in the culture of Lampung is *jarak kuto* (*Jatropha curcas*) which is usually planted as a hedgerows between each house and the *kayu labeu* or maja tree (*Aegle marmelos*) is planted at the edge of the yard. Based on these results, a recommendation concept of Lampungnese traditional home garden design that can support the architectural character of traditional houses is defined in figure 3.

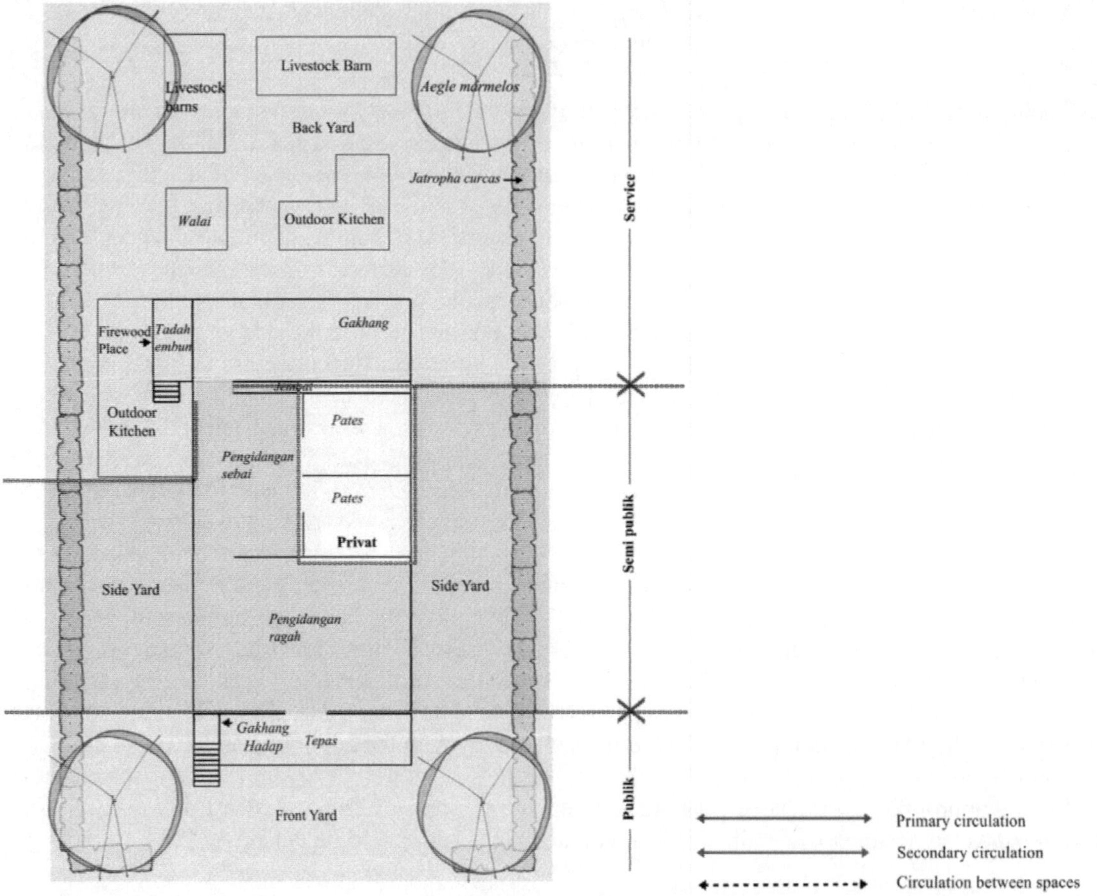

Figure 3. Concept plan of Lampungnese traditional home garden design based on culture

4. Conclusion

Traditional houses are usually in rectangular shape stilt house which spatial arrangements are influenced by kinship. House is usually divided into lower part of the house, centre part as the place of living, and higher part as the sacred place for god, while the yard is usually divided into front yard (*tengahbah/terambah/beruan*), side yard (*kebik/kakebik*), and backyard (*kudan/juyu/kebon*). The forming elements of Lampungnese traditional home garden are *gakhang hadap*, *walai*, outdoor kitchenette, firewood place, outdoor kitchen, livestock barns, as well as plants. The placement of each elements are based on eficiency to support the daily activities of the house owner. The concept is achieved by dividing the room into four by function, element placements, and the use of public space,

IOP Conf. Series: Earth and Environmental Science **91** (2017) 012024 doi:10.1088/1755-1315/91/1/012024

private space, semi-public space, and service space. Space both inside and outside the house are connected and can be accessed with a linear circulation. Plants that are used in Lampungnese traditional home garden are *jarak kuto* (*Jatropha curcas*) and *kayu labeu* (*Aegle marmelos*).

References

[1] Baran M, Yildirim M, Yilmaz A 2011 *Evaluation of Ecological Design Strategies in Traditional Houses in Diyarbakir Turkey* Journal of Cleaner Production 19(2011) 609-619

[2] Yildirim M, Turan G 2012 *Sustainable Development in Historic Areas: Adaptive Re-Use Challenges in Traditional Houses in Sanliurfa, Turkey* Habitat International 36(2012) 493-503

[3] Suharjanto G., 2011 *Membandingkan Istilah Arsitektur Tradisional Versus Arsitektur Vernakular: Studi Kasus Bangunan Minangkabau dan Bangunan Bali* Jurnal ComTech 2(2): 592-602.

[4] Rostiyati A 2013 *Tipologi Rumah Tradisional Kampung Wana di Lampung Timur* (Bandung: Balai Pelestarian Nilai Budaya Bandung)

[5] Juwita R *et al.* 2016 *Structural Test of Traditional Arfak House in Papua* Procedia Engineering 171(2017) 1542-1549

[6] Maningtyas RT, Gunawan A, and Munandar A 2013 *Kajian Desain Lanskap Permukiman Tradisional Madura* Jurnal Lanskap Indonesia 5(2)

[7] Ibrahim W, Nandang, 2011. *Arsitektur Tradisional Kenali Salah Satu Kearifan Lokal Daerah Lampung,* Jurnal Rekayasa. 15(1): 59-66

[8] Sulasman 2014 *Metodologi Penelitian Sejarah* (Bandung: Pustaka Setia)

[9] Creswell J W (2013) *Research design: Qualitative, quantitative, and mixed methods approaches* Sage publications

[10] [Depdikbud] Pusat Penelitian Sejarah dan Budaya, Proyek Penelitian dan Pencatatan Kebudayaan Daerah 1978 *Adat Istiadat Daerah Lampung* (Jakarta: Balai Pustaka)

[11] [Depdikbud] Bagian Proyek Pengkajian dan Pembinaan Nilai-Nilai Budaya Daerah Lampung 1997 *Adat Istiadat Daerah Lampung* (Jakarta: Balai Pustaka)

[12] Dinas Pendidikan Provinsi Lampung 2002 *Rumah Tradisional Lampung* (Bandar Lampung: Proyek Pelestarian dan Pemberdayaan Budaya Lampung pada Dinas Pendidikan Provinsi Lampung)

[13] [Depdikbud] Pusat Penelitian Arkeologi Bandung, Balai Arkeologi Bandung, 1996. *Laporan Penelitian Arkeologi Masa Islam (Kajian Etnoarkeologi Pemukiman) Situs Wana-Melinting Lampung Tengah*

[14] [Depdikbud] Kantor Wilayah Provinsi Lampung Bagian Proyek Inventarisasi dan Dokumentasi Kebudayaan Daerah 1987.*Arsitektur Tradisional Daerah Lampung* (Jakarta, Balai Pustaka)

[15] [Kemenbudpar] Direktorat Tradisi, Direktorat Jenderal Nilai Budaya Seni dan Film, 2011 *Arsitektur Tradisional Daerah Lampung* (Jakarta, Balai Pustaka)

[16] [Depdikbud] Direktorat Jenderal Kebudayaan, Direktorat Sejarah dan Nilai Tradisional, Bagian Proyek Inventarisasi dan Pembinaan 1992 *Dapur dan Alat-Alat Memasak Tradisional Daerah Lampung* (Jakarta, Balai Pustaka)

[17] [Depdikbud] Proyek Inventarisasi dan Dokumentasi Kebudayaan Daerah Lampung 1982 *Upacara Tradisional Daerah Lampung* (Bandar Lampung, Balai Pustaka)

[18] [Depdikbud] Direktorat Jenderal Kebudayaan, Direktorat Sejarah dan Nilai-Nilai Tradisional, Bagian Proyek Pengkajian dan Pembinaan Nilai-Nilai Budaya Daerah Lampung 1998 *Peranan Nilai-Nilai Tradisional Daerah Lampung Dalam Melestarikan Lingkungan Hidup* (Bandar Lampung, Balai Pustaka)

[19] Starke B, Simonds J O 2013 *Landscape Architecture: A Manual of Environmental Planning and Design* (New York: McGraw Hill Professional)

[20] Ingels JE 2004 *Landscaping Principles and Practices, 6th Edition* (New York: Thomson Delmar Learning)

[21] Crowe S 1981 *Garden Design* (Chicester: Packard Publishing Limited)

[22] Rutledge A 1985 *A Visual Approach to Park Design* (New York: John Wiley & Sons)

[23] Booth N K 1988 *Basic Elements of Landscape Architechtural Design* (New York: Waveland Press Inc).

Fitness Parks: A Comparative Study of the Components of Jakarta-Manila Parks and their Responsiveness to Support Physical Activities

Franklin S. Fontanoza Jr., Nappy L. Navarra, D.Engg.

Philippine Association of Landscape Architects

Email : fsfontanozajr@gmail.com, nappy.navarra@gmail.com

Abstract. Fitness has become more popular due to the cultural phenomenon that being fit can enhance one's perception of beauty. The sprouting of various outlets for physical activity such as bodybuilding gyms that cater to weightlifting, outdoor group dance classes, sports camps and cause-oriented marathons can be noticed in numerous parts of the world. But slowly its concept, that being fit is a mere physical representation of beauty, is shifting into a more health-oriented consciousness. Annual reports have shown that coronary heart disease is still in the top rank of the death causes in the world. This information has led more people to protect their health through several lifestyle improvements, with regular exercise being one of these methods to achieve health goals. Its numerous benefits range from the lowering of blood pressure, heightened learning capacity to the improvement of mood. The health-rooted awareness of the need for physical activities to support one's daily requirement has spread worldwide and has now been recognized by a lot of people.

Parks are usually designed with amenities such as playgrounds, pathways and wide open spaces where people from all walks of life convene, interact with each other and do various physical activities. With this in mind, the capacity of parks to host such activities has to be studied to determine which components do people who engage in active healthy lifestyles find highly attractive and usable. An analysis of such could lead to effective space programming of our local neighborhood parks making it more perceptive to the physical needs of the people. Two major sports complexes from South East Asia have been used as case studies to assess the responsiveness of the locals to the amenities offered in each complex to address health goals. The comparison revealed that the Gelora Bung Karno Complex in Jakarta, Indonesia has more activity-oriented amenities and longer operating hours, making it more receptive to meet the physical activity requirement.

Keywords: fitness parks, components, physical activity, health

1. Introduction

Ancient history has seen the primitive man engaging in physical activities as shown in his nomadic lifestyle of traversing long distances to move from one tribe to another on his bare feet. With the basic human need to provide food for himself and his family, he has developed a strong body to hunt, gather food and survive through harsh conditions. Over time, man started to build to empires and during those glorious eras, physical activities evolved and became part of his life. Athens and Sparta have seen a lot of vigorous training fit for the battlefield in their constant plight to claim territory over lands and the natural resources these possessed.

Ancient civilizations in China have promoted physical activities such as Cong Fu gymnastics, archery and dancing due to Confucius teachings promoting an active lifestyle. While in other parts of South East Asia, Sepak Takraw, a form of sport, started as early as the 15th century. In 1959, the South East Asian games or SEA games was born, scheduled every two years to bring South East Asian countries together.

Indonesia's fitness history started with tribal practices that dealt mostly with rituals and war folk dance. Archery became a sport in its classical times; football and badminton were introduced by the Dutch when the country was colonized. Badminton and basketball have become two of the most popular sports catering to local youths all over Indonesia.

The Philippines have almost had the same historical background with other South East Asian countries. Pre-Spanish period, the natives had their share of wandering forests for food and shelter, developing stronger limbs from running and climbing mountainous terrain. During the colonial era, several sports were introduced by the Spaniards and Americans such as Basketball, Soccer and Volleyball. Swimming has become a favorite pastime of children, most especially during warm summer months as they swam in rivers and the sea. The love for sports and the outdoors have been incorporated in the daily routine of Filipinos.

In the development of high-speed technology, urbanization became a global phenomenon and affected the entire world. Population increased immensely in cities, as rural-born families relocated and made settlements in urban areas, increasing the concentration of people in a limited city space. Transportation naturally developed after as people desired to move quicker from one place to another. Private cars and public utility vehicles were improved to ease the accessibility of a lot of places. This came with a price, resulting to pollution. With an expanding population, overcrowding became a problem as the open spaces originally allotted for recreation decreased in area and were converted to housing projects. People further increased their health risks as they became more inactive due to the comfort of vehicles at our disposal. Air pollution did not help either as it lessened the favorability of walking outdoors. The World Health Organization published an article through www.who.int stating that in 2012, ischemic heart disease, stroke, respiratory infections and lung disease topped the cause of death worldwide. The lack of physical activities has become a major health concern that our generation should start taking more seriously.

Exercise and other physical activities' importance cannot be overstated especially when it reaps a lot of health benefits. Researches have shown that exercising within a recommended intensity of 65 percent – 75 percent maximum heart rate for at least 30 minutes a day could lower blood pressure, make a person feel less anxious and release endorphins to create a more positive psychological state and enhanced mood. Exercise can limit excess weight gain by burning calories. Various physical activities such as sports and dance classes can also improve flexibility, stabilize heart rate and strengthen the immune system.

Outdoor spaces provide venues for physical activities to prevent the sedentary lifestyle that most of us experience in our lives. Being outdoors gives us a boost of Vitamin D which we all need to reduce pale skin and for us to absorb calcium. Exercising outdoors introduces the body to unpredictable factors such as wind resistance and hilly slopes. Communing with nature, plants in the gardens supply the much-required oxygen during heightened physical activity. On top of it all, exercising outdoors is free of charge and can be done in your own time and pace.

1.1. Problem Setting

21st century problems, including the health risks that we face today, have resulted from the urbanization of our cities. These have left us with limited open spaces which resulted from the negligence of proper implementation of responsive design or afterthoughts of poor city planning as damage control. Our open spaces have been affected to the point that the parks' capacity to provide livable leisure areas for growth and recreation has significantly diminished. Eventually, in the continued mismanagement of our lands, the negative effects would be felt by its users through the bodily manifestation of health risks.

Merging the physical need of the body to perform daily exercises and the desire of the human spirit to experience the bounty of nature should result to outdoor parks that incorporate spaces and amenities to house numerous physical activities. Neighborhood parks in Manila have been redesigned with jogging paths and several streets have included bike trails and improved sidewalks, to accommodate the growing requirement of Filipinos who engage themselves in physical activities. In Indonesia, empty lots are automatically planted with vegetation and kept maintained to be in utmost condition. Most of their parks have an area for fitness, equipped with outdoor equipment. Family members playfully use them, a strategy to introduce children to fitness at a young age.

But data regarding parks that should provide these amenities or at least have spaces allotted for exercise have yet to be gathered. The responsiveness of parks to cater to the daily physical activity requirement has yet to be assessed and checked whether the design of open spaces to promote fitness and healthy living are satisfactory or would need improvement.

This paper would try to address the following concerns:

1.1.1. How can a park become responsive to address the fitness needs of its users that could significantly lead to an increase in their physical activity level?

1.1.2. Which park amenities do patrons of the parks use most to meet their daily physical activity goals?

1.1.3. What park amenities attract people to use the park and exercise outdoors?

1.1.4. Why do patrons of the parks visit the specific parks?

1.2. Goals and Objectives

The main goal of this paper is to assess the responsiveness of the park to the physical activity requirement of its users. It tries to determine which park amenities do patrons deem attractive and useful based on the usability and accessibility of that particular amenity. It could also give initial information if they exercise in the park due to its proximity from their point of origin or if it is the unique features or specific amenities of the park that attracts them.

1.3. Scope and Delimitation

The two parks are chosen based on the following criteria: Relative size of a park that could house several park amenities, Availability of park patrons (runners and joggers) in the parks, Existence of at least a track oval or jogging paths, outdoor exercise equipment or sports facility that can be used by patrons to exercise, Proximity to residential properties and office buildings and Accessibility via roads and pedestrianized sidewalks along its perimeter.

Only two parks will be assessed in this study to have at least reasonable comparative data. One is the PSC PhilSports Complex in Pasig City in Manila, Philippines while the other is Gelora Bung Karno with Krida Loka in Jakarta, Indonesia. Gender and age bracket estimates will be observed for future reference. The level of fitness capacity of park users will not be included. The summer months of March and April will be considered when the weather is conducive and tolerable enough to do outdoor physical activities.

1.4. Significance of the Study

The study would benefit landscape architects, urban planners and the local government in planning and coming up with park designs that could effectively cater to the daily physical activity requirements of their users. Park management team would also benefit from the study using the information gathered to promote the well-designed park to a lot of guests and fitness enthusiasts which could eventually lead to an increase in their revenue. This could also serve as a supplementary study for researches that deal with fitness and health.

IOP Conf. Series: Earth and Environmental Science **91** (2017) 012025 doi:10.1088/1755-1315/91/1/012025

2. Methodology

The following methods were used.

2.1. Review of Related Literature

Public parks and recreation in the United States are studied in this article by the President's Council on Physical Fitness and Sports, posted in March 2008. Several researches in the article mentioned of the parks' ability to promote physical activity and recorded how much physical activity happens inside them. Park characteristics have been described and assessed based on what Americans find most attractive and have come up with the following considerations: park location (its proximity and walkability to residential properties and commercial buildings for access), park features (landscape elements such as trails, lawns, playgrounds and picnic groves), park programming (activities that the locals find interesting such as activity classes, cultural events and concerts, and sports competitions), park outreach (people's awareness of a park nearby through promotional activities and methods such as flyers, agencies and other social media) and park safety and maintenance (the need for safety and the quality of the landscape elements to be well maintained).

The paper written by Cohen, et. al. entitled *Parks and physical activity: Why are some parks used more than others?* studies how many people use the parks and determines which park feature the majority prefer to use. The authors used surveys and interviewed 51 park directors, park users and residents located around the parks. Direct observations have been made to determine which park features are more likely to be used during a seven-day observation period. Several information have been gathered such as which gender uses which type of outdoor feature more, the park's size capacity to house more organized activities that could lead to more park users and which park amenities are mostly used and neglected during those observation months. The study has concluded that focusing on park programming and promoting unique park features specific to the park would lead to an increased attendance of park users.

2.2. Site Criteria

The sites were chosen on the following parameters: Park location (proximity to home or workplace), Park amenities (jogging trails, basketball courts, outdoor gym equipment, etc.), Park programming (seasonal social activities such as sports events, zumba, yoga, etc.), Park outreach (availability of information about the park) and Park safety and maintenance.

2.3. On-site observation

Park patrons using the amenities were observed on site. The quantity was approximated during the chosen period of observation to have primary data on the number of people patronizing the amenities of the park.

Some park proponents were asked by the researcher about their usage of the park, like which amenity they use and enjoy most, how often they come to the complex and whom they usually come with. The security guard of the track oval was also asked about his general observations of the users of the site.

2.4. Site photos

Photos were taken to document the parks and their amenities to visually illustrate the on-site conditions. This could also help in documenting details that the researcher may have missed during the observation period.

3. Results and Discusion

The Gelora Bung Karno Stadium within the Gelora Senayan Complex, encompassing 284.2 hectares, is located in Central Jakarta, Indonesia. It is owned by the Football Association of Indonesia and has a

IOP Conf. Series: Earth and Environmental Science **91** (2017) 012025 doi:10.1088/1755-1315/91/1/012025

seating capacity of 88,000 on grass surface. Back then, it used to house 120,800 standing persons but has been reduced to 100,800 due to renovations made for the 2007 Asian Cup. Named after Indonesia's first president Sukarno, it mostly stages football matches which became a popular sport in the country and other international sporting activities such as the Asian Games, the SEA Games and the Tiger Cup. It was also the venue for the Pope John Paul II's visit wherein he said a Catholic mass and the centennial anniversary of Indonesian National Awakening day.

The Gelora Bung Karno Sports Complex has the following amenities: Gelora Bung Karno Stadium, the main stadium for football competitions, Madya Stadium, the smaller sized stadium for track and field events, Athletic Track Ovals (private and public), Outdoor gym equipment, Tennis Stadium (indoor and outdoor courts), Aquatics Stadium which has a capacity of 8,000 as well, Hockey field, Baseball field/ Softball field, Indoor gymnasiums for archery, wrestling, karate, silat and other martial arts sporting events, Senayan golf course, Lapangan A, B, C are several open air football training fields for the training of Indonesia's national football team, Al Bina mosque, Car circuit controlled by radio, and the Central Office of the National Sports Committee of Indonesia.

The Philippine Institute of Sports Complex or PhilSports Complex is a national sports development in the Philippines. The Philippine Sports Commission shares the total land area with the Department of Education. It is bound by Meralco Avenue, St. Paul Street, Molave Street, St. Martin Street and Capt. Javier Street. As stated in the Republic Act 6847 which gave birth to the Philippine Sports Commission, there was a need for Filipino athletes to have fully equipped facilities, complete with several sports amenities and dormitories to support them.

The PhilSports Arena has a seating capacity of 10,000 persons while the Football and Athletics Stadium can house up to 20,000 heads. Curfew hours are from 5:30 am to 10:00 pm.

The PhilSports Complex has the following amenities: Building offices, Dormitories, Karatedo gym on the roof deck, A dining hall, Amphitheatre, Fencing Hall, Meditation area and lagoon, Mini forest, PSC Gym – Basketball court, Bake shop, Swimming pool, Pool grandstand, Diving pool, Water reservoir, Multi-purpose area, Bleachers where they hold the dance/ sports training center and muay thai, Track oval, and the Grandstand.

3.1. Park location

The Gelora Bung Karno Sports Complex is bound by main roads and can be accessed through several roads. It has several entry points and can easily be accessed via public transportation and private vehicles. Most of the sports training facilities are open to public (Table 01).

The PhilSports Complex can be accessed through two entry points, one is for vehicular entry and the other one, is a one-way exit point which is also used for pedestrian access. It is a private complex since there are several office buildings and dormitories within the compound. Several facilities are open to public if not in use by the local athletes.

3.2. Park amenities

The Indonesian sports complex of Gelora Bung Karno has main courts but has an advantage with the presence of football fields since football has become a very popular sport in the country. Indonesia has national teams playing for various sports. The track oval is very busy on weekends, as runners and joggers flock the track oval. Outdoor fitness equipment is strategically located along the oval and freely used by many.

PhilSports Complex is also packed with runners and joggers on weekdays and weekends, although there is a certain period where the track oval is empty due to intense afternoon heat. The swimming pool is conducive for doing laps all year round, but mostly becomes occupied during the summer months since swimming lessons are popular for kids to learn in their vacation. Basketball athletes train in the indoor courts mostly for the governor's cup and commissioner's cup.

Most of the sport facilities inside Gelora Bung Karno Sports Complex and the PhilSports Complex have curfew. Depending on the event and usually rented out by users, the operational hours are relatively compliant to users' schedule. Gelora Bung Karno Complex has two athletic ovals: one is for the athletes, the other for the public. The public track oval opens early and closes at late hours. PhilSports Complex's track oval is open every day from 5:30 am to 9:00 pm. Usage of the track oval is free. The diving area is limited to athletes only as well as the private dormitories. The swimming pool is open from 8:00 am to 11:30 am and resumes by 1:00 pm up to 4:00 pm in the afternoon, Tuesdays thru Sundays. It is closed on Mondays for cleaning and maintenance. Pool management charges 45 Php per head in the entrance.

3.3. Park programming

Social activities and sporting events are set in the parks' calendar of events. Renovations ongoing, The Gelora Bung Karno Sports Complex is scheduled to host the SEA Games in the Main Stadium in 2018.

The PhilSports Complex has been a favorite venue of basketball tournaments. Mostly this sport is played in the complex.

The user-friendliness of the amenities attracts people to utilize the facilities. High volumes of users can be noted in most amenities of the Gelora Bung Karno Sports Complex. Amenities in PhilSports Complex, though without fee requirement, have a small patronage. This probably might be caused by the lack of information about the complex having those facilities to offer or maintenance issue.

3.4. Park outreach

The Gelora Bung Karno Sports Complex can be viewed online thru www.gelorabungkarno.co.id. The list of facilities, venues and events can be found in the site. It also has links to other social media.

The PhilSports Commission launched its website thru www.web.psc.gov.ph. It has been organized to view the facilities in the complex as well as venue rental fees for each facility depending on the type of event or training. The calendar of events can also be viewed and the schedule the amenities are open.

3.5. Park safety and maintenance

The parks are open to public but certain areas have limited access unless an event is scheduled in that area. The track oval of the Gelora Bung Karno Sports Complex is free to use as well as the outdoor gym equipment which are used by enthusiasts with higher intensity training goals other than just cardio exercises. The oval may be gated but can be accessed through opened gates. The Krida Loka which is the designated fitness jogging park has limited access since this amenity is gated and can only be accessed through its main gate on the opposite side of the main stadium.

Although PhilSports Complex is a gated development with restricted access, it has a relatively wide operating hours that can accommodate the public. Security guards are placed strategically within the complex. The most used facilities such as the swimming pool, the indoor basketball court and the track oval are well maintained but the outdoor amenities such as the lagoon, meditation area and amphitheatre are neglected.

Table 1. Comparison of Gelora Bung Karno Complex and PhilSports Complex.

	Gelora Bung Karno Complex	PhilSports Complex	Both
Park Location	• Several entry points • Most of the sports training facilities are accessible to the public	• Limited to two entry points for pedestrians • One entry point is a two-way entry-exit point while another is meant for exiting vehicles only	• Bound by main roads • Can be accessed by pedestrians through sidewalks • Can be accessed via public and private vehicles

Park Amenities	• Has the following amenities:
	  • Gelora Bung Karno Stadium, the main stadium for football competitions
	  • Madya Stadium, the smaller sized stadium for track and field events
	  • Athletic Track Ovals (private and public)
	  • Tennis Stadium (indoor and outdoor courts)
	  • Aquatics Stadium which has a capacity of 8,000 as well
	  • Hockey field
	  • Baseball field/ Softball field
	  • Indoor gymnasiums for archery, wrestling, karate, silat and other martial arts sporting events
	  • Senayan golf course
	  • Lapangan A, B, C are several open air football training fields for the training of Indonesia's national football team
	  • Al Bina mosque
	  • Car circuit controlled by radio
	  • Central Office of the National Sports Committee of Indonesia
	• Has football fields
	• Has outdoor gym equipment
	• Most of the facilities are open
	• The Stadiums are open to the public during events only
	• The Athletic track oval inside the stadium is used by athletes only, but the Athletic track oval outside is for public usage
	• The Aquatic stadium charges 15,000 during weekdays and 20,000 during weekends
		• Has the following amenities:
		  • Building offices
		  • Dormitories
		  • Karatedo gym on the roof deck
		  • A dining hall
		  • Amphitheatre
		  • Fencing Hall
		  • Meditation area
		  • Mini forest
		  • PSC Gym – Basketball court
		  • Bake shop
		  • Swimming pool
		  • Pool grandstand
		  • Diving pool
		  • Water reservoir
		  • Multi-purpose area
		  • Bleachers where they hold the dance/ sports training center and muaythai
		  • Track oval
		  • Grandstand
		  • Lagoon
		• Most of the amenities are open except for the Diving pool which is for athletes only and the dormitories and the dining hall which are for residents only
		• An outdoor volleyball court is currently being constructed
		• The Swimming pool charges 45Php per entry.
			• The main stadiums are open to public
			• The Athletic track ovals are open to public
			• The Aquatic stadium and swimming pools are open to the public
			• The Central offices are closed to the public unless you are an employee
			• Both sports complexes have operational hours

Park Programming	• Social activities and sporting events are posted in the calendar of events • The South East Asian Games are scheduled in 2018 • High volume of users can be noted using the amenities	• Basketball tournaments have been a favorite event in the sports complex • Moderate volume of users can be noted using the amenities	• Promote games and sporting activities
Park Outreach	• www.gelorabungkarno.co.id • Dual language both in Bahasa and in English • Links to other social media accounts	• www.web.psc.gov.ph • Rental fees per amenity can be viewed for easy reference • Contains links to other government agencies	• Both have browser-friendly websites • Has calendar of events
Park Safety and Maintenance	• The park amenities are functional and kept in good condition • Krida Loka, a jogging park, is gated and access to it is limited to one entry point • Krida Loka's maintenance needs to be improved	• Security guards are placed strategically in stations around the site • The facilities most used such as the swimming pool, the track oval and the basketball gyms are maintained, while the rest needs proper maintenance	

4. Conclusion

A well-designed park has the capacity to provide spaces for outdoor physical activities which the public can freely access and enjoy. It should have spaces for various recreational activities to meet the daily physical activity requirement of a person to grow strong and healthy. The benefits of exercise and doing sports in a space-programmed area are numerous as it significantly promotes overall health and well-being.

To achieve optimum health, exercise should be a part of one's daily routine. Thus, spaces to do exercise and other physical activities should be within the reach of every person. Public parks that are free and accessible can provide such spaces to house these activities. The responsiveness of our public parks to the physical daily requirement of a person can significantly lead to better space programming and in return could attract more park patrons to use the facilities to reach their fitness goals. Gelora Bung Karno Sports Complex in Central Jakarta, Indonesia has proved to be more responsive to locals' physical activity requirement based on the completeness of sports amenities and accessibility of most facilities within the complex. It positively responds to the call of the public for a venue where they can exercise and participate in various physical activities as recognized in the accessibility and free usability of the amenities that are kept in good condition, while still maintaining quality spaces for the local athletes to grow and train. Much can be learned and emulated by the Philippines to address the issue of inactivity by maximizing access to these facilities, and making a strong commitment to promote health and fitness through exercise, sports and other physical activities to the public through the local health department.

5. Acknowledgment

Many thanks to the patient professor and classmates of LArch 255 and Arch. Christian Valenzuela for accompanying the researcher in his data collection for this paper. Special thanks to Rina Dwica Desyana, Cici Nurfatimah for the warm accommodation in Bogor Agricultural University (IPB) and Agus Susanto for showing us around. Immense gratitude goes out to the family for funding the researcher's studies.

References

[1] Baran, P., Bocarro, J., Cosco, N., Edwards, M., Fang K., Floyd, M., Moore, R., Smith, W., and Suau, L. *Park-Based Physical Activity Among Children and Adolescents.* Retrieved from http://dx/doi.org/10.1016/j.amepre.2011.04.013

[2] Besenyi, G., Hipp, J., Hughey, S., Kaczynski, A., Schipperijn, J., Stanis, S., and Wilcox, S. *ParkIndex: Development of a standardized metric park access for research and planning.* Retrieved from: http://dx/doi.org/10.1016/j.ypmed.2016.02.012

[3] Cain, K., Carlson, J., Conway, T., Frank, L., Kerr, J., King, A., Saelens, B., and Sallis, J. *Interactions between psychosocial and built environment factors in explaining older adults' physical activity.* Retrieved from http://dx/doi.org/10.1016/j.ypmed.2011.10.004

[4] Cohen, D., Kaczynski, A., and Mowen, A. *The Potential of Parks and Recreation in Addressing Physical Activity and Fitness.* President's Council on Physical Fitness and Sports. Reseach Digest, Series 9, No.1. (2008).

[5] Cohen, D., Han, B., and McKenzie, T. *Quantifying the contribution of neighborhood parks to physical activity.* Retrieved from: http://dx/doi.org/10.1016/j.ypmed.2013.06.021

[6] Cohen, D., Derose, K., Marsh, T., Martinez, H., McKenzie, T., Setodji, C. and Williamson, S. *Parks and physical activity: Why are some parks used more than others?* Retrieved from: http://dx/doi.org/10.1016/j.ypmed.2009.08.020

[7] Floyd, M., Taylor, W., and Whitt-Glover, M. *Measurement of Park and Recreation Environments That Support Physical Activity in Low-Income Communities of Color.* Retrieved from http://dx/doi.org/10.1016/j.amepre.2009.010.009

Identifying Creative Urban Landscape Towards Creative Tourism in Bandung: *A Preliminary Study*

F Sinatra[1] , Kurniasih[1,2] and Indradjati[1,2]
[1)] Researcher of Urban Planning and Design Laboratory, SAPPK ITB, Indonesia
[2)] Lecturer of City and Regional Planning, SAPPK ITB, Indonesia

Email: sinatrafran88@gmail.com

Abstract. This article discusses about creative urban landscape towards creative tourism in Bandung. The objectives of this study are to identify and organize strategic plans to promote creative cluster and creative tourism. Three dimensions of creative space that are studied in this research are creative software landscape, creative hardware landscape, orgware. The results of this study suggests that creative spaces in Bandung need to reformulate creative clusters as some of the creative clusters have creative enclaves or sub creative clusters. There are eight main clusters and eighteen enclave clusters with each main and enclave clusters supported by creative software and hardware landscape. Both the creative software landscape and creative hardware landscape are designed to stem from the cluster theme to enhance tourists' experiences.

1. Introduction

Bandung has been a distinguished tourist destination in West Java for its beautiful landscape, scenery, and culture since the colonialism era. The attraction resulted in a growth of population in the last decade and making Bandung a newly metropolitan area with a dense population. Consequently, this has led to a significant increase of urban landscape and risk in tourism. Uncontrolled development has brought and impact to the quality of urban tourism or recreational activities negatively. The emergence of creative tourism was initially an effect due to the development of the creative class and creative economy. Creative communities, which were focused on business and home industries, have influenced the rise of creative economy. Distros were established as workshops for the products and were positively received by the market, which afterwards developed gradually and not limited to fashion.

Various innovations were discovered by the creative class to improve their product value. The idea of integrating shopping and creative tourism has brought a new experience to the guest. Visitors did not only buy the products but are also involved in the creative process of generating their desired designs. The success of creative tourism is to include the tourist in the activities. Santa Fe's working definition of creative tourism as a tourism activity that engages and offers an authentic experience to the visitor. The visitor could also participate in the learning process of the art, heritage, or unique traits of the destination and create a connection with the locals. In contrast with conventional tourism, the tourists are less involved with the process. This transformation shows that a shift of process has occurred between the tangible and intangible tourism resources.

The potentials of creative tourism in economic value has induced Bandung's local government to intervene, support, and enhance the quality by developing creative tourism attractions such as Kampoong wisata (urban tourist village) and festivals. Greg Richard [5] explained that modern development of creative tourism lies in the transformation of tangible elements to intangible elements of an area's culture for the tourist to consume. Three dimensions of creative development can be defined as creative hardware, which is represented by physical urban landscape such as infrastructure or space for creative production, consumption, and proconsumption; creative software, which is identified as the creation of creative atmosphere or ambience, fashion, quality of life, and perceived diversity; and orgware, which is related with sectors, industries, clusters, policies, and governance.

The objectives of this paper are to identify the means of development of the creative economy and creative industry into creative tourism, to explain the three dimension of creative tourism that have pushed Bandung city, and to provide suggestion for strategic plans which could be used to improve the completion of the creative tourism development in Bandung.

2. Method

This paper used qualitative and quantitative research. Data and information are collected by field observation and through deep interview with the representative from creative community and local government. This study is located in Bandung with eight urban development areas and 30 sub-districts.

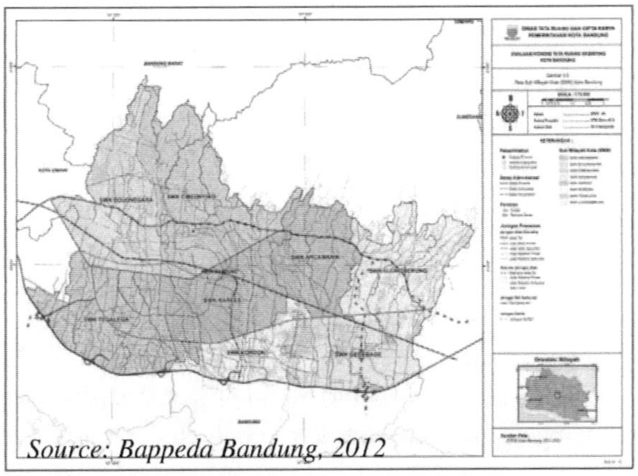

Source: Bappeda Bandung, 2012

Figure 1. Map of Bandung

3. Emerging of Creative Tourism in Bandung

In Indonesia, Initially, Creative tourism activities rise from gain creative economy and industries. Creative economy has been developing since 2005 through idea of President of Republic Indonesia. At that time, Indonesia has to increase craft industries and creativities. Officially, in 2011, creative economy was involved as a part of tourism activities and in the same year the Ministry of Tourism and Creative Economy was established. 2011 to 2025 is the years where central government has to accelerate this sector.

Table 1. Milestone of Creative Tourism Development

Creative Economy	*President: must increase craft and creative industry*	*Blue print of creative economy*	*platform of community of creative economy*		*Acceleration*
Year	**2005**	**2008**	**2010**	**2011**	**2025**
Creative Tourism				*Ministry of TourismdanCreative Economy was established*	*Acceleration*

Source: Kuswara, 2013 [8]

Bandung has been a tourist destination since the Dutch Colonialism in the 19th century. The natural and cultural aspects of Bandung are the most favorite attraction at that time. The contemporary tourism is starting to lose its popularity. The decrease of popularity has led to the tourism activity in the city to be unfavorable. Vast innovations were created in the development of tourism, which now has become a creative tourism. The development of the creative tourism in Bandung is also the result of the high number of productive cities and a number of campuses, which has contributed in innovation and activity.

In 2009, ministry of trade has classified 14 sub-sector of creative industries, which are: 1. Advertisement; 2. Architecture; 3. Art market; 4. Craft; 5. Design; 6. Fashion; 7. Video, Film, and Photography; 8. Interactive Games; 9 Music; 10. Theater; 11. Printing and Small Publishing; 12. Computer Services and Software; 13. Research and Development; and 14. Telecommunication and Radio. Table bellow describes the creative industry's sub-sectors in Bandung, with the addition of culinary sub-sector.

Table 2. Number of Creative Sector In Bandung

No	Creative Sector	Total
1	Advertisement	275
2	Art gallery	49
3	Craft	511
4	Design	17
5	Fashion	893
6	Film, Video, and Photography	297
7	Software and Interactive Entertainment	18
8	Music	156
9	Theater	1301
10	Architecture	90
11	Computer and software service	275
12	Television and Radio	38
13	Printing and small publishing	547
14	Research and Development	390
15	Culinary	532

Source: Bappeda Kota Bandung, 2008 in Tiara, 2016

4. Mapping Creative Urban Landscape

Creativity is an act that requires space to flourish creative ideas can be expressed within the space itself. In urban context, creative place is represented in urban landscape. There are three dimensions of urban creativities to achieve new development for creative urban development. They are creative hardware landscape, creative software landscape and orgware. The creative hardware, is represented by infrastructures, spaces or utility for creative production, consumption and proconsumtion; creative software landscape is associated with atmosphere or ambience, fashion, quality of life, perceived diversity, vibrancy, events, festivals; and the last part is creative orgware is the dimention that is described in policies, sectors, industries, clusters, and governance [4].

Bandung is an emerging creative city in the world, has been developing hardware, software and orgware towards creative city. In 2016, according recent survey identified some forms of urban creative landscape in Bandung:

- Creative software Landscape in Bandung is manifested in events or festivals are aimed to promote creative tourism. In 2015, there are 31 festivals that were organized by Milangkala Bandung Festival. Milangkala Bandung Festival aimed to integrate the event or festival in Bandung. That festivals: Bicycle rally night run 2015; Indonsia Art Voice Festival; Bandung Air Show; Konbawa Festival; Salman A Day Out Picnic; Music and Gimmick on The Street; MD Soul Session; Bandung Culture Folklore; Babakan Siliwangi Festival; Bandung Food Land; Baroque of Artliens; Bandung light Festival and so on.
- Creative Hardware Landscape is manifested in theme parks. Local government provided at least nine theme parks to encourage physic of creative places. They are:
  - Film Park. This theme park is the first film park in Indonesia, it is located in Taman Sari near with skateboard park. There are some facilities that are installed, such as: big videotron, good sound system, and unique tribune setting.
  - Skateboard Park. This park is located under Pasopati Flyover. Skateboard community in Bandung uses this space. This park also opened 24/7.
  - Fitness Park. This park is located in Iman Bonjol St. it has some facilities to workout such as: chest press, elliptical trainer, sit-up bench and cross trainer.

- Citarum Music Park. As the name implies is dedicate for the music communities in Bandung. This park is located in Belitung St. and most of the time the park is used by indie bands for their performance.
- Elderly Park. This park is dedicated for elderly people to get outdoor activities.
- Pet Park. This park is located in Cilaki St., which provide some facility for pet such as a toilet for pet and this site also as a meeting place for pet lover. Total area of pet park about $800m^2$. Beside has some facilities for pet, this park also provide jogging track, bench and so on.
- Photography Park. This park is located in Anggrek St. this park provides for photography community in Bandung.
- Persib Park is a park that designed for people who has interest in football. It name come from the famous football club in Bandung. Total area about 16x25 m.
- Superhero Park is located in corner of Bengawan and Anggrek St. There are some sculpture superhero figures such as Gatotkaca, Supermen, and Batman. This park is the best out door space for children.

• Creative Orgware. In this paper the main focus of orgware, it is associated with creative cluster. There are ten creative clusters that identified:
  - Creative Kampoong of Eco Bamboo, this creative kampoong is located in Dr. Setiabuhi St. this Kampoong promote edutourism relate with traditional games, and how to make angklung;
  - Creative Kampoong of Dago Pojok, this creative kampoong is located in Dago Pojok, this kampoong is initiated by Rahmat Jabaril. This kampoong has specialization for craft such as wayang, bamboo music instrument and so on;
  - Creative Kampoong of Cicadas, cicadas creative cluster is aimed for local acoustic development. Many traditional performance show on the open space, moreover this kampoong also has building for informal education to develop local creativity;
  - Old Town (Colonial Area), located in Asia-Africa St. this location has high historical value. This site also as a witness of international conference of countries from Asia and Africa;
  - Creative Kampoong of Barongsai (China Town). In this area, a tourist can visit monastery temple, and eat Chinese food;
  - Creative Kampoong of Cibuntu, this creative kampoong strives for the growth for Tofu home industries, located in Cibuntu;
  - Creative Kampoong of Cigondewah. This creative Kampoong is dedicated for development of local textile creativities;
  - Creative Kampoong of Cicukang, this kampoong aims to develop local creativities that relate with music instrument, wayang, marawis, jaipong dance and theater;
  - Creative Kampoong of Lewi Anyar, the other name of Lewi Anyar Kampoong is Langitan Kampoong. The main attraction for tourist activity is mural art. This site developed to gain the local economy, art and local creativity;
  - Creative Kampoong of Pasudan. This creative kampoong aims to promote local creative activities that relate with traditional dance, traditional music, theater, and fine art. Pasundan Kampoong is located in Balong Gede; and
  - Creative Kampoong of Belekok is located in Rancbayawak, Gedebage. Belekok is named after natural phenomenon of bird migration of *Araeola speciosa* or Belekok Bird (local name). The main tourist activity is bird watching.

Problems that are experiencing of Bandung to develop creative tourism relate with disintegration among creative dimensions (Hardware, software, and orgware). For instance, Creative Kampoong of Cicukang that has specialization in promoting local music instrument, Wayang, Marawis, and Jaipong dance does not supported by hardware and software creative landscape. It is highly recommended to develop creative cluster (orgware) that is integrated with others creative dimension.

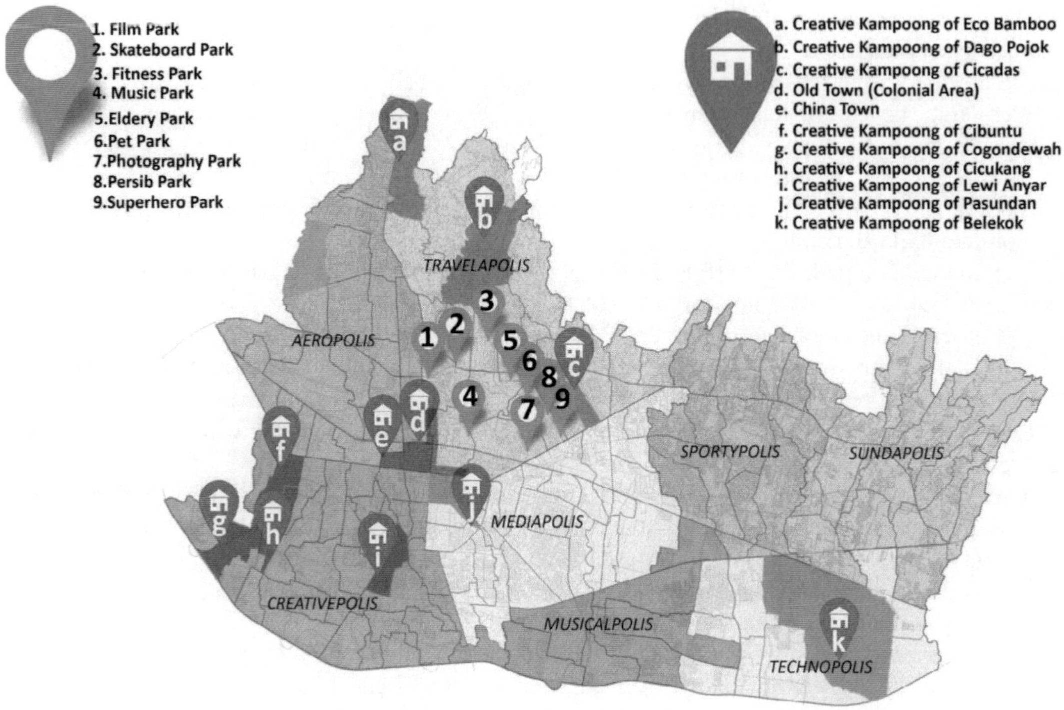

Figure 2. Inventories of Orgware and Creative Hardware Landscape in Bandung

5. Strategic Planning for Creative Urban Landscape Development

Creative urban landscape has been developing since 2011 when this city was stated as a part of world creative cities network together with 47 cities from 33 countries. By UNESCO, there are three obligation of this city that has to develop: first, Providing a thematic public park in Bandung dedicated to the UCCN, which will represent the Creative Cities in various artistic forms, as well as showcasing a diversity of urban cultures; second, Enhancing collaboration with other Creative Cities through creative festivals, joint events and workshops, notably by organizing an annual Asian-African Carnival with the aim of promoting underrepresented cities and South-South cooperation.

Moreover, in 2015, Bandung was issued the zoning regulation for eight urban development areas (SWK), each area has specific theme for development, and they are:

- SWK Ujung Berung as a Sundapolis: this urban area is planned to promote and conserve Sunda Culture
- SWK Tegallega: is developed as creativepolis that will accommodate some creative industries
- SWK Karees: will be developed as Mediapolis which will facilitate Information and communication technology (ICT) development in Bandung
- SWK Kordon: the development concept of this area is Musicalpolis, this concept attempts to gain the music as a creative industry.
- SWK Gedebage: Technopolis is the main concept development, this area is planned for new modern city in Bandung that consist new center of Bandung Municipality office.
- SWK Bojonegoro: this area is developed as a Aeropolis that will promote aerospace industry
- SWK Cibeunying: the concept of this area is Travelpolis that will be main area for tourism in Bandung
- SWK Arcamanik as a Sportypolis that will be center of sport facilities in Bandung

Creating the thematic SWK has advantages and disadvantages, on other hand, this policy will curb its own creativities, but creativities strongly need a place that has to organize systematically to be planned well. To deal with this problem enclave creative cluster is the best solution that could be taken. Table bellow shows the main creative cluster, enclave cluster (orgware) and general development guidelines for creative software landscape and creative hardware landscape.

Table 3. General development guidelines for orgware, creative software landscape and creative hardware landscape.

No	Creative Cluster (Orgware)		Creative Software Landscape	Creative Hardware Landscape
	Main Cluster	Sub-Cluster (Cluster Enclave)		
1	Areobiopolis	Aerocluster	Bandung air show	Provision of public space (park or street landscape) with aero theme
		China Town	Barongsai Festival	Provision of public space (park or street landscape) with china town theme
2	Travelapolis	C.K (Creative Kampoong) of Eco Bamboo (traditional game)	Traditional Game	Provision of public space (park or street landscape) for outdoor game
		C.K of Dago Pojok (craft,wayang)	Edutourism traditional craft exhibition	Provision of public space (park or street landscape) for craft exhibition
		Old Town (Colonial Area)	Asia-Africa Carnival	Provision of public space (park or street landscape) for Asia-Africa Carnival
		C.K of Cicadas (traditional music instrument)	Traditional instrumental music festival	Provision of public space (park or street landscape) to perform Traditional instrumental music
		Mixuse Cluster	Film, fashion, culinary, and so on	Provision public space with specific theme
3	Creativepolis	C.K of Cibuntu (Tofu Kampoong)	Bandung tofu festival	Provision workshop and public space for tofu festival
		C.K of Cogondewah (Textile)	Textile exhibition	Provision workshop and public space for textile exhibition
		C.K of Cicukang (instrumen, wayang, marawis, jaipong)	Traditional theater	Provision public space for theater
		C.K of Lewi Anyar (mural Art)	Mural art festival	Provision public space for mural art
4	Mediapolis	ICT Cluster	ICT exhibition	Provision workshop, hall and public space for ICT exhibition
		C.K of Pasundan (traditional dance, traditional music, theater, and fine art) Seni lukis	Painting exhibition	Provision workshop, hall and public space for Painting exhibition
5	Musicapolis	Music Cluster	Music and Gimmick on the Street Festival	Provision public space (park or street landscape) for Painting exhibition
6	Sportipolis	Sportcluster	Sport Festival	Provision sport facilities
7	Sundapolis	Sundacluster	Nyunda Festival	Provision public space (park or street landscape) with Sundaness theme
8	Technopolis	Technopolis cluster	smart technology exhibition	Provision workshop, hall and public space for smart technology exhibition
		C.K of Belekok	Bird watching	Provision bird watching track

6. Conclusion

The development of creative tourism was initiated from creative economy and creative industry by the creative class in Bandung. As part of the creative cities of the world, Bandung has developed its creative tourism through establishing creative tourism clusters policies. Clustering is the result of intervention and careful strategic planning from Bandung's local government to accommodate creative tourism focused to improve the experience and role of tourists. Other creative tourism development dimensions are the creative hardware landscape and creative software landscape. To this day, the three dimensions of creative tourism development are still not integrated and need proper measures to the reclustering creative areas. There are 8 recommended creative clusters with 18 sub clusters or enclave creative clusters in Bandung.

Reference

[1] Antonio Paolo Russo and Albert Arias Sans 2007 Tourism, Creativities and Development: Student Communities as Creative Landscape. E book: Routledge
[2] Giovanni Maciocco 2008 Urban and Landscape Perspectives; Springer, Italy
[3] Graeme Evans 2009 Creative Cities, Creative Space and Urban Policy. Journal: Urban Study
[4] Greg Richards and Julie Wilson 2007 Tourism, Creativities and Development: Creativities in Tourism Development. E book: Routledge
[5] Greg Richards and Wilson Julie 2006 Developing Creativity in Tourist Experiences: A Solution to The Serial Reproduction of Culture? Journal: Elsevier
[6] Iwan Kustiwan, Ibrahim Ukrin and Astri Aulia 2015 Identification of the Creative Capacity of Kampong's Community Towards Sustainable Kampong (Case Study: Cicadas and Pasundan Kampong, Bandung): A Preliminary Study
[7] Tiara Dwi Ade 2016 Bandung Creative Hub sebagai Penghubung Heritage Tema Connectivity. Undergraduate Thesis: Unikom, Bandung
[8] Ukus Kuswara 2013 Development of Tourism and National Creative Economy, *Ministry of Tourism and Creative Economy*
[9] Bappeda Kota Bandung 2012 Bandung General Spatial Planning for 2011-2031

Acknowledgments

We would like thanks to DIKTI that supported the financial of this research.

IOP Conf. Series: Earth and Environmental Science **91** (2017) 012027 doi:10.1088/1755-1315/91/1/012027

Places of Faith: A Reflection on Landscape of Manila Cathedral Plaza de Roma and Istiqlal Mosque Sacred Grounds of Jakarta

MM Pujalte[1] and N Navarra[2]
[1] Architect, MTLA, University of the Philippines
[2] Landscape Architect, Doctor of Engineering Ph.D, Manila Philippines

E-mail: pujaltemariamonica@gmail.com

Abstract: *Crossing boundaries of faith from Manila to Jakarta, this study is to classify the open spaces in their sacred grounds according to its characteristics, elements, use of space and hierarchy of importance in landscape design approach. The reflection of their religious landscape in preserving the traditional, and exploring the non-traditional aspect of their landscape design in global setting is carried out thru a spatial analysis for Plaza de Roma of Manila Cathedral and the sacred grounds of Istiqlal Mosque. The design framework would tackle: concepts, planning approach, functional symbolic values, and aesthetics used. The data and information are all examined based on observation, historical background, analyses, and literature content in determining spatial functions. Finally, when results are completed, this will give a better understanding on the importance of open areas in Manila and Jakarta's sacred spaces; paving way for a better sense of comfort in spiritual contemplation. This will also help reveal the commonalities in spiritual practices between Islam and Christianity, and the role of landscape in their religion and faith.*

I. Introduction

The concept of religious landscape should not be interchanged nor mistakenly assumed as cultic and polytheistic. It seems to be a matter of common knowledge that the concept of space are socially and culturally constructed and that this concepts are historically sensitive depending on the modern perspective and experience.[1] Religious landscape" has been a topic for the past twenty years. However, since the concept is not clearly stated and indistinct in nature, there is no established intellectual space to which it is derived. As defined by Matthew Dillon[2], religious landscape includes as well the itinerary a pilgrim is obliged to take or usually takes when approaching the sanctuary of his choice, the route which one has to take a specific place where sacrifices and other religious rites takes place. It includes the public or privately owned area surrounding a walled sanctuary not only with camping pilgrims, but also with fairs and markets, craftsmen and merchants who offer all kinds of services the pilgrims might need[3]. The space defines the temporary religious use of an often not sacred space; wherein the potential users are coming from different religious groups apart from Catholicism and Moslems and the practicing believers of these faiths that unite to experience the sacred ground.

II. Background of the Study

Places of worship, faith and religion hold a special position in the cultural and social aspects of a community. They create the design distinctiveness and hierarchy of importance in a certain structure on how these spaces are to be planned and oriented. However, the lack of spaces to work around that is sufficient to users; along with the design approach on the landscape reduce the efficiency of a sacred ground. With this, thc author would establish the relationship of the landscape in Mosque and Cathedral's open space on religious common grounds.

Historical Background Jakarta, Indonesia: Prayer (salat) is an indispensable religious ritual for Moslems which can be conducted anywhere except for the places they consider dirty such as tombs and slaughter houses. Masjid or Mosque is an Islamic religious building where many people with the same intention can pray together. Masjids throughout the world are built based on the axis facing Mecca. Indonesia boasts one of the most unique and monumental sacred grounds in Southeast Asia. Manila, Philippines: Philippines is the only Catholic Christian nation in the South East Asia. The Manila Metropolitan Cathedral Basilica or Manila Cathedral Basilica of the Immaculate Concepcion, is one of the oldest and most famous churches in Asia. With a storied history and legacy that spans over four centuries, the structure has seen the best and worst of time and humanity.

Research Problem With the geographic distribution of religious groups varying considerably, the global population of Manila and Jakarta in religious landscape therefore, influences the spatial configuration and functional efficiency of their places of worship. This study highlights the comparative scale of open spaces for Plaza de Roma and Mosque Courtyards (Sahn) in understanding their importance attached to their religious structure respectively. To clearly define space hierarchy, the different rituals for Catholicism and Moslems shall be magnified and analyzed.

Goals and Objectives In order to come up with the functionality and relevance of open spaces to sacred grounds this study aims to:

1. Identify the spatial sequence and uses of space for the religious structure of Istiqlal Mosque and Manila Cathedral;
2. To define and analyzed the historical influences on planning configuration and design elements of open spaces that makes it a sacred ground.
3. To provide spatial relationship of the natural and built environment for religious spaces of mosque and cathedral through site analysis and inventory.

Scope and Limitations The study is site specific to Istiqlal Mosque and Manila Cathedral's open space used for religious practices. For this study, the selected open spaces are those bounded within and adjacent to the interior and exterior constraint of these religious structures (courtyards and plaza).

Assumptions and Hypothesis Religious buildings that are monumental in scale symbolically represent spiritual dominance for a certain faith. The larger and grandeur it is designed and crafted, the more it embodies spiritual power over the other. For religious structures, it is assumed that even the smallest of space are considered sacred or blessed as long as it is attached to its main building. The orientation and size allotment of their spaces foretells the important use of the space. The sequential order and zoning of the spaces are relevant to the activities to be performed inside or outside the religious structure. Despite the changing environment, the physical characteristics of a religious structure is influenced and dictated by the traditional concept of design patterned to its historical and commemorative significance.

Significance of the Study The importance of this study is to understand the concept of religious landscape on the realms of Islamic and Catholicism's sacred grounds that reflects their culture and spiritual identity. The commonalities between two different faith and their religious practices can be a binding factor in resolving conflicts in the future. This will help define the attributes and integration of open spaces to their religious practices that is globally acceptable despite the changing conditions of climate, building materials and technological advancement used where sacred grounds are located (table 1).

IOP Conf. Series: Earth and Environmental Science **91** (2017) 012027 doi:10.1088/1755-1315/91/1/012027

Table 1. Open Spaces a Place of Worship and Sacred Grounds

CATHOLIC FAITH		ISLAMIC FAITH	
LOCATION:	Plaza del Roma, Manila Philippines	**LOCATION:**	Istiqlal Mosque, Jakarta, Indonesia
SACRED GROUNDS	Open space	**SACRED GROUNDS**	Courtyards
ACTIVITIES:	Liturgical Calendar	**ACTIVITIES:**	Political/ Civic/ Social/ Religious Celebrations
OBSERVATION, USER ANALYSIS , SPACE SYNTAX			
TANGIBLES:	Spatial sequence, Religious Activities, Spaces Required, Aesthetics		
INTANGIBLES:	Historical and traditional Background, Psychologica, Spritual and social perceptions, Modern Changes		

III. Concept and Framework

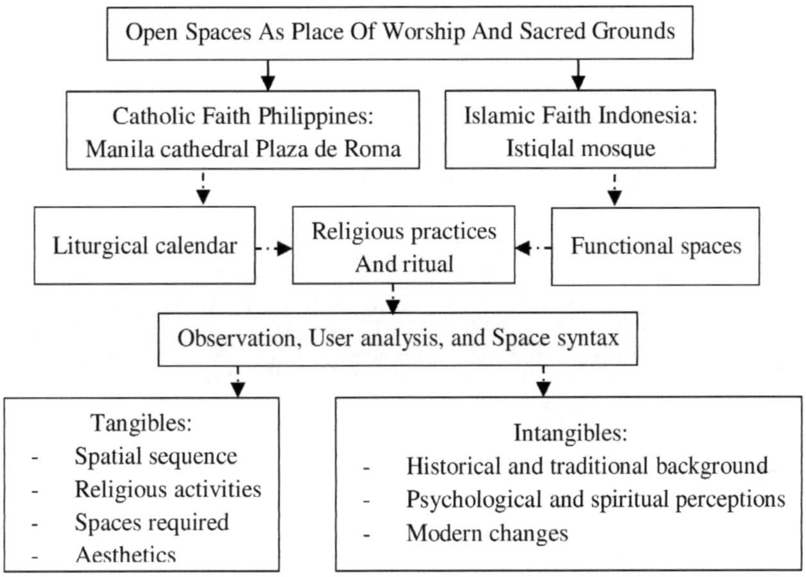

Figure 1. Concepts and Framework

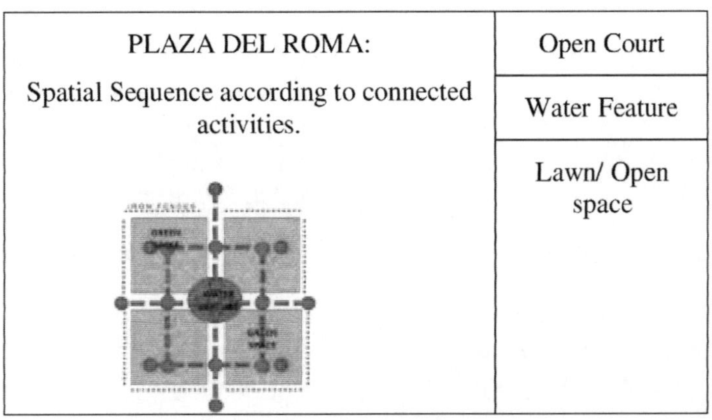

Figure 2. Space Syntax: Istiqlal Mosque

ISTIQLAL MOSQUE: Spatial Sequence according to connected activities. 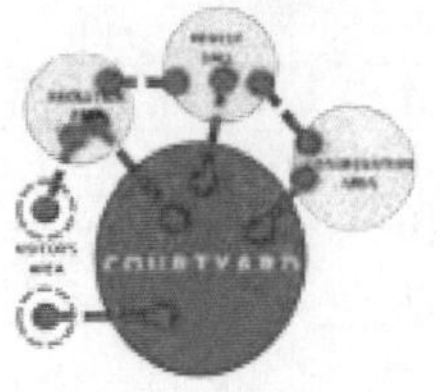	Visitor's Area Abolution Area Prayer hall Congregation Area Courtyard

Figure 3. Plaza del Roma

IV. Methods of Inventory

To come up with detailed understanding of the space hierarchy and its functional uses, this study provided a series of documentation in gathering and translating information in the form of photos, literature research, space syntax and comparative analysis.

V. Tables and Figures: Comparative Analysis

Table 2. Comparative Analysis of Plaza Del Roma and Istiqlal Mosque

PARAMETERS	ISTIQLAL COURT YARD	PLAZA DE ROMA	IN COMMON
Planning layout • Approach • Size • Concepts	• Seven entrances in the buildinggrandeur in scale; nine hectares lot • Square in plan; with surrounding columns and enclosed vast prayer halls with roofed areas • Paved flooring for indoor and outdoor spaces	• Considering its rectangular layout, it has four entrances accessible from North, East, West, South side of the lot. • Major plaza in Manila • Square in plan, eclosed by surrounding fences and hedges as perimeter planters; open green space	• Multiple accessibility to accommodate more users comming from different sides of the structure • Can accomodate larger numbers of users • Geometric in shape; formal symmetry.
Flexibility of the open space	The courtyard is attached to the main structure as worship area; prayer halls and vestibules were generally laid out in plan; space cn also be used apart from praying activities	Open space is adjacent to main structure where it is used for general functions	The space is flexible in use, depending on the time and season; both can be used as social, cvic, cultural activities
Physical efficiency of space	Despite irregularity of the site; the main area is configured in square plan to accommodate	Accessible to anyone at any given time of the day. The existing	The configuration of area s the same and the activities done

	more worshippers; courtyards are exposed to sunlight during the worshipping hours making it inconvenient for those using the space for prayers	trees and vegetation benefits the users as it is used for shading during hottest time of the day	except that the space where it is located is different. For courtyards, it is inside the mosque, while the plaza is outside the cathedral
Aesthetics • Motifs and patterns	Embellishments are intricate in design and influence by islamic art. Designs are reflective to their religious beliefs and practices	Dependet to the rulling power who govern, manage and maintain the area; dependent on the activities and celebrations to be held	For courtyards, it has definite look and fixed patterns; while plaza has a tendency to changes its appearance depending on the social activities
Historical importance	Built to conform President Sukarno'grand vision of a strong, multi faith state with the goverment as its center	Product of town planning system with reference to the laws of the indies where an open spaces is integral center of the goverment, civic, and secular structures	Apparently, the mosque is located near christian church in Jakarta where in the attched meaning is that it refrects to the world that religions could co-exist harmoniously with other

Spatial and Religious Experience

Table 02. Spatial and Religious Activities on Sacred Spaces

SPACE CONFIGURATION	MOSQUE COURTYARD	PLAZA DE ROMA
ENTRANCE	-seven point entrances that area are located adjacent to road networks structure's axis is oriented towards Mecca	-four main entrance point are coming from the north, east, west, south, axis of the area surrounded by civic, secular buildings, government building and ancillary road on each sides
PRAYER HALL	can accommodate 120,000 capacity five floors with each prayer hall accommodating 60,000 users	can accommodate 23,000 capacity one floor only
SIGNIFICANCE OF SPACE	Courtyard are important to mosque and cannot be seperated from the main structure as it serve as an extension praying area aside the prayers halls atop of each floors	The open space is used as town hall, convening areas for festive occassions and cultural activities
ACTIVITIES	Prayer area; education and teaching purposes; government forums	Prayer area on occassion; education and teaching purposes; for cultural and community activities, used as space for political forums; commercial space for selling merchandise to foreigners, locals and the like

FREQUENCY OF USE	**PRAYER AREA: five time a day FRIDAY: Whole day of prayer**; Id Al-Fitr and Id Al-Adha, Al Hijra (Islamic New year), Ramadan (Month od Fasting), "Id Al-Fitr- (Festival of Breaking the Fast), "Id Al-Adha	**Accessible any time of the day.** Liturgical Activities such as: Advent Season, Christmas Season, Season of Lent and Easter Season

VI. Results and Recommendation

The importance of courtyard in a Mosque is proportional to the importance of open space for Plaza de Roma. Using descriptive analysis; it is the most efficient technique to understand and provide substantial finding. It could be understood based on the literary research that the planning configuration of spaces in terms of scale and aesthetics has its bias on location, traditional systems used in design, materials used and even attachment to the laws and notable person; political or apolitical of the specific sites. In the architecture of faith of Moslems, the use of open spaces are for the purpose of accommodating larger number of faithful to pray and perform specific social, cultural, educational and even political events during off peak days and season. For the open space in Catholic Christian communities, particularly Plaza de Roma, the anticipation its space importance can be translated mostly during the liturgical celebrations. The sanctity of Plaza de Roma is not associated with religious practices rather on the civic use of space as community Park, on how it is adaptive to its changing use from social, commercial, political uses. There are similarities in the space intent between the sacred grounds of two different faiths. It could be seen in the early stages of planning that draws back to their historical significance, but a great difference on how these open spaces are being utilized when looking at a religious landscape point of view. Being the grandest among the grandest of all mosques in Indonesia, Istiqlal's exterior should be provided with the addition of greener landscape in order to mitigate the humidity in the surrounding. For Plaza de Roma, since it is the remaining green space at the heart of the important building of Intramuros, shall be restored without too much changes on the external features; to reflect and maintain the placeness of the area and wherein such changes can also add to the historic value of the area.

VII. References

[1] Alcock S and Scott E, 1993 *"Spaced- Out Sanctuaries: the Ritual Landscape of Roman Greece": Theoretical Roman Archaeology. First Conference Proceedings* pp. 155-165

[2] Dillon M, 1997 *Pilgrims and Pilgrimage in Ancient Greece* pp. 204-226

[3] Bresson A, 2000 discusses the question of fixed prices during religious festivals with markets, Lacité Marchande, Bordeaux pp. 174-178

[4] Mundigo A, Crouch D, 2004 *Laws of the Indies*, Proquest Information and Learning Company, Copyright Liverpool Press University.

[5] Al-Jaff, AJ 1989 . *Communication in Architecture: An Application of Space Syntax.* Department of Architecture, University of Technology, Baghdad, Iraq (M.Sc. Thesis).

[6] Mustafa AF, Hassan AS, 2013 , *Mosque Layout Design: An Analytical Study of Mosques Layouts in the early Ottoman Period*

[7] The Global Religious Landscape A Report on the Size and Distribution of the World's Major Religious Groups as of 2010 http://www.pewforum.org/2012/12/18/global-religious-landscape-exec/

[8] Jocano F, *Filipino Catholicism: A Case Study in Religious Change* http.//www.asj.upd.edu.ph

[9] Nadal CD, *Tracing the Urban Cultural Landscape Structure of Manila Sacred Space: A Case Study of Malate Church Ground, Thesis Paper in Master of Tropical Lanadscape Architecture*
[10] Galingan Z, *Muhon: A Journal of Architecure, Landscape Architecture and the Design Environment, Issue*. No. 4.

IOP Conf. Series: Earth and Environmental Science **91** (2017) 012028 doi:10.1088/1755-1315/91/1/012028

Landscape Character of Pongkor Mining Ecotourism Area

A Kusumoarto[1], A Gunawan[2], Machfud[3] and A Hikmat[4]

[1]Doctoral Student Program, Natural Resource and Environmental Management Science, School of Post-Graduate, Bogor Agricultural University, Jl. Raya Pajajaran, Bogor, Indonesia 16143
[2]Lecturer in Landscape Architecture Department, Bogor Agriculture University, Jl. Raya Darmaga, Bogor, Indonesia 16680
[3]Lecturer in AgroIndustrial Technology Department, Bogor Agriculture University, Jl. Raya Darmaga, Bogor, Indonesia 16680.
[4]Lecturer in Conservation of Forest and Ecotourism Department, Bogor Agriculture University, Jl. Raya Darmaga, Bogor, Indonesia 16680

E-mail: andri_anto@yahoo.com

Abstract. Pongkor Mining Ecotourism Area has a diverse landscape character as a potential landscape resources for the development of ecotourism destination. This area is part of the Mount of Botol Resort, Halimun Salak National Park (HSNP). This area also has a fairly high biodiversity. This study aims to identify and analysis the category of landscape character in the Pongkor Mining Ecotourism Area for the development of ecotourism destination. This study used a descriptive approach through field surveys and interviews, was carried out through two steps : 1) identify the landscape character, and 2) analysis of the landscape character. The results showed that in areas set aside for ecotourism destination in Pongkor Mining, landscape character category scattered forests, tailing ponds, river, plain, and the built environment. The Category of landscape character most dominant scattered in the area is forest, here is the river, plain, tailing ponds, the built environment, and plain. The landscape character in a natural environment most preferred for ecotourism activities. The landscape character that spread in the natural environment and the built environment is a potential that must be protected and modified such as elimination of incongruous element, accentuation of natural form, alteration of the natural form, intensification and enhanced visual quality intensively to be developed as a ecotourism destination area.

1. Introduction

The landscape character that formed from units of the landscape is a real pattern of elements that occur consistently in a particular type of landscape, which is created from a landscape formations formed from human activities that interact with land in a considerable period of time. In this period of continuous change, but the important thing is the damage or destruction of ecosystems through improper development [1]. The natural landscape character can be categorized as follows: mountain, dune, praire, swamp, lake, sea, streams, hill, canyon, forest, river, valley, pond, desert, and plain [2].

Human involvement in development who utilizing the natural resources, many lead changes in landscape character, causing damage and destruction of ecosystems [3] [4]. One alternative effort to protect the resources of the natural landscape character damage and maintain the aesthetic quality is through ecotourism. Ecotourism is one of the limited tourism activities that can be done in an

IOP Conf. Series: Earth and Environmental Science **91** (2017) 012028 doi:10.1088/1755-1315/91/1/012028

environment that has a natural landscape character, where its resources can also be utilized for economic purposes, as well as underground mining areas.

Pongkor Mining Ecotourism Area has a diverse landscape character as a potential landscape resource for development of ecotourism. Ecotourism activities based on landscape character in mining will provide an advantage in controlling the degradation of the environment, protect natural resources, education, conservation of biodiversity and protecting the cultural and economic benefits [5] [6] [7] [8] [9]. This is in accordance with the mandate SK. MENTAMBEM No. 1211/1995, that the development of mining areas should be able to provide social and economic benefits of society and the protection of ecosystems.

Ecotourism development based on landscape character requires a good assessment of natural and man-made landscape. This is because each area in mining area has some real differences landscape character that expected to provide protection of landscape character value. Based on this, it was necessary the identification and analyze of landscape character in the Pongkor Mining Ecotourism Area. This study aims to identify and analyze the category of landscape character in the Pongkor Mining Ecotourism Area for the development of ecotourism destination.

2. Methodology

The research location was in Pongkor Mining Ecotourism Area, which was part of the conservation area of UBPE Pongkor, PT. Aneka Tambang (figure 1) and located at an altitude ranging from 300 meters to 900 meters above sea level [10]. Research began in July 2016 until August 2016.

The method used in this research was descriptive method through field surveys and interviews. This study refer to the study approach by [11]. The study consisted of two phases: 1) identifying landscape character and 2) analysis of the landscape character category.

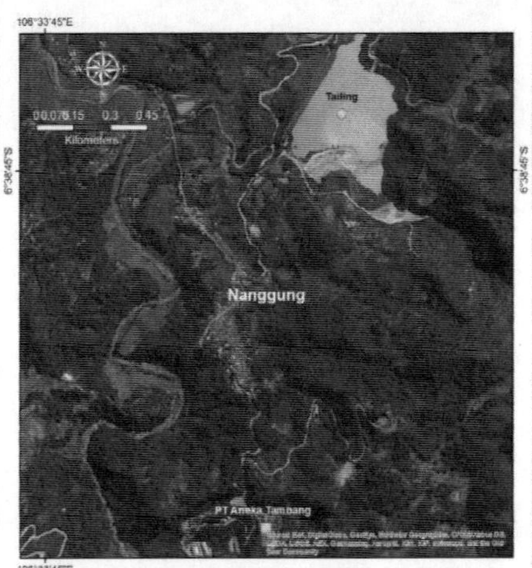

Figure 1. The Research Location
Source : ESRI (2016)

2.1. Research Data

Data collected in the form of primary data and secondary data. Primary data obtained from direct observation in the study site include:

a. biophysical data obtained through direct observation were a visual data, contours, slope, and ground check to landscape character category. The results of data processing base maps, thematic maps and satellite imagery that had vectorizationed into spatial database using ArcGIS 10.1 software.
b. observational data of landscape character elements. Checking the position of the elements in the region carried out by using a Global Positioning System (GPS). Scrutiny of the

landscape elements were directed to determine the key character to be the mainstay of area character as ecotourism destination.

2.2. Data Analysis

The analytical method that used is the landscape character analysis. Landscape character category analysis used as revealed by [2]. The Steps taken: 1) elevation and slope analysis, which consists of: (1) elevation analysis; (2) slope analysis; and 2) landscape character analysis.

2.2.1. Elevation and Slope Analysis

a. Elevation Analysis

Elevation analysis (altitude location of the earth's surface) to be done to create topographic maps and contour maps by Digital Elevation Method (DEM). Measures undertaken are: 1) created topographic data; 2) interpolated; 3) created a elevation digital map [12]. The formula for making interpolation as follows:

$$d/D = e/E$$

Where :

d = distance between the grid intersection toward a measuring point elevation obtained (meters); D = the distance between the grid intersections (meters); e = elevation change between the grid the same intersection and the intermediate point (meters); E = total elevation change between the grid intersections (meters).

b. Slope Analysis

Slope analysis will be grounded by a contour map (contour map) region [12]. Grade slope (slope categories) were analyzed based on the number of changes of DEM. Class slope of the land will be divided into several classes: 1) 0-3%, 2) 3-15%, 3) 15-25%, 4) 25-45%, and 5)> 45 %. Formulated into an area with a slope (slope) is different is as follows:

$$L = DE/S$$

Where :

L = length of the distance between the contour line; DE = contour intervals; S = the percentage of slope

c. Landscape Character Analysis

Landscape character maps derived from Landsat satellite imagery. Results obtained Landsat downloads the file containing metadata and files band, besides the image has geo-referenced on WGS 84 datum and projection UTM north orientation (product level 1). In the selection of the Landsat image, a reference used worldwide reference system (WRS) which is a catalog of Landsat data. Based on this, the coverage area of the study found only in one sheet (scene) that is on the path (path) 122 and the line (row) 65. Furthermore, the category processed before (pre-processing) which includes a correction cloud cover (cloud cover and clouds shadow mask), converted into a digital value and cutting reflectance image. Categories of landscape character based on [12] as follows: the mountain, stream, hill, canyon, forest, river, valley, tailings pond, and plain.

3. Result

3.1. General Condition

Location of the study are in Pongkor Mining Region. The research location is at 106°34'00"E - 106°35'00 E and 6°38'30 " S - 6°40'00 " S. This study was in the administrative region of Bantar Karet Village, Nanggung Sub-District, District of Bogor. There is a large river that flows in the western part.

Based on the elevation analysis through DEM obtained approximate location was at 300 - 900 meters above sea level (asl) (figure 2). In general, location of study scattered dominant elevations between 300 and 800 meters asl. DEM map showed this region had a fairly diverse elevation. To the east, and south increasingly visible elevation, reaches a height of ± 900-1000 meters asl.

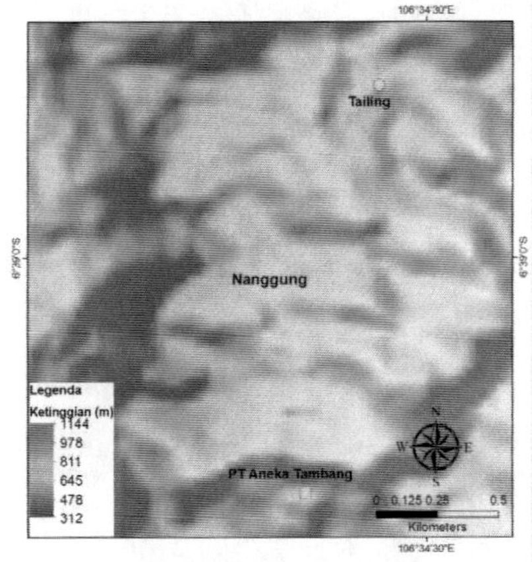

Figure 2. DEM Analysis
Source : data processed (2016)

Based on the slope analysis, there was a Several classes: 1) 0-3%, 2) 3-15%, 4) 15-25%, 5) 25-45%, and 6)> 45% (figure 3). The results of the analysis showed that the slope at locations scattered research slope from flat to steep. The slope of the land is dominated by 0-3% and 3-15%. Dominant steep slope was in the east and south.

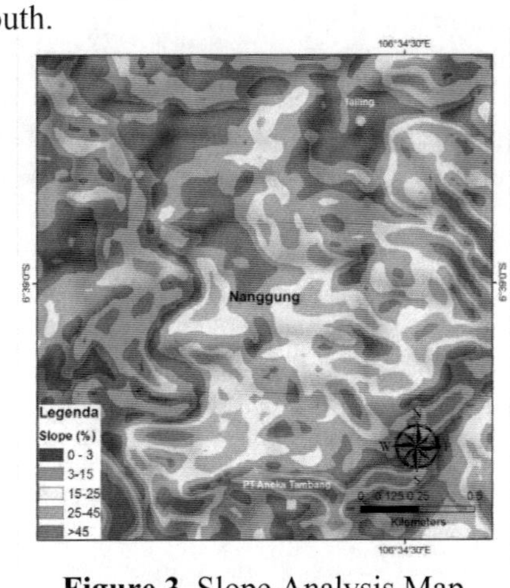

Figure 3. Slope Analysis Map
Source : data processed (2016)

3.2. Landscape Character Analysis
The results of study showed that the category of landscape character refers to [2] are forests, plain, rivers, tailings ponds and the built environment (figure 4). In general the landscape character was dominated by forests and plain. The analysis showed a change in the landscape character in some locations, especially in the location of the tailings ponds, on the edge of the river and the built environment (settlements scattered irregularly).

IOP Conf. Series: Earth and Environmental Science **91** (2017) 012028 doi:10.1088/1755-1315/91/1/012028

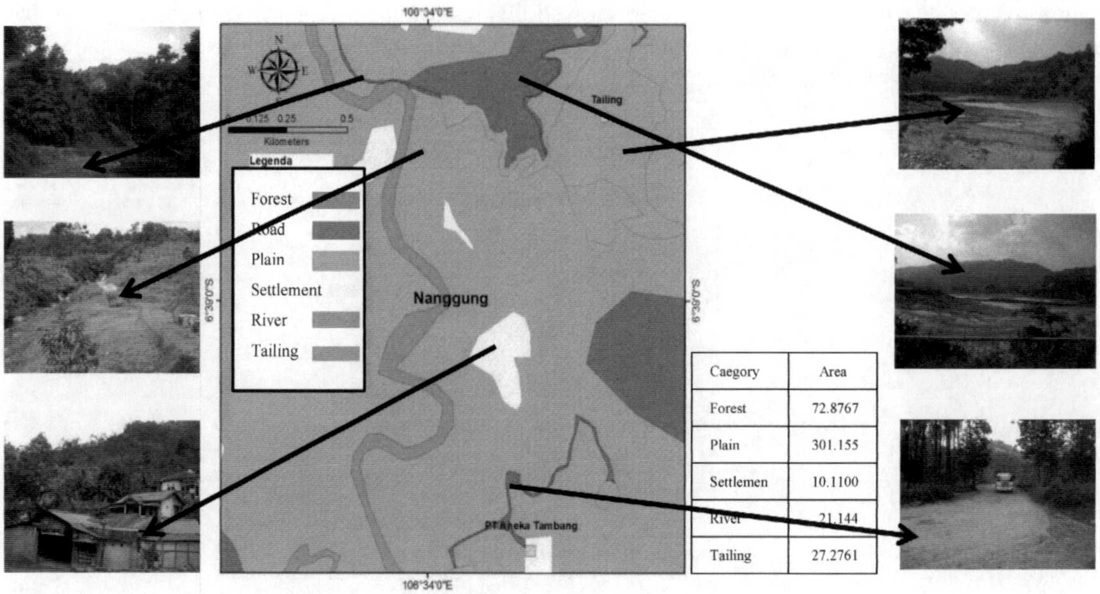

Figure 4. Landscape Character Category Analysis
Source : data processed (2016)

3.3. Landscape Character Protection and Modification

In the location of study, landscape character distributed forests, plains, tailings dams, rivers and built environment. The landscape character that spread in the natural environment and the built environment is a potential for development of ecotourism destination area. The river is the major feature of landscape character, whereas the forest, plain, tailings dam, and the built environment are the minor feature of landscape character. The landscape character must be protected and modified such as elimination of incongruous element, accentuation of natural form, destruction of the natural form, alteration of the natural form, intensification and enhanced visual quality intensively (table 1).

Table 1. Landscape Character Protection and Modification in Pongkor Mining Ecotourism Area

No	Landscape Character	Protected	elimination of incongruous element	accentuation of natural form	destruction of the natural form	alteration of the natural form	Intensification	enhanced visual quality intensively
1.	Minor Feature							
a.	Forests	v	v	v			v	V
b.	Plains	v	v	v			v	V
c.	Tailing dams	v	v			v	v	V
d.	Built Environment		v	v		v	v	V
2.	Mayor Feature							
a.	River	v	v	v			v	V

3.3.1. Protected The Landscape Character

Four types of landscape character in the location of study to be protected were forest, plains, tailing dams, and river. The development of ecotourism destination based on landscape character requires protection against natural character. This was because the natural landscape has a good diversity of biological resources. Protection also carried out on location that has hazards such as erosion, land use and land cover changed, water runoff from the tailing dams and river channel shape changed.

3.3.2. Elimination of incongruous element

Five types of landscape character in the location of study conducted incongruous element were forests, plains, tailing dams, built environment and river. The elimination of incongruous element usually effects an improvement [2] as well as the area that will be used for the development of ecotourism destination. Elimination of incongruous elements do if it conflicts with the natural landscape character. Some of these were the enclave (built environment) in the middle of the forest area. In addition, the colour, texture, form elements and the space where the elements were that the impression of unity and harmony with the natural landscape character. Bizarre landscape elements and damaged the unity and harmony in the area of built environment and the river eliminated, such as colour, texture, form the house and change the form of the channel shape of the river.

3.3.3. Accentuation of natural form

Four types of landscape character in location of study conducted accentuation of the natural form is forests, plains, built environment and the river. The accentuation of natural shape allows to achieve the optimal development of ecotourism destination. Landscape characters can be developed or improved by eliminating any negative elements and by the accentuating its positive qualities [2]. To improve the quality of the landscape needs to do accents in shaping the optimal character. The good view and their new experiences for tourists can be provided through the optimization of the landscape character accentuation through colour, texture, form and space.

3.3.4. Alteration of the natural form

Alteration of the natural form occurs on the landscape character of the tailing dams and built environmental. Construction of the tailing dams and some construction of homes in the enclave changing the natural landscape character. Such Alteration may be can cause damage the landscape character as the land becomes denuded, land cover change from natural landscapes into open land and undeveloped land. This can adversely impact the landscape or to the security landscape beneath it, for example in the form of erosion, landslides and flooding. Restrictions expansion of the built environment to the natural location area needed to be done and used it as part of the landscape elements that can be utilized for the development of ecotourism destination area. The changing natural landscape character should be an effort to optimize the function while maintaining balance and preservation of existing landscape.

3.3.5. Intensification

Landscape character may be intensified by strengthening the landscape character. Three natural landscape character types in the location of study that could be strengthened are forests, plains and river. Intensification was site dramatized. What is meant is the elements of the landscape in a natural landscape character presented following the landscape character profile as the original and did not alter or eliminate them. Intensification can also be done on the built environment and tailings dam in a way site negated. What is meant is giving the boundary with the elements of the natural landscape, constructing building construction preventing erosion and landslides, constructing of retaining wall to prevent damage tailing dams and the water runoff. Intensification of the built environment by balancing the natural landscape elements both of the colour, weight and distance and set the setting pattern and extract the full potential site.

3.3.6. Enhanced visual quality intensively

Five type of landscape character in the location of study could be improved visual quality intensively to develop the area as an ecotourism destination. All of the action plan will be developed on the landscape should give priority to the achievement of the attribute of beauty. The landscape elements that presented must have suitability and beauty of the landscape character. The protection of the natural landscape character such as forests and river give the desired unity and harmony. In plain areas, landscape character profiles maintained and intensified form of character. The landscape elements presented to give unity and harmony with the surrounding environment. The visual quality of the built environment can be improved through the planning and design that provides good value for unity and harmony such as colour, shape, balance, weight and distance to the surrounding environment. Contrast that are not predominantly be presented within the limits still achieve suitability, unity and harmony with the environment in which these elements are awakened.

4. Conclusion

Pongkor Mining Ecotourism Area has a diverse landscape character as a potential landscape resources for the development of ecotourism destination. The development of ecotourism destination based on landscape character requires a good assessment of natural and man-made landscape. The category of landscape character are forests, plain, rivers, tailings ponds and the built environment, was dominated by forests and plain. The river is the major feature of landscape character, whereas the forest, plain, tailings dam, and the built environment are the minor feature of landscape character. The landscape character must be protected and modified such as elimination of incongruous element, accentuation of natural form, alteration of the natural form, intensification and enhanced visual quality intensively

Acknowledgements

Researchers would like to thank to PT. Aneka Tambang which have helped to completion of the study.

References

[1] Western A 2014 *The Landscape Character Of Derbyshire* (English: Derbyshire Country Council).
[2] Starke B W and Simonds J O 2013 *Landscape Architecture : A Manual of Environmental Planning and Design* (United Stated of America (ID) : McGraw-Hill Education LLC).
[3] Zonneveld I S and Forman R T T 1989 *Changing Landcape : An Ecological Perspective* (New York (ID): Springer-Verlag Inc).
[4] Waterman T 2009 *The Fundamental Landscape Architecture* (London (UK): AVA Publ).
[5] Meletis Z A and Campbell L M 2007 Call It Consumption ! Re-Conceptualizing Ecotourism as Consumption and Consumtive *Geo. Compass* vol 1 eds 4 pp 850-870 doi:10.1111/j.1749-8198.2007.00048.x.
[6] Blamford A, Beresford J, Green J, Naidoo R, Walpole M and Manica A 2009 A Global Perspective on Trends in Nature-Based Tourism. *Plos Bio.* vol 7 eds 2 pp 1-6 doi: 10.371/journal.pbio1000144.
[7] Cobbinah P B 2013 Contextualising the meaning of ecotourism *Tour. Management Perspectives* vol 16 eds 3-4 pp 179-189 doi: 10.106/ j.tmp.2015.07.015.
[8] Dologlou N and Katsoni V 2016 Ecotourism in Protected Area, A Literatur Review. *Ecotour. Paper Series* vol 38 eds 3 pp 1-20.
[9] Wang L E, Zhong L, Zhang Y and Zhou B 2014 Ecotourism Environmental Protection Measures and Their Effects on Protected Area in China *Sustainability* vol 6 eds 3 pp 6781-6798 doi: 10.3390/su6106781.
[10] PT. Aneka Tambang 2013 *Laporan Hasil Verifikasi Lapangan-Proper 2013* (Jawa Barat: PT. Aneka Tambang UBPN)
[11] [SNH dan TCA] Scottish Natural Heritage dan The Countryside Agency 2002 *Landscape Character Assessment : Techniques and Criteria for Judging Capacity and Sensitivity* http://www.countryside.gov.uk/cci/guidance and http://www.snh.org.uk/strategy/LCA. Diunduh tanggal 3 Juni 2014 pp 20.
[12] Strom S, Nathan K and Woland J 2009 *Site Engineering for Landscape Architects* (New Jersey: John Wiley & Sons, Inc).

Landscape Potential Analysis for Ecotourism Destination in the Resort Ii Salak Mountain, Halimun-Salak National Park

A Kusumoarto[1], A Gunawan[2] and G R Nurazizah[3]

[1]Lecturer in Study Program of Architecture, Faculty of Technology, Mathematic, and Natural Science, Indraprasta University, PGRI, Jakarta
[2]Lecturer in Department of Landscape Architecture, Bogor Agriculture University
[3]Lecturer in Study Program of Management Resort & Leisure, Universitas Pendidikan Indonesia

Email : andri_anto@yahoo.com

Abstract. The Resort II *Salak* Mountain has variety of landscape potential for created as ecotourism destination, especially the potential of the waterfall (*curug*) and sulphur crater (*Kawah Ratu*). The aim of this study was to identify and analyze the potential resources of the landscape to be created as ecotourism destination, Resort II *Salak* Mountain. This research was conducted through two phases: 1) identification of the attractions location that have potential resources for ecotourism destination, and 2) analysis of the level of potential resource of the landscape in each location using Analysis of Tourist Attraction Operational Destination (ATAOD). The study showed Resort II *Salak* Mountain has many ecotourism objects which have been used for ecotourism activities, such as hot spring baths, *Curug Cigamea, Curug Ngumpet, Curug Seribu, Curug Pangeran, Curug Muara, Curug Cihurang, Kawah Ratu,* camping ground, *Curug Kondang* and *Curug Alami*. The location of all waterfalls –*curug,* spread widely in the core zone for ecotourism. In the other hand, camping ground is located in the business zone, while *Kawah Ratu* is located in the natural forest, which is included in the buffer zone of Halimun-Salak National Park (HSNP). The result showed that the ecotourism objects with the highest potential value are *Kawah Ratu, Curug Seribu, Curug Muara, Curug Kondang* and *Curug Ngumpet.*
Keywords: *potential of ecotourism destination, The Resort II Salak Mountain, Halimun-Salak National Park (HSNP), Analysis of Tourist Attraction Operational Destination (ATAOD)*

1. Introduction

Tourism is one sector of development that can be utilized as one of the main of economic activities of a country [1] [2]. Tourism activities that performed in the national park area must be based ecotourism [3] [4], for stabilizing the ecology sustainability. Ecotourism is essentially a form of tourism that is responsible for the preservation of the area's natural, economic benefits, maintain the cultural integrity of local communities [5] [6] [7] [8], and give a new experience for the visitors [9].

The development of tourism activities in the region will influence and encourage the development of other sectors, particularly in terms of expanding employment, business opportunities, and poverty alleviation [10] [11]. Ecotourism is important to do because the industry is moving in non-extractive and non-consumptive [3] [12], based on environmental sustainability [13] [14], while still providing economic benefits to the local community [15] [16].

The resort II *Salak* Mountain, *Halimun Salak* National Park (HSNP) is an area of natural environment which have some potential of Nature Attractions (NA). The presence of some of these objects can give a new experience to the tourist. HNSP has unspoiled ecosystem representation and has undergone degradation, modification and or assisted, has unique natural attractions, rare, and beautiful as well as landscape and natural potential that can be used as Tourist Attraction Destination (TAD). Unfortunately, in the resort area, not all the potential landscape can be identified and analyzed to be developed as an ecotourism destination.

.Based on this background, the study conducted landscape analysis of resource potential for ecotourism in the Resort II *Salak* Mountain. The aim of this study was to identify and analyze the potential resources of the landscape to be created as ecotourism destination, Resort II *Salak* Mountain.

2. Methodology

The research location was in Resort II *Salak* Mountain, HSNP, Bogor, Indonesia. This research was conducted in October 2015 until January 2016. The study was conducted by descriptive method. Collecting data in this study conducted by field observation and crosschecking/exploration that used to collect data related to the identification of the potential and the initial condition of the resort area. Two steps of analysis performed are mapping using Geographic Information System (GIS) [17] [18] and assessment of tourism potential using Analysis of Tourist Attraction Operational Destination (ATAOD) instrument [19].

2.1. Research Variables

The obtained data were then processed based on the criteria that have been determined to obtain guidelines for the analysis of tourist attraction operational destination. ATAOD : after the entire The TAD in Resort II *Salak* Mountain was identified, the potential measurements was done with measurements of variables that can be seen in table 1.

Table 1. TAD Potential and Research Variables

TAD Potential	Research Variables
Natural Phenomena	Scarcity
Green Open Space	Beauty
Flora	Sensitivity
Fauna	Seasionality
Culture	Accessibility
	Social Function

2.2. Method of Data Collecting

The mapping data obtained through digitization, scanning and conversion of Landsat TM imagery. It intended to form the basis of spatial data that processed by the GIS. To determine the distribution of objects of ecotourism was done through several steps: 1) determination of the research area; 2) identification of the location; 3) inventory of biophysical data location; 5) interpretation of the data; and 6) mapping the distribution of objects of ecotourism.

The population in assessing ATAOD were the government, tourism entrepreneurs and community leaders who came from a village in the Resort II Salak Mountain. Respondents who took by purposive sampling method were some key persons that was considered to know and fully understand any potential that exists in. Respondents were then given instrument ATAOD to be filled and assessed the potential of each TAD in the region.

2.3. Method of Data Analysis

In the assessment process, it should be understood that the character can be data assessment quantitative and qualitative data. In this study, the assessment process was to structure a value that is easy and commonly used, the scoring system. This scoring system was using a Likert Scale that moves from 1 to 5. The scoring average range results can be seen in table 2.

Table 2. Scoring Average Range

Score	Interpretation Value of Attractions
4.21 – 5.00	Very High
3.41 – 4.20	High
2.61 – 3.40	Moderate
1.81 – 2.60	Low
1.00 – 1.80	Very Low

3. Result

3.1. General Condition

The region was located at: longitude 106° 36 '30' '- 106° 45' 55" E and latitude 6° 31 '0' - 6° 47' 15" N. Total area is 760.36 ha. This research administrative area was in *Pamijahan* district that passed by three major rivers, namely *Cikuluwung* River tipped in *Kawah Ratu*, *Cigamea* River tipped from the buffer area of the foot of *Salak* Mountain, and *Ciapus* River.

3.2. Ecotourism Attractions Destination in Resort II Salak Mountain, HSNP

In the Resort II Salak Mountain spread some attractions which was still running. These objects were 1) Hot Spring Baths (*pemandian air panas*), 2) *Curug Cigamea*, 3) *Curug Ngumpet*, 4) *Curug Seribu*, 5) *Curug Pangeran*, 6) *Curug Muara*, 7) *Curug Cihurang*, 8) *Kawah Ratu*, 9) camping ground, 10) *Curug Kondang* and 11) *Curug Alami*. The distribution of these attractions can be seen in figure 1.

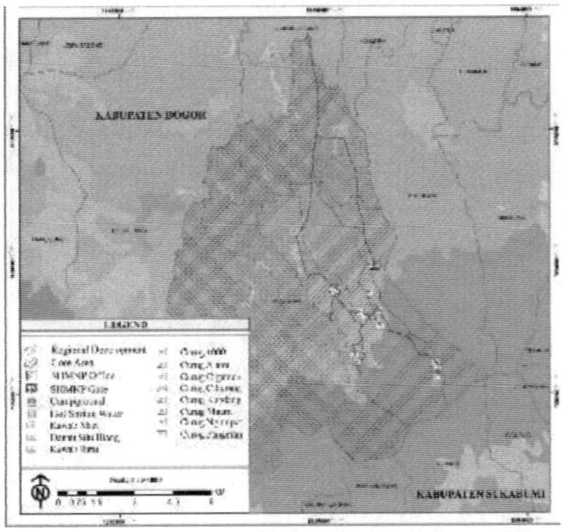

Figure 1. Location of Tourist Attractions Destination in research location

To reach the area there were several possible access can be developed. Assessing from the way of the status paths to reach this location is divided into three groups, namely: *Jakarta-Parung-Semplak–Ciampea* is the Provincial Road. *Ciampea-Cibungbulang-Pamijahan* is the District Road. *Cibungbulang-Pamijahan*-Central of The District is the Sub District of the Village Road (figure 2). The distance to the location of attractions can be seen in table 3 and table 4.

Table 3. The Mileage to HSNP Region

No.	The Path	The Meliage
1.	Via cikampak-HSNP	4 km + 6.2 km
2.	Via cibatok-HSNP	4 km + 7.3 km

IOP Conf. Series: Earth and Environmental Science **91** (2017) 012029 doi:10.1088/1755-1315/91/1/012029

Table. 4. The Meliage from The Gate to The Ecotourism Objects

No.	Name of The Ecotourism Objects	The Distance from The Gate	The Size of The Ecotourism Objects (ha)
1.	Hot Spring Baths	404 meter	0.719918
2.	Curug Cigamea	410 meter	0.442131
3.	Curug Kondang	243 meter	0.033039
4.	Curug Alami	218 meter	0.010225
5.	Curug Pangeran	256 meter	0.041065
6.	Curug Ngumpet	69 meter	0.021013
7.	Curug Cihurang	185 meter	0.100413
8.	Curug Muara	867 meter	0.121262
9.	Curug Seribu	1294 meter	0.083196
10.	Camping Ground	75 meter	0.167236
11.	Kawah Ratu	3776 meter	8.374262

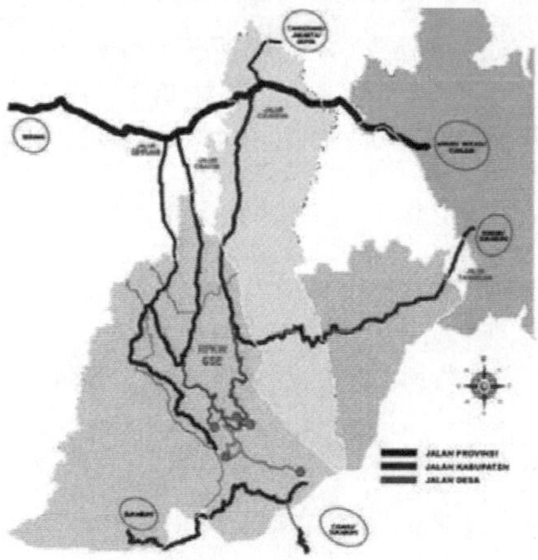

Figure 2. Accessibility and Class of Road to The Resort II *Salak* Mountain

3.3. The Potential Value of Ecotourism Attractions in the Resort II Salak Mountain

The obtained data were then processed based on the criteria that have been determined to obtain guidance of tourism analysis. The analysis of the potential of tourist attraction used to mapped the internal forces that have in some attraction in order to become a visitor attraction. In this analysis, all attractions in Resort II *Salak* Mountain researched every potential category.

3.3.1. Natural Phenomena Potential Value

From the analysis of the potential can be seen that *Kawah Ratu* and *Curug Kondang* has potential natural phenomena assessment scores, were the highest than other tourist attractions that exist in the Resort II *Salak* Mountain. As for his scores can be interpreted high (4:00 and 4:41). As for the details, the assessment of potential value can be seen in figure 3.

3.3.2. Green Open Space (GOP) Potential Value

From the green open spaces function analysis, *Kawah Ratu*, *Curug Seribu*, and *Curug Muara* got the highest ratings, with a score (5:29, very good). As for details, the assessment of potential value shown in figure 4.

IOP Conf. Series: Earth and Environmental Science **91** (2017) 012029 doi:10.1088/1755-1315/91/1/012029

3.3.3. Flora Potential Value

The highest potential value of flora in *Kawah Ratu*, *Curug Seribu*, and *Curug Muara* with a score of 4.00 (good). The green open space condition that has a composition of shape, colour, and the pristine of green open space, unspoiled, and the process dynamics of green open space which highly integrated with the natural potential of the waterfall (*curug*). Thus creating an overall landscape beautiful and reconcile (figure 5).

3.3.4. Fauna Potential Value

The Resort II *Salak* Mountain has potential variety of fauna. Based on analysis of potential fauna in this region, the potential value of this fauna, *Curug Cigamea* get the highest score (a score of 3:29, is quite high), followed premises *Curug Ngumpet*, *Curug Kondang*, and Camping Ground (a score of 2.71, high enough) (figure 6).

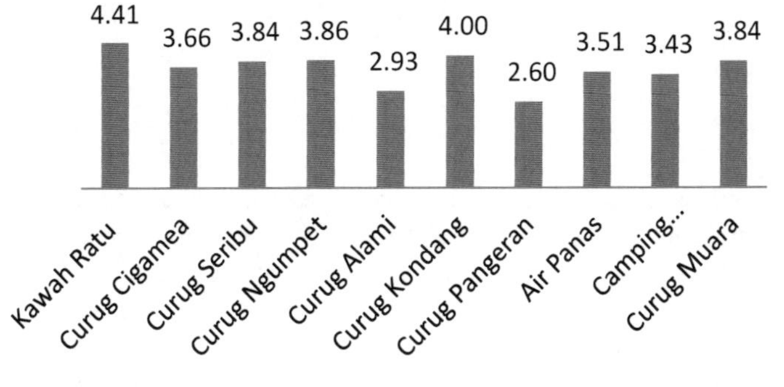

Figure 3. Natural Phenomena Potential Value Analysis
Source : data processed, 2016

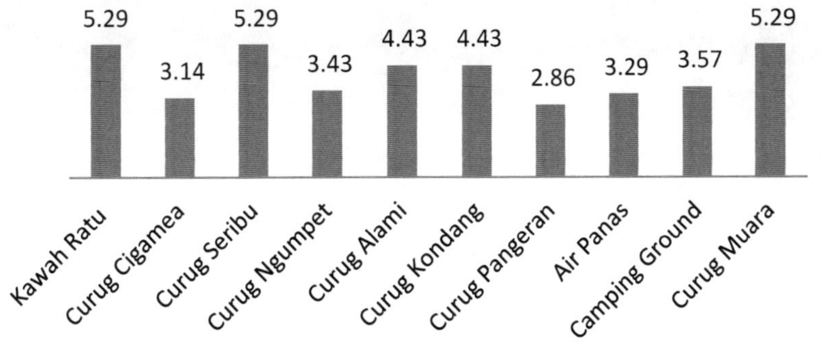

Figure 4. Green Open Space Potential Value Analysis
Source : data processed, 2016

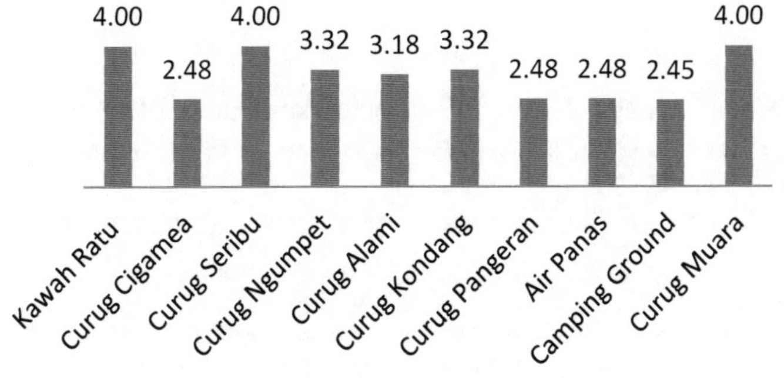

Figure 5. Flora Potential Value Analysis
Source : data processed, 2016

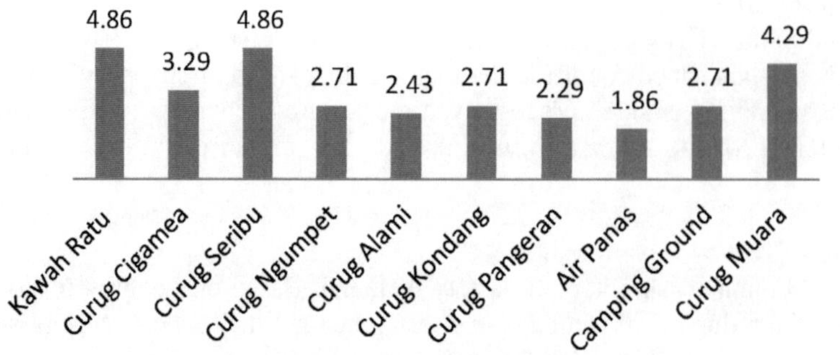

Figure 6. Fauna Potential Value Analysis
Source : data processed, 2016

3.3.4. Attractiveness Analysis of Potential Ecotourism Destination Attractiveness in Resort II Salak Mountain

The results of the attractiveness analysis potential of ecotourism indicated the priority of TAD that has the highest potential were *Kawah Ratu, Curug Seribu, Curug Muara, Curug Kondang* and *Curug Ngumpet* (figure 7). These TAD have some potential attractiveness to be the area of ecotourism development because of its physical characteristics were still easy to planning and design based on carrying capacity. In addition the number of tourists who visit are not too many (mass tourism) as well as and the merchant only were in very small amounts (minimizing conflicts of relocation).

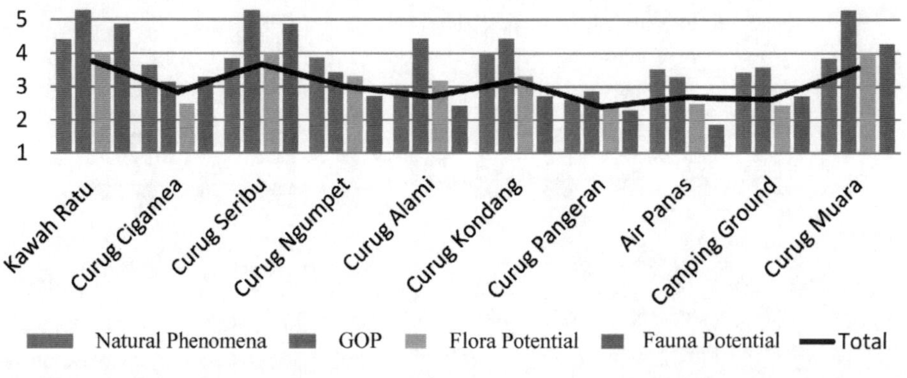

Figure 7. The Total Ecotourism Attractiveness Potential Value in Resort II *Salak* Mountain
Source : data processed, 2016

3.3.5. Culture

Around Resort II *Salak* Mountain has cultural tourism potential (material and immaterial heritage). A wealth of heritage material in the form of art studios are including studio art sculptures, *wayang golek*, *jaipongan* dance, *pencak silat*, *degung*, *marawis degungan* and others, if explored in the past. For immaterial heritage was rather hard to identify because of the lack of documentation related to.

4. Conclusion

The Resort II Salak Mountain has the landscape resources that have the potential for development of ecotourism, such as Hot Springs Baths, *Curug Cigamea, Curug Ngumpet, Curug Seribu, Curug Pangeran, Curug Muara, Curug Cihurang, Kawah Ratu*, Campgrounds, *Curug Kondang*, and *Curug*

Alami. TAD which has the highest potential are *Kawah Ratu*, *Curug Seribu*, *Curug Muara*, *Curug Kondang* and *Curug Ngumpet. Curug Ngumpet was a* shortest distance from the gate.

The five potential TAD to be the area of ecotourism development. They have some physical characteristics were still easy to planning and design based on the carrying capacity, the number of tourists who visit are not too many (mass tourism) as well as the merchant only slightly (minimizing conflicts of relocation).

References

[1] Gössling S 2000 Sustainable tourism development in developing countries: Some aspects of energy use *J. Sustain. Tour.* vol 8 eds 5 pp 410-425.
[2] Mitchell R E and Reid D G 2001 Community integration: Island tourism in Peru *A. Tour. Research* vol 28 eds 1 pp 113-139.
[3] Goodwin H 1996 In pursuit of ecotourism *Biodiversity & Conservation* vol 5 eds 3 pp 277-291.
[4] Scheyvens R 1999 Ecotourism and the empowerment of local communities *Tour. Management* vol 20 eds 2 pp 245-249.
[5] Ceballos-Lascurain H 1991 Tourism, Ecotourism, and Protected Areas Parks. *J Sustain. Tour.* vol 2 pp 31-35.
[6] Carter E and Lowman G 1994 *Ecotourism : A Sustainable Option* (New York : John Willey & Sons).
[7] Honey M 1999 *Ecotourism and Sustainable Development : How Owns Paradise* (Washington DC: Island Press).
[8] Bjork P 2000 Ecotourism From a Conceptual Prespective , an Extended Definition of Unique Tourism Form *Intl. J. Tour Research* vol 2 pp 189-202.
[9] Gunn C A 1994 *Tourism Planning Basics, Concepts, Cases. Third Edition* (London : Taylor & Francis Ltd).
[10] Lordkipanidze M, Brezet H and Backman M 2005 The entrepreneurship factor in sustainable tourism development *J. Cleaner Product.* vol 13 eds 8 pp 787-798.
[11] Chok S, Macbeth J and Warren C 2007 Tourism as a tool for poverty alleviation: A critical analysis of 'pro-poor tourism'and implications for sustainability *Current issues in Tour.* Vol 10 eds 2-3 pp 144-165.
[12] Sirakaya E, Sasidharan V and Sönmez S 1999 Redefining ecotourism: The need for a supply-side view. *J. Travel Research* vol 38 eds 2 pp 168-172.
[13] Kiss A 2004 Is community-based ecotourism a good use of biodiversity conservation funds?. *Trends in ecology and evolution* vol 19 eds 5 pp 232-237.
[14] Valentine P S 1993 Ecotourism and nature conservation: A definition with some recent developments in Micronesia *Tour. Management* vol 14 eds 2 pp 107-115.
[15] Ross S and Wall G 1999 Ecotourism: towards congruence between theory and practice. *Tour. Management* vol 20 eds 1 pp 123-132.
[16] Wunder S 2000 Ecotourism and economic incentives—an empirical approach. *Ecologic. econom.* vol 32 eds 3 pp 465-479.
[17] Dawod G M 2013 Suitability Analysis for Tourism Infrastuctures Utiizing Multi-Criteria GIS: A Case Study in Al-Hada City, Saudi Arabia. *Intl. J. Geo and Geosciences*. vol 4 eds 2 pp 313-324.
[18] Akten M 2013 Possibility to Employ AHP as a Multi-Citeria Decision Making Method in Landscape Planning Initiatives P 87-104. In M Özyavuz (ed). *Advances in Landscape Architecture*. (Croatia: InTech Publ).
[19] [Ditjen PHKA] Direktorat Jenderal Perlindungan Hutan dan Konservasi Alam 2003 *Analisis Daerah Operasi Obyek dan Daya Tarik Wisata Alam. (ADO-DTWA)* (Jakarta: Departemen Kehutanan).

Promoting Productive Urban Green Open Space Towards Food Security: *Case Study Taman Sari, Bandung*

M. Ridwan[1], Fran Sinatra[2] and Petrus Natalivan[3]
[1)] Student of City and Regional Planning, SAPPK ITB, Indonesia
[2)] Researcher of Urban Planning and Design Laboratory, SAPPK ITB, Indonesia
[3)] Lecturer of City and Regional Planning, SAPPK ITB, Indonesia

Email: sinatrafran88@gmail.com

Abstract. The common trend of urban population has been growing significantly in Indonesia for decades, are affected by urban green space conversion. Generally, this area is utilized for urban infrastructures and residences. Furthermore, urban area has grown uncontrollably that could enhance the phenomenon of urban sprawl. The conversion of green urban area and agricultural area will significantly decrease urban food security and quality of urban environment. This problem becomes a serious issue for urban sustainability. Bandung is a city with dense population where there are many poor inhabitants. Families living in poverty are subjected to food insecurity caused by the rise of food prices. Based on the urgency of urban food security and urban environment quality the local government has to achieve comprehensive solutions. This research aims to formulate the policy of productive green open space towards food security for poor people in Bandung. This research not only examines the role played by productive green open space to supply food for the urban poor but also how to govern urban areas sustainably and ensure food security. This research uses descriptive explanatory methodology that describes and explains how to generate policy and strategic planning for edible landscape to promote urban food security. Taman Sari is the location of this research, this area is a populous area that has amount of poor people and has a quite worse quality of urban environment. This study shows that urban green open space has the potential to be utilized as an urban farming land, which poor inhabitants could be main actors to manage urban agriculture to provide their food. Meanwhile, local government could contribute to subsidize the financial of urban farming activities.

1. Introduction

A city is a large scale of settlement with a non-agricultural main activity. The characteristics, which stated defines the dependence of a city to its hinterland due to food supply. In spite of that, the rapid urbanization that is taking place goes together with a rapid growth of urban poverty and food insecurity. The World Bank (2000) estimates that approximately 50% of the poor live in urban area, while there were only 25% in 1988. Most cities in developing countries have great difficulties to cope with this development and are unable to provide services to the poor (Armar-Klemesu, 2000). Food security has always been a key facet for people living in cities. However, supplying food in the urban areas may have obstacles in production and distribution that leads to unaffordability for the urban poor.

The sustainability of one city is coherence between modernization and sustainable food system. The urban strategy by supplying food in cities and merge the society and food system to build a healthy and sustainable urban lifestyle. Food resilience in a city could solved by bringing food back to the city.

In this case, by empowering urban agriculture. This urban design model focuses on public health, community development, and ecological stability. As for applying urban agriculture requires an available open space in a city.

In most cities in Indonesia, urbanization has resulted in converting available green open spaces. Generally, the land conversion is an effort to provide public housing and supporting facilities. These conversions will decrease the quality of urban environment and food security significantly. Green open spaces in city is a land capital to provide urban farming. Though, one of the purposes of open spaces in cities is for agriculture (Minister of Home Affairs Regulation, 2007).
Nevertheless, the empowerment of green open spaces as an urban farm has yet productive in supporting the basic need of urban societies. Crops haven't been planted in available urban spaces. By attempting a sustainable urban food system needs supports from both the state government and local societies. In spite of that, the implementation of urban farming would require land extension from the existing green open spaces. Applying urban farming activity could stimulate city independency in supplying food so that creates a food resilient city. This study is to identify the impact by applying urban agriculture in fulfilling the basic needs of urban societies.

2. Area Description: Urban Slums in Tamansari, Bandung

Bandung is the largest city in West Java Provence. Since 1985, Bandung has a common trend of rapid urbanization which cause sprawling urban areas to the east side of the city and Cimahi on the west. The urban growth that is not well-maintained leads to a significant land conversion from open spaces to public housing. While the population increases, the public facilities are yet able to support the people's needs, especially the poor. One of the locations of poor neighbourhood is Tamansari Neighbourhood which located around Cikapundung River.

Figure 1. Map of Tamansari Neighborhood, Bandung City

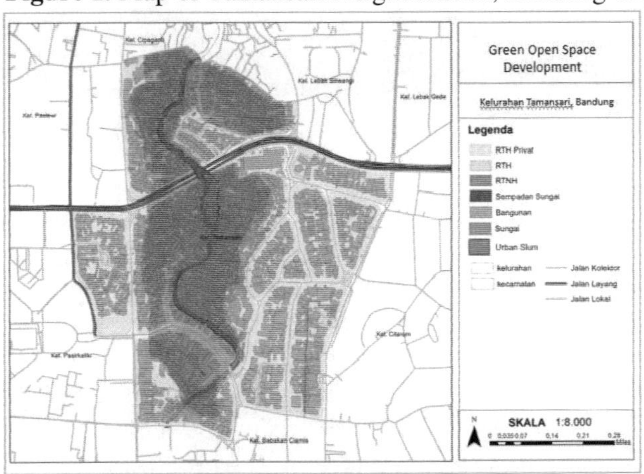

The picture above shows Tamansari Neighborhood which located in Cibeunying District, Bandung. According to the city's zoning regulation, Cibeunying District is a housing area built with supporting facilities. Most area of Cibeunying is for ecological conservation and cultural heritage. But in fact, rapid urbanization that is taking place in this northern Bandung area emerges urban slums in some particular area and unplanned land use conversions. Tamansari Neighborhood has a large slum area surrounding Cikapundung River. The slum is shown in red in the picture above. The slum are has taken almost half the area of Tamansari.

The area that marked in red on the picture above is a populous slum area that has amount of poor people and has quite a bad quality of urban environment. In spite of it, a typical unplanned urban slum area does not facilitate the residents with a decent utilities and other needs. One of those other needs is open space for local society to interact daily. The map above also shows green open space distribution on Tamansari Neighborhood. There is only minimum space that is used for green open space in the whole Tamansari, none in the riverbank slum area.

3. Land Potential for Agriculture

Urban agriculture may take place in locations in the inner city area, suburbian area, and even the peri-urban area. The activities may take place on private owned space such as homestead, a land away from the residence, private building yard (owned, leased) or on public land such as parks, conservation areas, along roads, streams and railways. Some practices may even use semi-public land such as schoolyards, grounds of schools and hospitals, either the buildings are still operating or not.

Meanwhile, Tamansari Neighborhood has many private open green space along building yards and homesteads. Most of the private open green space occurred on the eastern side of Tamansari where lies a residential and some commercial areas that regulated with a low building base coefficient. Those area has many private yards and some public parks. But on the other hand, urban agriculture targets food resilience for the urban poor. While then space availability to implement an urban farm mostly lies on the prosperous part of the neighborhood.

Table 1. Built Area Allocation

Development	Area (Ha)	%
Building	50.19	49.3
Building Yard and Homestead (Private Owned)	11.058	10.9
Green Open Space + Riverbank Area (Public Owned)	1.14	1.1
Public Space	0.06	0.1
River	2.09	2.1
Roads	11.45	11.2
Alley	25.802	25.3
TOTAL AREA	***101.79***	***100.0***

The table above shows the land distribution for particular use across Tamansari. According to the Ministry of Home Affairs regulation (2007), the minimum requirement of green open space existence in a city 30% of the city area. From the 30% area, consists both public owned and private owned which took 20% public owned and 10% private owned. From the table above can be seen that the private owned green open space has already fulfill the requirement. While the public owned green open space is falling behind the minimum requirement by 18.9% less. Tamansari only have public green space as 1.1% of the whole neighborhood area, which consists of a few small parks.

The concept of productive green open space is empowering green spaces as an urban farmland, which produces crops and other types of plantation. The most suitable type land use to implement an urban farm is green open spaces (RUAF, 1999). However, the total area of public green space in Tamansari is only 1.14 Hectares. From which is the only available public land to be developed as an urban farmland. Therefore, only small part of the neighborhood that could be utilized as an urban farm to produce crops for the populous slums in Tamansari.

4. Cultivation of Land Allocation

The can author specify local residents' needs by taking several sample respondents to determine their vegetative food preference. These respondents are people that lives in the slums surrounding Cikapundung River. According to the data collected, most inhabitants preferred vegetables (consists of spinach, kale, onion, garlic, chili, tomato) as 48% of the total consumption. Followed by spices (22%), fruits (13%), nuts (12%), and tuber (6%). From then, the author can specify societies' preference and make weighing in land allocation of certain commodities to be produced.

The existing available land to develop an urban agriculture in Tamansari is only 1.14 hectares. Which means that the author can only calculate the agricultural productivity based on the land available. The land allocation for preferred crops are 703 m2 for tuber, 5,875 m2 for vegetables, 492 m2 for nuts, 1,583 m2 for fruits, and 2,744 m2 for spices. Every type of commodity has a different productivities and harvest terms. By a year period, the green open space land could only produce 73.2 tons that consists of 4.5 tons of tuber, 54.3 tons of vegetables, 0.5 tons of nuts, 4.3 tons of fruits, and 9.4 tons of spices. Due to land limitation, harvesting intensification would be needed in order to fulfil food needs.

5. Design Strategies for Urban Agriculture

A framework of urban design strategies can be used to aid in creating urban agricultural landscapes that promote ecological biodiversity and social sustainability in urban environments no matter what the scale, type, or location of the landscape. Incorporating a systems thinking approach that considers appropriate methodologies and tools, creativity and innovations, combined with the design principles that follow, urban agricultural landscape can flourish and aid in the evolution of the sustainable city. By building an integrated systems thinking model for urban agricultural landscapes, there are several principles that promote a more integrated approach.

A framework of urban design strategies can be used to aid in creating urban agricultural landscapes that promote ecological biodiversity and social sustainability in urban environments no matter what the scale, type, or location of the landscape. Incorporating a systems thinking approach that considers appropriate methodologies and tools, creativity and innovations, combined with the design principles that follow, urban agricultural landscape can flourish and aid in the evolution of the sustainable city. By building an integrated systems thinking model for urban agricultural landscapes, there are several principles that promote a more integrated approach.

- **Promote biodiversity;** The importance of biodiversity in urban environment is being a part of ecological services that make life livable on cities (National Wildlife Federation). It could also provide us with an array of foods and materials and it contributes to the economy. Without biodiversity stability, an ecosystem can be jeopardized.
- **Increase food security and food safety;** Should focus on the utilization of foods through proper nutrition, preparation, and feeding practices, and the stability of these conditions over time (Maxwell and Smith 1992; FAO 2006).
- **Incorporate education and outreach for awareness;** The teaching and outreach of the food landscape and food system ecological, social, and economic benefits to both private and public sectors. This part of the system includes on-site education, training and mentoring local societies as well as the outreach associated with the marketing and branding of the urban agriculture development.
- **Be climate adaptive for environmental resilience;** Agricultural landscapes, on their own and/or when combined with buildings and other urban infrastructures, can increase the ability to climate control and aid environmental resilience. As we lose permeable surfaces that support vegetation and soil life to paved surfaces of streets and buildings.
- **Maximize water accessibility, availability, and quality;** All vegetation requires water. The amount of water needed depends on the crop and other aspects of the layout. According to the Governmental Regulation (2011) the requirement of developing city area as an urban farm is gaining water source, both by irrigation and rainwater. Furthermore, building a drainage system is also important in managing agricultural water waste.
- **Provide for soil health;** Through past urban development and conventional framing practices, the health of the soil has been compromised and degraded. Through urban agriculture landscapes integration of ecological based practices, the soil can be revitalized adding to food vitality and also the soil's ability to reduce erosion in urban environment.
- **Develop a systems network that is both regionally and locally appropriate;** Connectivity of the various natural and built resource systems provides for a more integrated resource management structure that current conventional agriculture and current city planning systems do

not consider. The benefits and value of a regional system add to the ecological, social, and economic sustainability of the city and the outer city boundaries.

- **Protect and increase human health benefits;** This human health aspect consists from psychological, biological, and physical. Urban agriculture landscape will prove fruitful to increase health and well-being benefits in a city. Adding plants for both nutritional and medicinal needs, community gardens that provide for social interaction, and educational opportunities, all add up to a healthier environment for people and ecosystems.
- **Provide for the connection for people with nature to the enrichment of both;** The Nature Deficit Disorder is a description of the human costs of our alienation from nature (Richard Louv in *Last Child in the Woods*). Therapeutic landscape design has statistics that show that landscapes help to heal and soothe patients as well as people walking in urban environments. Agricultural landscapes offer an excellent chance to engage not just experience nature in the city.
- **Foster community, placemaking, and social resilience;** Placemaking is both an idea and a tool that considers the context of creating communities that are desirable places to live. It is based on the idea that places are complex and dynamic systems. Some of these tools are asset mapping, visualization, or predictive modelling (Boyd, 2002). One tool is shaping a community's identity and activity pattern. The other is identifying the many aspects of food and agriculture that are central to fostering human health and happiness.
- **Promote sustainable economic benefits and opportunities;** Poor people generally spend a substantial part of their income, approximately 50%. Growing the relatively expensive vegetables therefore saves money as well as on bartering of produce. Selling products brings income. The municipality and sectoral organizations can play a crucial role in stimulating micro-enterprise development related to urban agriculture.
- **Increase the treatment of waste as a resource for plantation compost;** Thinking of waste as a resource, or in a way to produce food, is an important part of rethinking waste streams in a city (Bill McDonough in *Cradle to Cradle*). Upcycling a material to be reused in a new and improved way is one method to address some waste products. Designing for deconstruction and repurposing those materials to another use is another way to address waste. Green can be turned into soil and compost.

Due to the cross cutting and multi-dimensional nature of urban agriculture, policy development and action planning on urban agriculture should involve various sectors and disciplines: agriculture, health, waste management, community development, parks and nature management, among others. Moreover, urban farmers have to be involved in the planning process. An important aspect of strategic urban planning is related to the participation of the urban poor themselves in the analysis of the situation, in the definition of priorities and in action planning and implementation. Such consultative processes will make the outcomes of policy development and action planning not only robust and comprehensive, but also accepted and sustainable.

By implementing municipal strategies for sustainable urban agriculture, urban policy makers can substantially contribute to the development of safe and sustainable urban agriculture by:

- Creating a conducive policy environment and formal acceptance of urban agriculture as an urban land use;
- Enhancing access to vacant open urban spaces and the security of agricultural land use;
- Enhancing the productivity and economic viability of urban agriculture by improving access of urban farmers to training, technical advice, and credit and supporting the establishment and strengthening of urban farmer organizations; and
- Taking measures that prevent/reduce health and environmental risks associated with urban agriculture.

In the case of Tamansari, the available land to develop an urban agriculture and fulfilling the urban poor's needs are very limited. Therefore, the government should make an inventory of vacant land and

IOP Conf. Series: Earth and Environmental Science **91** (2017) 012030 doi:10.1088/1755-1315/91/1/012030

by adding an extension by making land acquisition to provide urban farmlands. This practice should gain city-planning objectives of acquiring more green open spaces and creating a better urban activity pattern. Furthermore, the urban agriculture development should require the involvement of municipality, private parties, and local communities to achieve environmental, social, and economic goals.

6. Conclusion

In order to actualize a food resilient city, urban agriculture plays a major role. It has an impact on human health, local markets, and city inclusivity. This development concept is addressed to the urban poor where food affordability is questioned. Due to its regional characteristics, Tamansari Neighborhood has a number of urban slums whereas many poor inhabitants live. Indeed, the current spatial planning in Tamansari does not support this landscape development. But a productive green open space is much needed to increase public interaction and to provide economic opportunities. On the other hand, productive green open space could also bring environmental sustainability in the city.
Developing a productive green open space, in other words food landscape, needed both good landscape design and policy planning. The agricultural productivity in the vacant land in Tamansari is far beyond necessity. Therefore, policy planners should consider zoning reform by expanding urban land for agriculture in order to gain a food resilient city. While the food landscape itself should be maintained by local community to regain ecological biodiversity and social sustainability.

Acknowledgments

We would like to thank Center for Research and Community Cervices Institut Teknologi Bandung for the support to the research activities.

References

[1] Phillips, April. *Designing Urban Agriculture.* 2011. New Jersey: Wiley.
[2] National Wildlife Federation. 2016. *Wildlife Conservation.* www.nwf.org/wildlife/wildlife-conservation/biodiversity. 30 October 2016.
[3] Upton, Joanna B., Jennifer Denno Cisse and Christopher B. Barrett. 2015. *Food Security as Resilience: Reconciling Definition and Measurement.* Cornell University.
[4] UAF Foundation. 2016. *Urban Food Policies and Strategies.* www.ruaf.org/topics/urban-food-policies-and-strategies. 12 September 2016.
[5] Boyd, Susan, and Roy Chan. 2002. *Placemaking: Tools for Community Action.* Washington DC: CONCERN, Inc.
[6] Agrotani. 2016. *Agrotani Kategori Pertanian.* www.agrotani.com/a/pertanian. 20 September 2016
[7] Butler, L., and D.M. Moronek. 2002. *Urban Agriculture Communities: Opportunities for Common Ground.* Ames: Council for Agricultural Science and Technology.
[8] American Society of Landscapes Architect. *How to Expand Urban Agriculture.* dirt.asla.org/2010/01/28/how-to-expand-urban-agriculture/. 31 October 2016
[9] Center for Ecoliteracy. 2008. *Big Ideas: Linking Food, Culture, Health, and the Environment.* Berkeley: Learning in the Real world.
[10] Louv, Richard. 2008. *Last Child in the Woods: Saving Our Children from Nature Deficit Disorder.* Chapel Hill: Algonquin Books.
[11] De la Salle, Janine, and Mark Holland, eds. 2010. *Agricultural Urbanism: Hand for Building Sustainable Food Systems in 21st Century Cities.* Winnipeg: Green Frigate Books.
[12] Puriandi, Fandy. 2012. *Studi Bentuk Pertanian Kota oleh Komunitas Berkebun di Kota Bandung sebagai Masukan Pengembangan Pertanian Kota di Kawasan Perkotaan.* Bandung: Institut Teknologi Bandung.

IOP Conf. Series: Earth and Environmental Science **91** (2017) 012031 doi:10.1088/1755-1315/91/1/012031

Landscape Characteristics of Oriental Honey Buzzards Wintering in Western Part of Flores Island Based on Satellite-Tracking Data

Syartinilia[1,3], G H Al Farisi[1] and H Higuchi[2]

[1] Department of Landscape Architecture, Faculty of Agriculture, Bogor Agricultural University, Kampus IPB Darmaga, Bogor 16680, Indonesia
[2] Graduate School of Media and Governance, Keio University, Endo 5322, Fujisawa, Kanagawa 252-0882, Japan

syartinilia@yahoo.com

Abstract. Oriental Honey Buzzards (OHBs, *Pernis ptilorhynchus*) are migratory raptor that has been satellite-tracked since 2003. Some islands in Indonesia which are used for wintering habitat are Flores and Borneo. However, both islands have different characteristics of climate and land cover. The objectives of this research were to analyze the landscape characteristic of the OHBs wintering habitat in western Flores, and to subsequently compare landscape characteristic of the OHBs wintering habitat in Borneo. Landscape habitat characteristics were analyzed using Principal Component Analysis (PCA) combined with GIS and then compared to the previous study in Borneo Island. The result showed that the first of six principal components explained 79.14% and 77.59% of the observed variation in landscape characteristics of both core and edge habitats, subsequently. Habitat selection by OHBs at wintering site was influenced by the availability of thermal wind and food. Savannah was identified as the main landscape characteristic that was different between wintering habitat in Flores and Borneo. Savannah is well-known as a habitat for many species of amphibians, reptiles, and small mammals so that it can be a hunting area that provide alternative feed for OHBs.

1. Introduction

Migration is the globally animals movement caused by change of seasons, significantly. The ultimate goal of migration is to find food sources in order to survive. Migratory raptors need to have at least three different environments: breeding grounds, wintering grounds, and stopover grounds [1]. In general, most raptors in Asia are breeding in northern Asia and wintering in the southern Asia. Indonesia is an important wintering area for several species of migratory raptors that breed in the Eastern Palearctic [2]. From 69 raptors recorded in Indonesia, 39 species are migratory raptor and 26 species have categorized as intercontinentally migration (long-distance migration) [3].

Oriental Honey Buzzards (OHBs, *Pernis ptilorhynchus*) are the migratory raptors, which have wintering sites in many places in Indonesia. Satellite tracking (ARGOS) has been successfully used to track the OHB since 2003 in order to observe the migration route of OHBs. Based on the ARGOS satellite-tracking data, distribution of wintering habitat in Indonesia spreads over several islands in western part of Indonesia such as Sumatra, Java, Borneo and eastern part of Indonesia such as Flores and Kupang [4]. Borneo and Flores Islands are wintering habitat of OHBs, whereas both islands have a different climate condition and land cover. Although different, OHBs still prefer both islands as their wintering habitat.

Recently, landscape characteristics of wintering sites in Borneo consist of core and edge habitat, where both characteristics are essential to be studied as a single entity of OHBs habitats [4]. Therefore, the study on the characteristics of wintering habitat in western part of Flores Island and comparing to landscape characteristics found in Borneo is important for understanding the ecology of wintering habitat of OHBs. Result of this study can be use as baseline information for providing a recommendation on the management of wintering habitat to maintain the existence of OHBs.

2. Methods

2.1. Location and Time

The research location is located in the western part of Flores Island, precisely in regencies of West Manggarai, Central Manggarai, and East Manggarai (Figure 1). This study was conducted from March 2015 to June 2016.

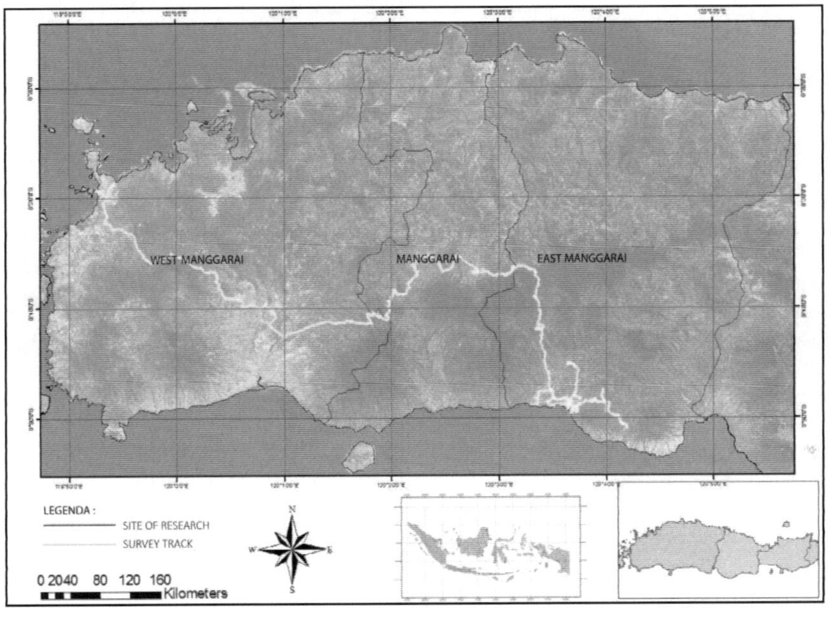

Figure 1. Study Site Map

2.2. Delineation analysis of core and edge habitat of OHBs

The Satellite tracking data used in this study wasfrom ARGOS data in 2006-2010. This study used a data sample from two individuals of OHBs which their tracks were recorded by the ARGOS satellite. The numbers of Platform Transmitter Terminal (PTT) as identity of the two individuals were 84427 and 66554. Recorded data from the satellite tracking of two individuals were 241 locations. In this analysis, kernel-density estimation (KDE) was used. This method is often used to analyze the spatial distribution. Using GIS, this method will create a line based on the intensity of use by calculating the average of the influence of point data. Kernel-density estimation method was selected because this method is often used to study a habitat. Delineation of wintering habitat of OHBs was produced using Fixed Kernel_95% (FK_95%) for edge habitat and Fixed Kernel_50% (FK_50%) for core habitat [4]. The aim of this analysis was to determine the core and edge habitats of OHBs based on point data of satellite tracking. In this discussion, the core is defined as an area and known to have the high intensity of the presence of OHBs, indicated from satellite tracking data recorded in the area.On the other hand, the edge is habitat that surrounds the core and known to have the low intensity of the presence of OHBs, indicated by satellite tracking data recorded in the area. This analysis used the extension Hawth's tools on a menu of fixed-kernel density estimator.

2.3. Analysis of Landscape Characteristics of Wintering Habitat

In this analysis, making a basic map that would be used as an environment variable was conducted (Table 1). Environmental variables were required in principal components analysis to determine the characteristics of the wintering habitat of OHBs. Basic maps used for this study were the land cover, slope, and elevation. Land cover map was produced by supervised classification method. Imagery map used was Landsat 8 on September 14th 2014, with a resolution of 30 x 30. Ground truth check was done for obtaining the data for training area and predicting the accuracy of land cover map. Land cover classification in the western part of Flores Island was divided into eight classes: water body, forest, savannah, paddy field, shrub, build up area, plantation and bare land.

Table 1. Environmental variables

No.	Environmental variables	Abbreviation	Data source
1	Nearest distance to elevation of 0-300 m.a.s.l.	DE1	Extracted from ASTER GDEM that made as euclidean distance map
2	Nearest distance to elevation of 300-500 m.a.s.l.	DE2	
3	Nearest distance to elevation of 500-700 m.a.s.l.	DE3	
4	Nearest distance to elevation of 700-1 000 m.a.s.l.	DE4	
5	Nearest distance to elevation of>1000m.a.s.l.	DE5	
6	Nearest distance to slope of 0-3 %	DS1	Extracted from ASTER GDEM that made as euclidean distance map
7	Nearest distance to slope of 3-8 %	DS2	
8	Nearest distance to slope of 8-15 %	DS3	
9	Nearest distance to slope of 15-25 %	DS4	
10	Nearest distance to slope of 25-40 %	DS5	
11	Nearest distance to slope of> 40 %	DS6	
12	Nearest distance to water bodies	DWB	Extracted from land cover map that made as euclidean distance map
13	Nearest distance to forests	DFR	
14	Nearest dista nce to savannah	DSV	
15	Nearest distance to paddy fields	DPF	
16	Nearest distance to shrubs	DSR	
17	Nearest distance to build-up area	DBA	
18	Nearest distance to plantation	DPT	
19	Nearest distance to bare lands	DBL	

Land cover, elevation and slope maps had been obtained were then converted into a binary map, which had two values (0 and 1). Value of 1 was for the environmental variable that will be used as a distance map, and value of 0 was for the environmental variable to be removed. Furthermore, raster binary map was converted into vector maps. Then the vector map was converted into a Euclidean distance map that provided the nearest distance information of each pixel into a specific type of environmental variables.

Principal Component Analysis (PCA) was used to integrate 19 environmental variables and identify landscape characteristics of core and edge habitats used by OHBs. Subsequently, PCA was carried out with Varimax rotation with the Kaiser normalization to extract the principal component that characterize core and edge habitats. The analysis was performed using SPSS to get the importance level of environmental variables that became a principal component of wintering habitat of OHBs in western part of Flores Island.

2.4. Comparative Analysis of Habitat Characteristics

Characteristics comparison of the habitats in the western Flores and Borneo was obtained through descriptive comparison. The characteristics comparison of habitats was conducted on each environment variable in the core and edge habitats at each location. The previous study was used for wintering habitat characteristics of OHBs in Borneo [4].

3. Results and Discussions

3.1. Distribution of Wintering Habitat of OHBs

The distribution of wintering habitat of OHBs included core habitat with an area between 0.06 to 40.29 km^2 with a total area of 165.86 km^2 and edge habitat with an area between 1.34 to 178.66 km^2 with a total area of 785.66 km^2 (Figure 2). Core and edge habitats of OHBs were predominantly distributed in the southern part of the western Flores. Highest distribution of core habitat was in East Manggarai regency (52.5%) and Central Manggarai regency (25.8%) and lastly West Manggarai regency (21.7%). On the other hand, highest distribution of edge habitat was in East Manggarai regency (50.24%) and West Manggarai regency (26.16%) and the last Central Manggarai regency (23.6%).

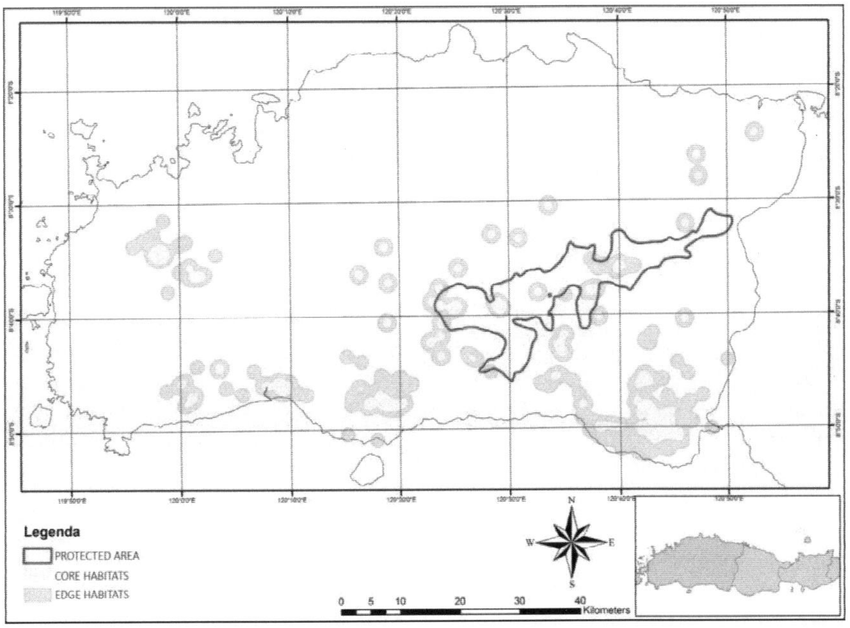

Figure 2. Distribution of core and edge habitats, and overlaid with protected area.

Delineation of core and edge habitats resulted in the largest core habitat in forest area (43.22%), shrubs (36.16%), savannah (8.96%), and the garden (7.29%) if it was overlaid with land cover map. Largest edge habitat was in the area of forest (36.31%), shrub (33.22%), savannah (14:43%) and plantation (9.16%). It was indicated that the formation and types of land cover exist both in core and edge habitats were similar, but the only difference was the percentage proportion of area of each land cover type. Forests still had the highest proportion of both core and edge habitats.

Comparing the core wintering habitat to existing protected area boundaries showed that only 21.13 km^2 (12.7 %) of the core habitat is protected; the remaining area extends far outside the protected areas. Dominant core and edge habitats in East Manggarai regency were in protected area, while the rest were in the Central Manggarai regency. Core and edge habitats were not found in the protected area of West Manggarai. Overall, it can be concluded that more than 80% of the core and edge habitats area were outside protected areas. It should be a concern of local government and managers of protected areas in order to maintain sustainability habitat of OHBs in the Flores Island.

3.2. Characteristics of Wintering Habitat of OHBs in western part of Flores

The first of six principal components explained 79.14% and 77.60% of the observed variation in landscape characteristics of core and edge habitats, respectively (Tables 2 and 3). Based on these results, we focused on the first three principal components in each core and edge habitat, which explained about 54.09% and 50.15% of data variance, respectively.

Table 2. Principal component analysis result for core habitat

Environmental variable	Principal Component (PC)					
	1	2	3	4	5	6
DFR	0.914					
DS6	0.895					
DE4	0.854					
DE3	0.767					
DS5	0.732					
DPT						
DE5		0.855				
DBL		0.787				
DWB		-0.755				
DE2			0.760			
DE1			0.721			
DSV			0.718			
DPF				0.825		
DSR				0.688		
DBA						
DS4					0.716	
DS1						
DS2						0.877
DS3						
Eigenvalue	4.721	2.973	2.584	1.690	1.609	1.460
Variance (%)	24.845	15.649	13.600	8.894	8.467	7.682
Cumulative (%)	24.845	40.494	54.094	62.988	71.455	79.137

The first principal component of the core habitat (PC1) was interpreted as characteristics associated with forest (DFR), slope more than 25% (DS5, DS6), elevation of 500-1000 m a.s.l (DS3, DS4). The second characteristic was associated with an elevation of over 1000 m a.s.l and bare land and water body (DBL, DWB). The third component was associated with the savannah (DSV) and elevation of 0-500 m a.s.l. (DE1, DE2). Unlike the core habitat, forest and elevation were not included in the first principal component of the edge habitat, only the slope of 0-15%. The second component was associated with the elevation of 0-700 m a.s.l (DE1, DE2, DE3) and savannah (DSV), while the third main component in edge habitat was forest (DSF), elevation of 700-1000 m a.s.l (DE4) and slope more than 40% (DS6).

OHBs tend to prefer their wintering habitat based on the abundance of food and the presence of other predators [4]. Forest became the first factor in the core habitat and became the third factor in edge habitat. It indicated that main food source for OHBs was available in forest. Food is a major driving factor for the bird in selecting their habitat during nonbreeding season. Habitat preference of OHBs was significantly related to the existence of a nest of honey bee colonies. *Apis dorsata* was most bee usually found in the study site. *Apis dorsata* was categorized as giant rock bee because it is larger than other honey bees. *Apis dorsata* usually establishes colonies in areas that have sufficient availability of food, especially in the forests area and plantations around forest in the area of core habitat. During the survey, there were three locations of trees beehives of *Apis dorsata*. Based on interviews with local residents, the local name of the tree is Mbete.

A difference in the characteristics of the core and the edge habitat was savannah that turns into a hunting area for OHBs in edge habitat. It was indicated from the savannah that was in the second principal component in edge habitat (PC2). Savannah provided many alternatives food for OHBs such as reptiles (lizards and snakes) and small mammals (mice). It showed that the food was potentially available in the both habitats, but the difference was only in terms of quality, the potential of bees as the main food and honey was more commonly found in the core, while edge provided alternative food such as reptiles, amphibians, and small mammals.

Beside food, core and edge habitats preference was also based on the fulfillment of the thermal wind to fly. Slope and elevation are the physical variables that shape the topography of an area which associated with the availability of thermal wind [4]. The combination of slope and elevation variations can cause thermal wind used by OHBs in soaring and gliding. Variations of slope and elevation became important variables to OHBs in soaring. Solar radiation and variation of landforms may influence the thermal [5]. Either in the core or edge, there was a potential for wind thermals, but the core habitat had higher thermal wind potential than the edge habitat. It was evidenced by the first principal component of the core habitat that had a variation of steep slope with a slope of > 25% (DS5, DS6) and the first principal component of the edge habitat which had a relatively flat slope variation with a slope of 0-15% (DS1-DS3).

Table 3. Principal component analysis result for edge habitat

Environmental variable	Principal Component (PC)					
	1	2	3	4	5	6
DS1	0.841					
DS2	0.799					
DS3	0.777					
DBL						
DE2		0.916				
DE1		0.794				
DSV		0.715				
DE3		0.649				
DFR			0.897			
DE4			0.673			
DS6			0.635			
DPT						
DBA				0.880		
DWA				-0.683		
DE5				0.621		
DPF					0.861	
DSR						
DS5						0.790
DS4						0.782
Eigenvalue	3.699	2.977	2.853	2.198	1.537	1.480
Variance (%)	19.466	15.669	15.015	11.567	8.088	7.790
Cumulative	19.466	35.135	50.150	61.717	69.805	77.595

3.3 Characteristics comparison of wintering habitat of OHBs in western Flores and Borneo

Comparison of wintering habitat characteristics of OHBs in Borneo with Western Flores either in the core or edge showed similarities in environment variables used as the main reason in selecting wintering habitat. The variables as the main reason in selecting wintering habitat were the forest, variations of both slope and elevation. Landscape characteristics of core and edge habitat in Borneo were highly influenced by the availability of thermal winds and food. However, the more frequent presence of OHBs in core habitats indicates that habitat selection is basically influenced by increased thermal winds associated with particular landform characteristics [4].

The similar characteristic found in core and edge habitats in both islands was on the existence of forests. It was concluded that, in the hunting area requirement, OHBs require a combination of forest and vegetated areas. It suggests that variation in land cover is an important factor that makes the dietary needs of OHBs can be met throughout their wintering habitat. Variation of land cover as landscape attributes caused heterogeneity that has an important role in the ecological system. Heterogeneity can describe the spatial complexity of a landscape mosaic [6].

The difference characteristic found in core and edge habitats in both islands was on mosaic of landscape that was affected by the abundance of food that could be hunting area for OHBs. The results of this study found that the savannah became the distinguishing characteristic landscape of wintering habitat in western part of Flores Island with Borneo Island. According to the "Honey Buzzard", wintering habitat is in rainforest or savannah forest [7]. The presence of savannah is very important, because the savannah can provide alternative foods such as amphibians (frogs), reptiles (lizards and snakes) and small mammals (mice) that play important roles to establish the body fat when existence of bees was less. Hunting areas on Borneo had a mosaic landscape composed of natural land (forests) and modified land by human activity (developed land, paddy fields, and gardens).

Wintering habitat of OHBs in Borneo has similarities to breeding habitat that is landscape mosaic of Satoyama in Japan. Satoyama comes lexically from the word of "sato" that means habitable areas and "yama" that means mountain. As developed in the agricultural system in Asia, this system is a complete mix between the natural environment and human settlement areas. These results have similarities with the results of previous studies [4], [8], [9] in West Java that represent wintering habitat characteristics of OHBs in western part of Indonesia. Characteristics of wintering habitat of OHBs in western Indonesia tend to be similar with the mosaic landscape of native habitat of OHBs in Japan i.e. Satoyama landscape mosaic where savannah as land cover does not exist in Japan.

4. Conclusions

This study was successfully delineated the core and edge habitat of OHBs. Fixed kernel-density successfully identified 165.86 km^2 of core area, while the edge area was 178.661 km^2. Core and edge habitats of OHBs were predominantly distributed in the southern part of the western Flores Island. Principal component analysis revealed that the characteristics of wintering habitat of OHBs were influenced by both the thermal wind and the availability of food. Thermal winds and the food were equally present in both habitats, core and edge, which distinguished only in terms of quality. Potential of food and higher thermal winds occurred in the core habitat so that visiting frequency of OHBs was higher in the core than in the edge. The similar characteristic found in core and edge habitats in Borneo and Flores Islands was on the existence of forests. Savannah became the main characteristics that distinguished landscape of wintering habitat of OHBs in Flores and Borneo.

Acknowledgments

The authors would like to thank the Ministry of Research, Technology and Higher Education of Republic Indonesia under “Penelitian Unggulan Perguruan Tinggi” grant in fiscal year of 2016.

References

[1] Bildstein KL 2006 Migrating raptor of the world: their conservation and ecology. Cornell University, New York

[2] Germi F 2005 Raptor migration in east Bali, Indonesia: observations from a bottleneck watch site. Forktail 21:93–98

[3] Sukmantoro, W 2007 Migrasi Burung Pemangsa Asia dasarburung.wordpress.com [12 April 2014]

[4] Syartinilia, Makalew ADN, Mulyani YA, Higuchi H Landscape characteristics derived from satellite-tracking data of wintering habitats used by oriental honey buzzards in Borneo 2015 *Landscape Ecological Engineering* **11**(1): 61-71. doi : 10.1007/s11355-013-0237-4.

[5] Careau V, Therrien J-P, Porras P, Thomas D, Bildstein K Soaring and Gliding Flight of Migrating Broad-winged hawks: Behavior in The Nearctic and Neotropics compared 2006 *The Wilson Journal of Ornithology* **118**(4):471–477.

[6] Farina A 2006 *Principles and Methods in Landscape Ecology* (Netherlands: Springer)

[7] Gamauf A, Hearing E Molecular phylogeny and biogeography of Honey-buzzards (Genera Pernis and Henicopernis) 2004 *Blackwell. J. Zool. Syst. Evol Research* **42**:145–153

[8] Pradani RC 2013 *Pengelolaan Lanskap Habitat Musim Dingin Sikep Madu Asia (Pernis Ptilorhynchus) Berbasis Satellite Tracking Di Kawasan Talaga Bodas, Jawa Barat [skripsi]* (Bogor: Institut Pertanian Bogor)

[9] Ameliawati P, Syartinilia, Mulyani A Y, Highuci H Karateristik Habitat Musim Dingin Sikep Madu Asia (Pernis Ptylorhynchus) Di Talaga Bodas, Jawa Barat Berbasis Data Satellite Tracking 2013 *Jurnal Lanskap Indonesia* **5**(2):37-42.

Ecological Design of Fernery based on Bioregion Classification System in Ecopark Cibinong Science Center Botanic Gardens, Indonesia

S Nafar[1] and A Gunawan[2]

[1]Student, Dept. of Landscape Architecture, Bogor Agric. Univ.
[2]Lecturer, Dept. of Landscape Architecture, Bogor Agric. Univ.

E-mail: sholihin.nafard@gmail.com

Abstract. Indonesia as mega biodiversity country has a wide variety of ferns. However, the natural habitats of ferns are currently degrading, particularly in lowlands due to the increasing level of urban-sprawl and industrial zones development. Therefore, Ecology Park (Ecopark) Cibinong Science Center-Botanic Gardens as an ex-situ conservation area is expected to be the best location to conserve the lowland ferns. The purpose of this study is to design a fernery through an ecological landscape design process. The main concept is The Journey of Fern, this concept aiming on providing users experiences in fernery by associating conservational, educational, and recreational aspects. Ecological landscape design as general is applied by the principal of reduce, reuse, and recycle (3R). Bioregion classification system is applied by grouping the plants based on the characteristics of light, water, soil, air, and temperature. The design concept is inspired by the morphology of fern and its growth patterns which is transformed into organic and geometric forms. The result of this study is a design of fernery which consist of welcome area, recreation area, service area, and conservation education area as the main area that providing 66 species of ferns.

1. Introduction

The Total amounts of ferns diversity in Indonesia is no less than 2,000 species. Papua and Borneo Island is estimated to have around 2,000 and 1,000 species of ferns. However, during the last five decades, the natural habitat of ferns has been degraded due to the massive land changing for urban, industrial, and agricultural [6]. One way to remain the sustainability of ferns diversity is through conservation. The Centre for Plant Conservation – Bogor Botanical Garden as government institution has various of ex-situ area, one of them is Ecology Park (Ecopark), Cibinong Science Center-Botanic Gardens (CSC-BG). Ecopark as ex-situ conservation areas have urgency in preserving the diversity of ferns. The urgency arises because the current natural habitat in the lowlands are more likely to be more degraded than the highlands.

Conservation of ferns in Ecopark conducted through the development of garden collection. Aside from being a conservation area, a garden collection can also be used as facilities of research, education, recreation, and environmental services, corresponding to the functions of the botanical garden in the law of Indonesian Institute of Science (LIPI) No. 1 Year 2014. Therefore, the purpose of this study is to design a fernery in Ecopark through an ecological approach by assessing the site characteristic.

2. Methodology

2.1. location and time

The study is located at Jln. Prof. Dr. A. Tisna Amidjaja, Ecology park, Cibinong Science Center-Botanic Gardens, Cibinong District, Bogor Regency, West Java Province, Indoesia. Geographically, the site of study is at coordinates 6o29'26 LS – 6o29'28" LS and 106o51'5" BT – 106o51'7" BT (Figure 1). The study was conducted in April 2015 until September 2016.

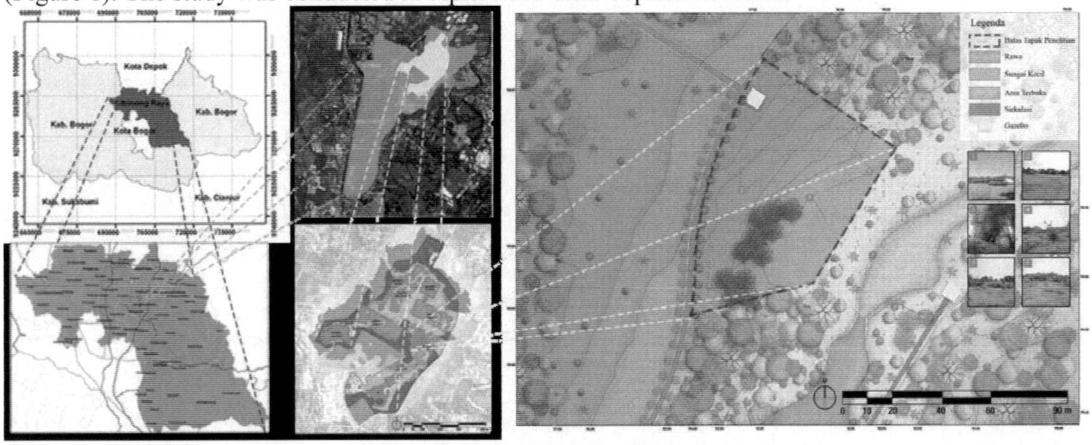

Figure 1. Site location

2.2. Methodology

2.2.1. Preparation

The preparation phase consists of setting the goal and preparing the tools and materials. Tools and materials on this study consists of hardware (GPS, measure tape) and Software such as Autodesk AutoCad and Trimble SketchUp.

2.2.2. Site Inventory

Site Inventory phase is to collect the data of biophysical, visual, socio-cultural, and regulation aspects which consist of primary and secondary data. The primary data is obtained through surveys, interviews, and questionnaires. Meanwhile, the secondary data is obtained through books, maps, journal, and other related references.

2.2.3.Analysis and Synthesis

Analysis and synthesis performed to determine the most suitable areas for the use of the site and score the feasibility of the planned use of the site. Analysis and synthesis focused on three aspects, that are the potential and the constraints of the site; the characteristics, perceptions, and preferences of the visitor; and the carrying capacity of the site. The analysis-synthesis on potentials and constraints, of the site, and also on characteristics, perceptions, and preferences of the visitor is conducted by descriptive method. Meanwhile, the analysis-synthesis on the carrying capacity of the site is performed by descriptive and mathematical methods. The analysis of carrying capacity is focused on the sensitivity of ferns. The analysis is conducted by giving the score of 1-3 on each indicator that have been modified to the ferns (Table 1). Furthermore, the value of every indicator summed together and being classified into three levels of sensitivity, that are the low sensitivity (score 5-8), moderate (9-11), and high (score 12-15).

Table 1. Rate sensitivity of ferns

Indicator	Parameter	Score

Status	IUCN*: least concern	1
	IUCN*: near threatened & vulnerable	2
	IUCN*: endangered & critically endangered	3
Type	seasonal plants	1
	annual plants	2
	perennial plants	3
Roots and Rhizome	tap roots	1
	fibrous roots	2
	aquatic, epiphytes, & lithophytes	3
Stem and rachis	Sturdy stem, prickly, and/or toxic	1
	Pseudo stem and/or easy to germinate	2
	Succulent	3
Leaf and lamina	The appearance of shape and colour is not attractive	1
	The appearance of shape and colour is quite attractive	2
	The appearance of shape and colour is very attractive	3

*IUCN (International Union for the Conservation of Nature)
Source: Ruhiyat (2008) with modification

2.2.4.Design

The design phase begins with a concept diagram. The concept diagram consists of the main idea, design concept, and the concept of development. The result developed through schematic design concept in the form of blocks and scenario planning phase. Then, the design is detailed through design development consisting of site plans and illustration.

3. Result and discussion

3.1. General Condition

Ecopark is a public park that located in CSC-BG, LIPI. Ecopark is built as an extension of Bogor Botanical Garden that have a common purpose, namely as a means of education, research, and ecotourism. Currently, Ecopark has a garden collection grouped by bioregion, covering Sumatera Bioregion; Java and Bali; Borneo; Lesser Sunda Island; Celebes; The Mollucas, and Papua. In addition, there is a lake known as Dora Lake, information center, rest facilities, and parking facilities. However, there are also empty land which has not been used and developed by the manager so that not all areas can be used by visitors.

Site study located in Ecopark with 4,720 m2 area dominated by grass. There are wild bamboo groves in the south and southwest arches and a gazebo in the northwestern part of the site. The site is bordered with Kalimantan bioregion on the north and northeast, the outlet of Dora Lake in the east, a path on the west, and Papua Bioregion in the southern part of the site. Currently the site is included in the area that has not been utilized by the manager of Ecopark.

3.2. Analysis and Synthesis

3.2.1. Analysis-Synthesis the Potentials and Constraints of the Site

The site is located near from the center of Bogor Regency administration, allowing visitors, especially from Bogor City, Depok City, and Jakarta City to visit the site. However, the site has less attraction for visitors because of the absence of the object recreation and education so that the necessary additional recreational and educational attractions that can attract visitors.

Judging from the slope of the land, the site has a slope ranging from 0-15%. The of 1-10% is category of ideal slope to serve a variety of functions without the need for grading room. Meanwhile, the slope of 10-15% is considered too steep for variety land use and easily eroded [1]. Therefore, engineering footprint through excavation, embankment, and retaining wall into solutions to minimize and avoid the problem of erosion.

Soil at the site is latosol that manifold with high clay content is not good for the growth of ferns for texture is hard and the lack of aeration. The problems addressed by the addition of organic mulch on planting area. On the other hand, clayey soil has good soil strength. That conditions allow the construction of small to medium scale landscape can be built without special treatment of the soil.

The quality of water included in the category quite well through the identification of physical, chemical, and biological functions of water. However, there is garbage in some parts of the water body so that it can have negative effects on water quality. Therefore, the application of the bulkhead in spam filter is required to facilitate garbage collection. Review of aspects of drainage, the absence of drainage channels with clay soil become obstacles footprint. Stormwater management becomes the means used to overcome these obstacles [7]. Temperature and humidity monthly average in Bogor in 2014 was 25.93 oC and 82.83 %. The ideal temperature for tropical ferns range 21-27 oC with humidity between 60-80 % [3]. The comfortable temperature for human in tropical regions are ranged at 22.80-25.80 oC with humidity between 40-75 % [5]. Based on that, the temperature is ideal for fern habitat and human comfort. Instead, the humidity is less than ideal for fern habitat and also human comfort. The humidity is reduced through modifications to the wind, by arranging landscape elements and pursed tightly thereby creating Windstream [8]. Judging from the sun's radiation, the site conditions are still comfortable to visit because of the solar radiation absorbed by vegetation tended lawns with 0.25 albedo values and broadleaf vegetation albedo values between 0.15 to 0.18 [10].

The site is dominated by vegetation and reed (*Imperata cylindrica*) makes visibility horizontally and vertically extensive, but also creates the impression of monotony. The monotony can be reduced through adding landscape elements. While from an ecological standpoint, vegetation plays in climatic amelioration, improving water quality, and affect wildlife.

The element of landscape dominated vegetation present a natural character and provide a positive visual impression. Objects outside the footprint of Dora Lake, Mount Halimun-Salak and Mount Gede-Pangrango be borrowing landscape to the site. This potential needs to be improved through the addition of landscape elements that support. However, there are also bad view due to litter and vandalism. These obstacles are overcome through the addition of garbage bin.

3.2.2. Analysis-Synthesis the Characteristics, Perceptions, and Preferences of Visitors

Based on data form the questionnaire, respondents consisted of men (32.26 %) and women (67.74 %). Almost half of them live in Bogor City (48.38 %). Distribution of the age of the respondents was dominated by the age of 16-20 years (54.84 %). Most of the respondents were students (64.52 %) and amounted to 45.16 % of respondent receive allowance < Rp 1,000,000 per month. Judging from visitor perception, most respondents know Ecopark as an education-tourism area (31.43 %). However, the activity of dominant recreational activities just like enjoying the scenery (27.27 %) and photographed (25.00 %). Most Respondents found hot climate (54.55 %) and dirty park (24.24 %) as a major problem at the site.

Based on the preferences of the fernery function, the majority of respondent want the facilities that support leisure activities such as bench (36.17 %), gazebo (21.28 %), and water feature (14.89 %). Meanwhile, the preference for educational facilities such as interpretation boards that giving information of the fernery (47.22 %). Judging from the preference for fernery design, most respondents wanted organic pattern (51.61 %) or a blend of organic and geometric patterns (41.94 %). Elements of landscape vegetation (93.55 %) was selected as the main elements compared to building and pavements. Meanwhile, the use of natural landscape material (64.52 %) such as stone and wood is preferred as the main material than the manmade materials such as glass and metal.

3.3. Design

3.3.1. Basic Concept

In general, the planning and design of the site in this study raised the concept of The Journey of Fern, which is a concept of the park harmonize aspects of conservation, education, and recreation, through grouping ferns based classification proximity habitat/bioregion/ecoregion (Figure 2a). This concept was formulated on the basis of the unity of the theme in Ecopark and also consistent with respondents high interest to the implementation of the fernery based on bioregion classification.

3.3.2. Design Concept
The design concept of the site is inspired by the morphology of the ferns. Ferns morphological transformation into an organic form is the basis for the design (Figure 2b). This form gives an dynamic, cheerful, soft, and grows on the landscape. In addition, the symmetrical shape will also applied to the design. Symmetrical horizontal lines provide a relaxed and calm atmosphere. Meanwhile, a symmetrical zig-zag lines give passionate and dynamics atmosphere [10].

(a) (b)
Figure 2. (a) basic concept (b) design concept

3.3.3. Concept Development
The concept of space, activities, and facilities of the site consists of four main area namely welcome area, recreation area, conservation and education area, as well as service area (Table 2). The concept of circulation and accessibility of the site refers to the concept of universal design [2]. Universal design is an approach to design that provided circulation for visitors, including persons with disabilities. However, not all circulation at the site is designed using a standard disability circulation. Therefore, the concept of circulation and accessibility at the site is divide into three types, two of them, the primary circulation and secondary circulation is used for all visitors, meanwhile, the tertiary circulation is used to be specific visitors.

The concept of vegetation refers to the basic concept of the site. Fern planting plan carried out systematically through bioregion proximity grouping to simplify the process of data collection and documentation of vegetation (Table 3). However, other plants outside of fern is also a consideration in the design. The concept of hydrology and drainage refers to the goal of garden design. This aspect has a role in influencing the microclimate in each collection area. In addition, the development of this concept is also done through an ecological approach that is by creating a sustainable water management system. However, the site also as a public park need to consider this aspect of the aesthetic and recreational function in addition to the ecological function. Therefore, the concept of hydrology and drainage footprint consist of three types, namely as stormwater management, water feature, and litter trap. The visual concept refers to the basic concepts, so that the natural character of the landscape elements preserved and presented at the site. The visual concept consists of themes related to landscape macro in nature. Application of theme park is used to distinguish the collection of ferns and provide a different experience for visitors. Technically, the themes raised include landscape mountains, craters, forests, rivers, valleys, swamps, lakes, islands, meadows, and cliffs. As spatial, those concept is merged to block plan and being detailed on site plan (Figure 3).

Table 2. The concept of space, activities, and facilities of the site

Space	Sub space	Activities	Facilities	Area (m^2)	Percentage (%)

Welcome area		Marking spot	Gate, signage, information gate	90	1,9
		Meeting point	Bench, plaza		
Recreational area	Nature recreation area	Gathering, photo	Bench, gazebo	1400	29,7
		Viewing landscape	Promade, water feature		
	Multifunctional recreation area	Gathering, photo, picnic	Bench, gazebo, lawn, oudor auditorium, plaza		
Conservational & Educational Area	Collection gardens	Marking spot	Interpretation board, signage	3200	67,8
		Observe	Fern collection		
	Indoor conservation	Research, breeding	Green house		
Service Area	Waste management area	Waste management	Garbage bin, litter trap	30	0,6
	Water management area	Water purification	Vegetated swales, bioretention swales, bioretention basins/ rain gardens		
Total				4720	100

Table 3. Ferns classification through bioregion closeness and vegetation sensitiveness

No	Zone	Amount of species
1	*Open/ direct sun plant - Terrestrial – Normal land*	10
2	*Open/ direct sun plant - Terrestrial – Moist land*	8
3	*Open/ direct sun plant - Epiphyte – Normal land*	2
4	*Open/ direct sun plant - Epiphyte – Moist land*	2
5	*Shaded plant - Terrestrial – Normal land*	15
6	*Shaded plant - Terrestrial – Moist land*	12
7	*Shaded plant - Epiphyte – Normal land*	10
8	*Shaded plant - Epiphyte – Moist land*	7

No	Vegetation sensitiveness	Amount of species
1	Low	19
2	Moderate	40
3	High	7

(a) (b)

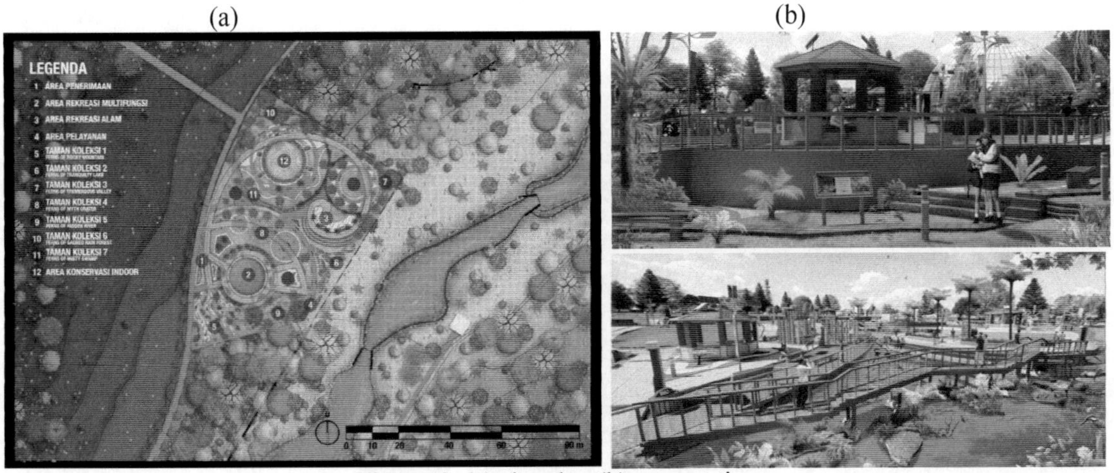

Figure 3. (a) site plan (b) perspective

4. Conclusion

Based on research, the site has a characteristic lowlands with high rainfall throughout the year. Site conditions in accordance with ferns lowland habitat. The number of species of ferns have been identified with 66 species are classified into seven types of garden collections based on proximity bioregion and three categories of sensitivity by the sensitivity analysis of vegetation. The site as exsitu conservation has the function to preserve a diverse collection of plants. The site as a public park to accommodate visitors in a variety of outdoor activities. The fernery design uses the concept of The Journey of Ferns that links the aspects of conservation, recreation, and education nature through ecological approach and grouping ferns based bioregion. The pattern of fern garden design collection comes from the transformation of morphology and growth patterns of fern into organic and geometric shapes. Ecological design of the garden fern collections applied through stormwater management, litter trap, environmentally friendly energy, and composting organic waste. There are four main area on this design, that are welcome area, recreation area, service area, and conservation-education area as the main area. The conservation-education area consist of ferneries with the various kind of visual themes, those are fernery with mountainous, lake, valley, crater, river, rain forest, swamp theme, and an indoor conservation as the core of the ferneries with island theme.

References

[1] Booth NK 1983 *Basic Elements of Landscape Architectural Design* Illinois (US): Waveland Press

[2] Harris CW and Dines NT 1998 *Time Saver Standards for Landscape Architecture: Design and Contruction Data* 2nd ed New York (US): McGraw-Hill.

[3] Hoshizaki BJ and Moran RC *2001 Fern Grower's Manual* Hong Kong (HK): Timber Press, Inc.

[4] Kato M 1992 *Cytotaxonomic and Reproductive-Biological Atlas of Asian Pteridophytes* Proc Sem Asian Pterid. **2**: 5-8.

[5] Mandagi AUG 2013 *Perancangan Taman Edukasi Lingkungan untuk Anak-Anak di Situ Cikaret, Kecamatan Cibinong, Kabupaten Bogor* [Skripsi]. Bogor (ID): Institut Pertanian Bogor.

[6] Mehltreter K 2010 *Fern Conservation*. Di dalam: Mehltreter K, Walker LR, Sharpe JM, editor. Fern Ecology. New York (US): Cambridge University Press.

[7] [PUB] Public Utilities Board 2011 *ABC Waters Design Guidelines* 2nd ed. Singapore (SG): PUB.

[8] Reed S 2010 *Energy-Wise Landscape Design: A New Approach for Your Home and Garden* Gabriola Island (CA): New Society Publishers.

[9] Ruhiyat Y 2008 *Studi Daya Dukung Biofisik Kawasan Rekreasi Kebun Raya Bogor* [skripsi]. Bogor (ID): Institut Pertanian Bogor.

[10] Rushayati SB, Alikodra HS, Dahlan EN and Purnomo H 2011 *Pengembangan Ruang Terbuka Hijau berdasarkan Distribusi Suhu Permukaan di Kabupaten Bandung* Forum Geografi **25 (1)**: 17-26.

IOP Conf. Series: Earth and Environmental Science **91** (2017) 012033 doi:10.1088/1755-1315/91/1/012033

Riparian landscape management in the midstream of Ciliwung River as supporting Water Sensitive Cities program with priority of productive landscape

T U Z Noviandi[1], R L Kaswanto[2], H S Arifin[2]

[1]Research Scholar, Department of Landscape Architecture, Bogor Agricultural University, West Java, Indonesia.
[2]Lecturer, Department of Landscape Architecture, Bogor Agricultural University, West Java, Indonesia.

E-mail: zelannoviandi@gmail.com

Abstract. Nowadays, Ciliwung River is facing problem of the settlement occupation in its riparian zones. This phenomenon caused ecological damage in riparian, so it can aggravate the disaster of annual flooding in Jakarta. As an effort to control this catastrophe, riparian landscape management of Ciliwung River is needed. Based on its topography, Ciliwung River is divided into three segments, there are the upstream, the midstream, and the downstream. Data shows that riparian in the midstream is the largest area, it covers more than 60% of the total riparian area. This segment is very important to be managed in order to reduce runoff towards the downstream. The method used was comparing many standards to get the ideal riparian width in the midstream, which is 50 m for urban areas and 100 m for outside the urban areas. Next method was analyzing spatially to get riparian landscape characteristic of Ciliwung River. The result showed that 37.11% of riparian zones in the midstream had occupied by settlement. Analysis of riparian function and utilization had held by using Analytical Hierarchy Process. Priority of riparian function in the midstream of Ciliwung River is production. This can be realized with the plan of community garden or inland fisheries. Riparian landscape management in the midstream aims to support the food consumption diversification, and maximize the function of water catchment and water retention in order to support the program of Water Sensitive Cities.

1. Introduction

Ciliwung River flows from the upstream in Puncak, Bogor District, and ends at the Bay of Jakarta. From the total lenght of 117 km, Ciliwung River is divided into three segments, there are the upstream, the midstream, and the downstream. During rainy season, this river is often blamed as the cause of the flooding in Jakarta. Flooding actually occurs due to reduced of water catchment area, that can be achieved with the existence of green open space, one of which is riparian zones [6] [24].

The phenomenon of settlement occupation in Ciliwung River may destroy land and vegetation formation [16]. As a result, water retention capability in the riparian was diminished, and lead to increase runoff volume that can be cause floods in downstream [2] [5]. Because of the problems in riparian of Ciliwung River, many strategies to protect riparian zones is needed. The strategy must be integrated from the upstream to downstream of Ciliwung River. But in this study, the focus of study

IOP Conf. Series: Earth and Environmental Science **91** (2017) 012033 doi:10.1088/1755-1315/91/1/012033

site is on the midstream riparian of Ciliwung River. Based on the area of riparian zones, the midstream is the largest riparian zones, and it covers more than 60% of the total riparian area. Ciliwung River in the midstream through Bogor City, Bogor District, Depok City, and part of South Jakarta. With a declivious slope, this segment is very important location to be managed as water catchment and water basin area. The cover of vegetation, both natural and cultivated, are still dominating in this segment. This potential is very important to be developed for community around riparian.

Several approaches had been done to achieve the goal of riparian landscape management. Analysis of the width and landscape characteristics of riparian is necessary to know the general condition of the Ciliwung River riparian in the midstream. Determining the function and use of riparian landscapes had obtained by Analytical Hierarchy Process (AHP). The approach to get riparian landscape management in the midstream of Ciliwung River is the river restoration concept, which aims to restore the river and riparian zones in natural conditions and protection after the restoration [12] [15] [23] [26]. This concept is also an effort to support the program of Water Sensitive Cities (WSC), where the WSC program aims to realize a city or region to be adaptive with climate change and population pressure on urban water [25].

2. Method

2.1. Study site

The reseacrh was conducted from April until September 2015 in the midstream riparian of Ciliwung River along 67.52 km of the total Ciliwung River length 117 km (Figure 1), with the distance between the two river valleys is 35.83 km. The midstream riparian of Ciliwung River is bordered by Bendung Katulampa in Bogor City to Rawajati in South Jakarta.

2.2. Analysis of riparian width and landscape characteristic in the midstream of Ciliwung River

Analysis of the Ciliwung River riparian width in midstream was conducted by comparing the riparian width standards. There are three standards of riparian width that compared (Table 1), that is (1) the standard of riparian width with flood protection function, (2) the standard of riparian width based on the Decree of the Minister of Public Works No 28 of 2015 and the Indonesian Government Regulation No. 38 of 2011, and (3) the standard of riparian width based on morphology, ecology, and hydraulic of river flood [17].

Table 1. Comparison of the riparian width standard

Function	Sources	Explanation	Riparian width
Review of Literature and International Journals			
Flood protection	[11]	Improve meanders	Two canopy trees
	[7] [9]	100-year flood protection	50 - 90 m
	[14]	Reducing the water level increase	30 m
Regulation regarding Riparian Width Standard			
Flood protection and slope stabilization	Decree of the Minister of Public Works No 28 of 2015; Indonesian Government Regulation No. 38 of 2011	Embankment; urban area	3 m
		No embankment; urban area	10 - 30 m
		Embankment; outside urban area	5 m
		No embankment; outside urban area	50 - 100 m
Standard of riparian width based on morphology, ecology, and hydraulic of river flood			
Flood protection	[17]	Urban area	10 - 50 m
		Urban fringe	30 - 75 m
		Outside urban area	50 - 100 m

Riparian land cover analysis was carried out spatially. The method used was to classify the land cover of riparian landscape into two categories: developed and undeveloped land (Figure 2). The developed land on the riparian includes settlements, public facilities, and public space with non-permeable materials domination. The undeveloped land on the riparian includes riparian vegetation, either natural or cultivated vegetation. Field observation and ground truth check are performed to find out existing land cover in the riparian area of Ciliwung River. Field observation conducted at several locations, such as (1) Sempur, (2) Kedung Halang, (3) Bojong Gede (4) Tirta Jaya (5) Tugu (6) Lenteng Agung and (7) Tanjung Barat.

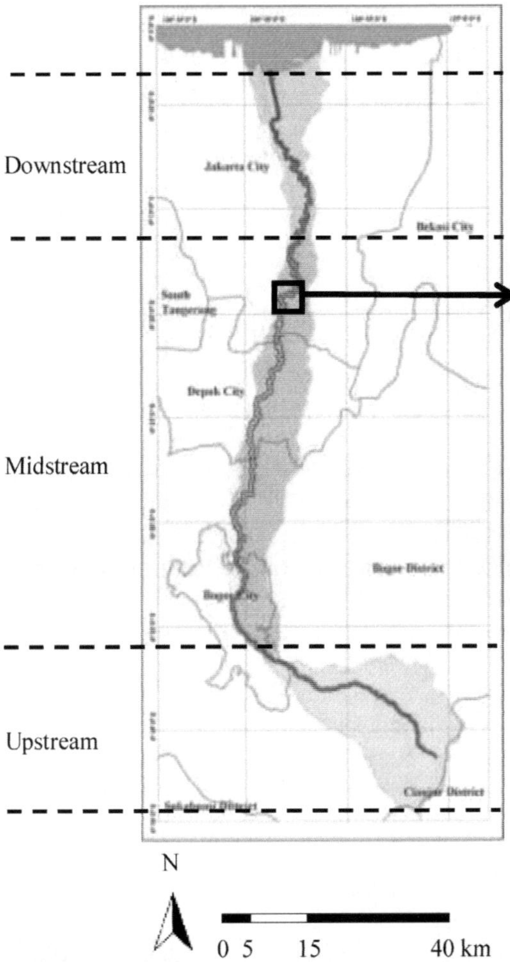

Figure 1. Study site in the riparian midstream of Ciliwung River

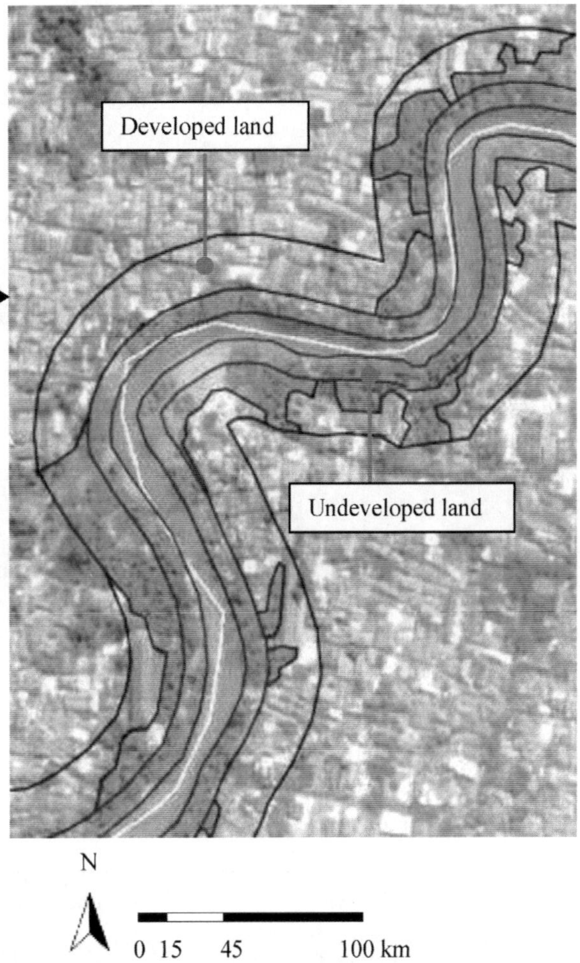

Figure 2. Analysis of riparian land cover at Lenteng Agung, South Jakarta

2.3. Analysis of riparian landscape function and utilization in the midstream of Ciliwung River

The process to determine the function and utilization of riparian used Analytical Hierarchy Process (AHP). This method is a structural technique to analyze complex decisions [21]. Experts who were respondents in this study consisted of (1) experts on Civil Engineering and Water Resources Management, (2) experts on Social and Community-based Management of Riparian Landscape, and (3) experts on Watershed. The experts faced with prioritiy choices of functions and utilization for all segments of the Ciliwung River riparian. The function of riparian is adjusted based on ecosystem services by Arifin and Nakagoshi [3], which is function of (1) water resources management (WRM),

(2) production (Pr), (3) biodiversity conservation (BC), (4) carbon sequestration and stock (CSS), and (5) landscape aesthetic (LA). Utilization of riparian is obtained from the derivative function of the riparian landscape.

2.4. Riparian landscape management in the midstream of Ciliwung River

Riparian landscape management in the midstream was conducted from the result of riparian width and landscape characteristics analysis were combined with the results of the riparian landscape function and utilization. Research approach that used in arrangement of the riparian landscape management was the concept of river restoration with application of eco-engineering [18] as an alternative to the river bank reinforcement with natural materials that not to damage riparian ecosystems. Restoration of river systems is centred on increasing the biodiversity of the system. These results as well as a supporter of the WSC program, that aimed to realize the city become more adaptive to water, both during the rainy season and dry season.

3. Results

3.1. Riparian width determination in the midstream of Ciliwung River Riparian

Based on the region, Ciliwung River riparian is divided into two kinds, that is riparian in urban areas (Bogor City, Depok City, and South Jakarta) and in the outside of urban areas (Bogor District). The width of riparian Ciliwung River in the midstream were determined by considering the three standards, that is (1) the standard width of riparian river with flood protection function, (2) the standard width of riparian based on the Decree of the Minister of Public Works No 28 of 2015 and the Government Regulation No. 38 of 2011, and (3) the standard width of riparian based on morphology, ecology, and hydraulic of river flood [17]. In consideration on runoff during the rainy season and erosion risk on the riverbank or riparian, the determination of the riparian width refers to the analysis, which is 50 m for riparian in the urban areas and 100 m for riparian in the outside of urban areas. This determination are not much different to riparian width standard by Maryono [17]. Riparian width in urban areas is still narrower than in the outside of urban areas, because land demands for development in the urban areas are relatively higher than in the outside of urban areas.

3.2. Riparian landscape characteristic in the midstream of Ciliwung River

The land use of Ciliwung River riparian in the midstream quite varied. Riparian that had been occupied by settlement dominantly can be found in urban areas, like Bogor City (Figure 3), Depok city and South Jakarta. Whereas, riparian land use in form of productive landscape, such as inland fisheries, community garden, and bamboo formation (Figure 4), can be found a little bit in urban areas and dominantly in the outside of urban areas.

Figure 3. Settlement occupation in the midstream riparian of Ciliwung River

Figure 4. Bamboo formation in the midstream riparian of Ciliwung River

Developed land occupation in this segment is 312.41 ha, or 37.11% of total riparian area in the midstream. The remaining area of 529.55 ha is still vegetate area, either natural vegetation such as bamboo, or cultivated vegetation such as mixed graden. Ciliwung River in this segment also has a quite high sinuosity value of 1.88 and in the category of meander river. The river that classified as meander indicates that the slope of the riverbed in this segment is also relatively ramps and water velocity in this segment is 5 m/s averagely.

3.3. Riparian landscape function and utilization priority in the midsteream of Ciliwung River

Based on sensitivity analysis of AHP, in the midstream riparian of Ciliwung River, the function priority is Production (Pr), with a value of 30.6% (Figure 5). The next priority in midstream are Carbon Sequestration and Stock (CSS) 27.7%, Biodiversity Conservation (BC) 25.7%, Water Resource Management (WRM) 23.1%, and Landscape Aesthetics (LA) 21.6%. Although the function of production is a priority in the midstream riparian of Ciliwung River, other functions can also be applied to this segment because the value of priorities were not too significant. The utilization priority in the midstream riparian of Ciliwung River was obtained from the function of production. The utilization for production that already exists in this segment are inland fisheries and community garden. So these utilization will be managed and develop to make sustainable landscape riparian in the midstream of Ciliwung River.

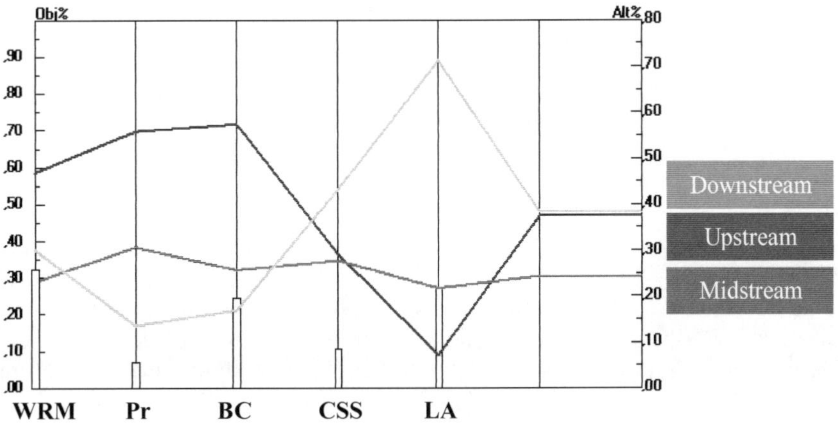

Figure 5. Sensitivity graph of AHP to determine the function of Ciliwung River riparian

3.4. Riparian landscape management in the midsteream of Ciliwung River

The approach that used in the concept of Ciliwung River riparian landscape management is the river restoration concept, which is a concept that aims to restore the river in its natural condition. Therefore, developed land such as settlement in the riparian zone must be released and returned as a buffer zone. Strengthening the river banks must no longer with retaining wall or concrete, but by using natural materials (eco-engineering), such as formation of rock, soil and vegetation on the riparian.

Function of production can be realized by making productive community garden and inland fisheries. This garden can be managed by community and the harvest can be used for daily purposes or resold to generate income. Function of water resources management (WRM) is also a very important function, especially in reducing runoff. In the midstream, WRM function is directed at water retention, with the initiation of water retention pond on the outer curved edge of the river (Figure 6). The pond can hold excess water during rainy season and utilized for specific needs, like inland fisheries or waterfront landscape recreation.

4. Discussion

Riparian is a transition zone between terrestrial and aquatic ecosystem which the condition of biotic and abiotic strongly influenced by stream water and ground water [4] [10] [20]. Rapid development that occupies riparian zones had caused damage the ecological function [13] [16] [22]. Therefore, this zones should be protected and free from any massive development (Decree of Minister of Public Works No. 28 of 2015 and Government Regulation No. 38 of 2011). The phenomenon of developed land occupation in riparian zone is exacerbating the impact of floods in Jakarta and it is one of environmental damage indicator [1] [8].

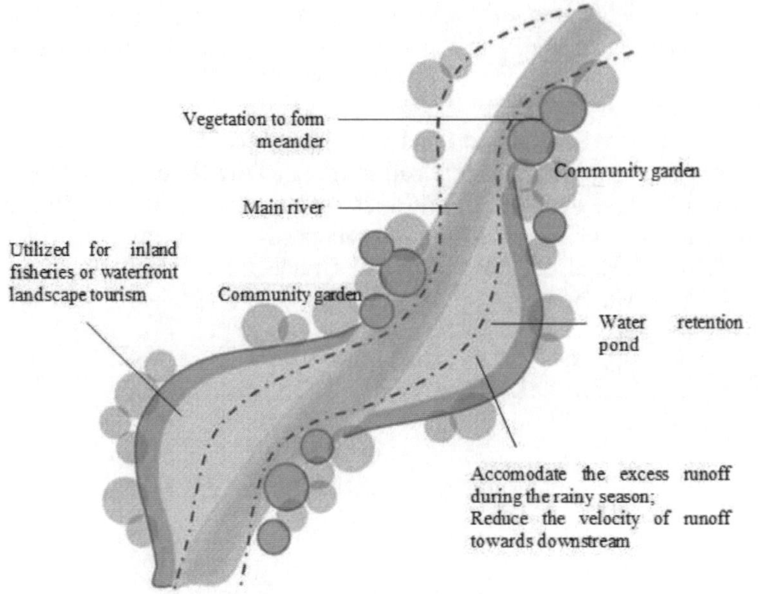

Figure 6. Water retention pond concept on the midstream of Ciliwung River

River management concepts currently applied in Indonesia are still oriented to conventional hydraulic that seeks to discharge water as quickly as possible as toward downstream. Conventional hydraulic concept can even lead to worse flooding effect, because the runoff can not infiltrate to the riparian effectively, then flows even faster and finally accumulated as flooding in the downstream. The bad impact of the conventional hydraulics concept, brings new concepts of river restoration. River restoration aims to restore the river and its riparian zone to its natural conditions, including protection and ecological restoration in areas that have been disturbed by anthropogenic activities [12] [15] [23] [26].

Ciliwung River restoration concept can be combined with a kind of different utilization on each segment. In the midstream, the priority of production function is quite appropriate because there are existing riparian utilization for productive community garden and inland fisheries. However, the existence of productive community garden and inland fisheries have not been well managed and is found only in a few villages. Furthermore, the concept of riparian landscape management for water retention is the initiation of riverside pond. This concept aims to capture the excess runoff during the rainy season. All the concept that had been generated in this study, are the effort to support the Water Sensitive Cities (WSC) program. Water sensitive cities is a road map of a cities to be adaftive for climate and population pressure on urban water [25]. Three principles set the foundation for this vision of a Water Sensitive City, that is (1) citis as water supply catchment, (2) cities providing ecosystem services, and (3) cities comprising water sensitive communities [19].

5. Conclusion

The width of Ciliwung River riparian in the midstream that aims to control the flood, are 50 m for riparian in urban areas and 100 m for riparian in the outside of urban areas. Developed land has occupied 37.11% of Ciliwung River riparian in the midstream, and the rest is undeveloped area that covered 62.89% of total riparian zones in the midstream of Ciliwung River.

The function priority of Ciliwung River riparian in the midstream that was obtained from AHP is for production. The priority utilization for production are productive community garden and inland fisheries. Concept of riparian landscape management in the midstream would be directed as water retention with the initiation of water retention pond on the river bank to increase water storage volume, so it can reduce runoff during rainy season.

Acknowledgements
This research is part of multiyears research funded by Ministry of Research, Technology and Higher Education of the Republic of Indonesia and The Australia-Indonesia Centre (AIC) Water Sensitive Cities (WSC), with major research entitled "Analysis of Green Water and Blue Water Availability in Sustainable Landscape Management of Ciliwung Watershed". The authors would also like to thank *Balai Besar Wilayah Sungai Ciliwung Cisadane* (BBWSCC) and *Balai Pengelolaan DAS Citarum Ciliwung* for their data about existing condition of Ciliwung River.

References
[1] Albano R, Sole A, Sdao F, Giosa L, Cantisani A, Pascale S. 2014. A Systemic Approach to Evaluate the Flood Vulnerability for an Urban Study Case in Southern Italy. *J Water Resource and Protection*. 6:351-362.
[2] Arifasihati Y, Kaswanto. 2016. Analysis of Land Use and Cover Changes in Ciliwung and Cisadane Watershed in three Decades. Procedia Environmental Sciences 33:465-469.
[3] Arifin HS, Nakagoshi N. 2011. Landscape ecology and urban biodiversity in tropical Indonesian cities. *J Landscape and Ecological Engineering*. 7(1):33-43.
[4] Arroyo AL, Johansen K, Armston J, Phinn S. 2010. Integration of LiDAR and QuickBird imagery for mapping riparian biophysical parameters and land cover types in Australian tropical savannas. *J Forest and Management*. 259:598-606.
[5] Barbosa AE, Fernandes JN, David LM. 2012. Key Issues for Sustainable Urban Stormwater Management. *J Water Research*. 46(20):6787-6798.
[6] Bertoldi W, Drake NA, Gurnell AM. 2011. Interactions between river flows andcolonizing vegetation on a braided river: exploring spatial and temporal dynamics in riparian vegetation cover using satellite data. *J Earth Surf. Process. Land.* 36(11):1474-1486.
[7] Bertulli JA. 1981. Influence of forested wetland on a southern Ontario Watershed. Proceeding of The Ontario Wetlands Conference. p. 33-47.
[8] [BPDAS] Balai Pengelolaan DAS Citarum-Ciliwung. 2013. Rencana Pengelolaan DAS Terpadu DAS Ciliwung. Bogor (ID): Departemen Kehutanan.
[9] Castelle AJ, Johnson AW, Conolly C. 1994. Wetlands and stream buffer size requirements [review]. *J Environmental Quality*. 23:878-882.
[10] Clerici N, Paracchini ML, Maes J. 2014. Land-cover change dynamics and insights into ecosystem services in European stream riparian zones. *J Ecohydrology and Hydrobiology*. 14(2):107-120.
[11] Divelbiss CF. 1994. A Review of Selected Functions of Riparian Buffer Zones and Widths Associated With Them. In Rivers Without Boundaries Conference, American Rivers Management Society, (April 1994):1-7.
[12] Helfield J, Engström J, Michel J, Nilsson C, Jansson R. 2012. Effects of river restoration on riparian biodiversity in secondary channels of the Pite River, Sweden. *J Environ. Manage*. 49(1):130-141.
[13] Li S, Gu S, Tan X, Zhang Q. 2009. Water quality in the upper Han River basin, China: the impacts of land use/land cover in riparian buffer zone. *J Hazard.Mater.* 165(1-3):317-324.
[14] Lynch JA, Corbett ES. 1990. Evaluation of best management practices for controlling nonpoint pollution from silvicultural operations. *J The American Water Resources Association*. 26:41-52.
[15] Kaswanto. 2015. Land Suitability for Agrotourism Through Agriculture, Tourism, Beautification and Amenity (ATBA) Method. Procedia Environmental Sciences 24:35-38.
[16] Kaswanto, Utami FNH. 2016. The Disparity of Watershed Development between Northern and Southern Region of Java Island. Procedia Environmental Sciences 33:21-26.
[17] Maryono A. 2009. Kajian lebar sempadan sungai: studi kasus sungai-sungai di provinsi Daerah Istimewa Yogyakarta. *J Dinamika Teknik Sipil*. 9(1):56-66.
[18] Maryono A. 2014. Menangani Banjir, Kekeringan, dan Lingkungan. Yogyakarta (ID): Gadjah Mada University Press.
[19] [MONASH UNIVERSITY]. 2016. What is a Water Sensitive City. Publising. http://www. waterforliveability.org.au/ ?page_id=1706. Accessed 30 October 2016.

[20] Naiman RJ, Decamps H, McClain M. 2005. Riparia - Ecology, Conservation, and Management of Streamside Communities. Academic Press, 448 pp. ISBN 9780126633153.
[21] Saaty TL. 2008. Decision Making with The Analytic Hierarchy Process. *Int J Services Sciences.* 1(1):83-98.
[22] Sliva L, Williams DD. 2001. Buffer zone versus whole catchment approaches to studying land use impact on river water quality. *J Water Res.* 35(14):3462-3472.
[23] Stella JC, Rodríguez-González PM, Dufour S, Bendix J. 2013. Riparian vegetation research in Mediterranean-climate regions: common patterns, ecological processes, and considerations for management. *J Hydrobiologia.* 719(1):291-315.
[24] Stevaux JC, Corradini FA, Aquino S. 2012. Connectivity processes and riparian vegetation of the upper Paraná River, Brazil. *J South American Earth Sciences.* 46:113-121.
[25] Wong, T. 2015. Water Sensitive Cities: a road map for cities adaptation to climate and population pressures on urban water. Publishing. https://www.monash.edu/__data/assets/pdf_file/0003/293691/msi-seminar_10-05-25_cwsc_presentation.pdf. Accessed 30 October 2016.
[26] Xia J, Zhai X, Zeng S, Zhang Y. 2014. Systematic solutions and modeling on eco-water and its allocation applied to urban river restoration: case study in Beijing, China. *J Ecohydrology & Hydrobiology.* 14:39-54.

The Concept of Historical Landscape Design at Watugong Archaeological Site Area in Malang City

I Setyabudi[1], R Alfian[2], W R Hastutiningtyas[3]

[1]Lecturer of Architectural Design, Tribhuwana Tunggadewi University, Malang, Indonesia
[2]Lecturer of Architectural Planning, Tribhuwana Tunggadewi University, Malang, Indonesia
[3]Junior Lecturer, Tribhuwana Tunggadewi University, Malang, Indonesia

E-mail: isetyabudi.st@gmail.com

Abstract. Malang city has the high historical value. It showed by many archaeological situses found, such as: The Dutch Colonial Building until kingdom era on classical history period. Generally, it could be seen at urban affairs like government building even the ancient house. But the last kingdom archaeological site only found at the village. The oldest archaeological site in Malang city was found on Kanjuruhan Kingdom, which is concentrated in Tlogomas. The Watu Gong Hamlet that was located in Tlogomas Political District had an archaeological site. It was a big stone in which the stone looked like a traditional music instrument, it was called Gong. As the archaeological site in megalithicum, before the Hindu and Budha came in from India, that stone was predicted as the foundation structure of the big building. The Watu Gong Hamlet was located in Tlogomas archeological site area, also Merjosari and Karang Besuki. Three of them are the archeological sites for Kanjuruhan Kingdom at eigth century, until Kahuripan Kingdom around the eleventh century, as the heir of the Ancient Mataram Kingdom. The urban government has a program to improve the village required to their region potential and it was possible to revitalize the Tlogomas village, so that the historical character can be seen well. The modernity of a hamlet has impact on the local identity blured. In which, they did not think about economic only and it can be minimized, also the hamlet will be a characterized tourism object. The revitalization purposed to continuing the past, then it's connected to present. It's performed as corridored garden planning. The landscape development appropriated to promote about the characters of Kacapiring flower, Rose, Jasmine and Puring. They are the special plants from Kanjuruhan Kingdom, beside the other furniture street model. This research was descriptive explorative and discussed about the concept with architecture design approach, started from data collecting, precedent study, programming until the developing concept of the historical landscape. The historical landscape concept for Tlogomas at present is recommended to Watu Gong as the village tourism object.

1. Introduction

Historical sites in architecture context provides evidence there is a settlement that has been developed culture and its values beneficial to society lessons for decades afterward. Long time before the Dutch colonial period are still largely intact, many historical relics that exponentially degraded. The basic difference of this problem is technology. Relics of the colonial era has a younger age and they are created in the needs of the user's physical (tangible), while the kingdom era are already old and had a philosophical function, generally places of worship (intangible), and material from wood or stone.

Many relics of history that has been razed to the ground which should be protected by law cultural heritage.

Great nation was born because of their history, culture and civilization. The culture and civilization is marked by the presence of the site. Malang is inseparable from the history although the city is growing rapidly in the Dutch colonial era, starting from Kanjuruhan kingdom history. There are many sites spread across the city and district of Malang that is still preserved. No exception to the existing site in Tlogomas.

RW 03 (citizen association 03) in Tlogomas, there Kanjuruhan kingdom heritage sites in the 7th century in the form of stone resembling the shape of a gong. Originally the stone relics of a foundational structure base with a sacred building where the brahmins meditate, study and train their inner power. The stone looks a music instrument has about 80 cm diameter, 40 cm height and a protrusion with 15 cm diameter and 20 cm high. This stone was known as a foundation of wooden structures perforated at the bottom, so if an earthquake occurs only vibration without damaging the structures thereon. The number of 13 pieces and stones collected at the pavilion site. The foundation structure supported a very large building on it by stone pedestals, serves as a religious and government place. In addition to the stone, the stone also found that perforated for dispensing medicine and gathering food and a place to get cleaned up. There is also a red-brick thick under the courtyard houses, with a lengthened position pattern forming ground floor.

Tlogomas was originally a part of the Kanjuruhan kingdom, the area covers Dinoyo, Merjosari and Tlogomas. Watugong hamlet is a region located in Tlogomas since 1987 and entered Dau subdistrict, Malang district, but at Dutch period, Watugong include in section Dau Malang district in accordance with the Staatblad No. 19 of 1819. Tlogomas name was found after 1930, originally place on there were Ngelo, Karuman, Guyangan, Pelandungan/Kelandungan, Merjoyo, Merjosari, Dinoyo and Ketawanggede, so they need to call in one name. Tlogomas formed when people often find objects of gold (emas/mas: in bahasa) near the water source (telaga/tlogo: in bahasa). (Source: Tlogomas village profile).

According to interviews with village elders Mr. Bambang, the research team so Malang in 1984 estimated that a relic of megalithic sites Watugong that when cultured cobblestone community as a means of worship prior to the entry of Hindu-Buddhist religion of India. Watugong sites located among a row of neighborhood Dinoyo archaeological sites, Tlogomas, Merjosari and Karang Besuki. The fourth area is a legacy of the Kanjuruhan Kingdom at 8th century until Kahuripan at 11th century as the heir of the Ancient Mataram kingdom who moved to East Java. The distance between Candi Badut and Karang Besuki temple with Watugong site not far away, there are also Dinoyo inscriptions which tells purposes of the king for granting brahmin with a big house. It can be concluded that the Watugong area is an existing hostel building sacred pastors to glorify resi Agastya.

From the brief history described Watugong, so it region is very important in Malang city development. As for now Watugong condition is more advanced, but not all segments of society understand the importance of history then someday their potential condition will be degraded. Based on the cooperation with city government and villages, the village will be developed which indicate the existence of their identity which manifested in village planning based on local wisdom. Their local potential in the form of historical sites make public RW 03 as the hometown of the matter of history and be an icon of 'past in present' in a tourist village planning. Main criteria for the development of the region, among others refers to: improvement of environmental quality, public service, service to the tourists who visit, the preservation of Watugong heritage sites, the development of talent and creativity of the citizens to realize RW.03 Tlogomas as ' Watugong Tourism Village ' and became new destinations in Malang, and turning the wheels of the economy.

The planning is not only related to the architectural form of the building (hardscape) but also the landscape (softscape) which refers to the history. Plant as the local nature of the Kanjuruhan kingdom are kacapiring (*Gardenia* sp), rose, jasmine and puring (*Codiaeum* sp.) planned as 'dresscode' of Watugong region. Urgency of these activities is to increase public awareness through history and provide a venue to accommodate community activities. Village tourism can serve as a means of education, culinary, and outbound, so main problem in this article is 'how is design concept of landscape history of Watugong archaeological site?' In order to create friendly environmental for rural tourism.

2. Methodology

The method of analysis in this research use descriptive-qualitative with explorative approach and architectural design method according Hakim [1] and Setyabudi [2] theory the context of landscape architecture. Hakim and Setyabudi, explains that the method of designing the architecture starts from the determination of the project, the object of study, inventory footprint, programing, concept, pre-design, and design development. In this article only discussed the results of the research that comes from the results of programing to design development. The tools used for data analysis using computer graphics software that SketchUp, AutoCAD, and Lumion.

Data is divided into primary and secondary data. The primary data of the physical data location (photo, measuring results and sketches), data from interviews with officers and supervising field space, and location mapping. The data obtained from field observations with camera tools, stationery, recorder and GPS. The usefulness of data for qualitative analysis in the form of the potential and constraints of the site, analysis functions and needs, as well as analysis of the object. Secondary data from literature on the potential and constraints Watugong village derived from the data Tlogomas village. The location selected in the village Watugong because it has the potential of local wisdom undeveloped and can be projected to the village for future travel. Here the object of study is presented in the map (googlemaps search: 7 ° 56'20.0 "S 112 ° 35'51.1" E) and photos.

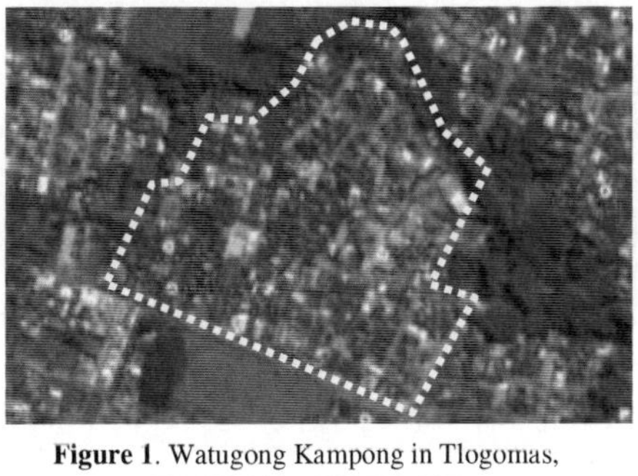

Figure 1. Watugong Kampong in Tlogomas, 7°56'20.0"S 112°35'51.1"E

Figure 2. Watugong archaeological site (it called pendopo) contains stones like traditional music instrument (gamelan/gong)

3. Result and Analysis

Tlogomas as the entrance area of Malang from the west direction (Kediri and Jombang), so it has a strategic role in the development of the city. As a buffer zone, Tlogomas grew rapidly into an area that is heterogeneous. Since the 90s Tlogomas with a definite step move towards a positive development since the establishment of several universities that are in the region of Tlogomas. The area is more developed and grown into urban communities after Malang Government move to the center bus stops and public transportation from Dinoyo to Landungsari where administratively included in the Tlogomas.

Geographically, Tlogomas has an area of 167.59 km^2, located of 444 m height above sea level, with 24^0C max. temperature and of 18^0C min. temperature. The maximum rainfall ever minimum 2,71 mm and 2,31 mm that has ever happened. People diversity who settled in region Tlogomas raises the potential of different with the other region. Tlogomas has 13,450 resident which prural conditions with many kind of excellent products. Tlogomas also has quantity of 9 RW(s) (citizen association) and 49 RT (s) (neighborhood association). Total amount resident by sex is 8002 men and 5,448 women.

Results ground inventories and interviews with community leaders indicate that the planned location located along the main road corridors (Kanjuruhan street) and the alley leading to the site, so the focus of this research is the landscape planning of road corridors by considering the history of both hardscape and softscape elements. Here is a photo existing conditions as explanatory.

Figure 3. Existing condition at Kampong Watugong toward archaeological site area

Preliminary studies conducted in-depth interviews to community leaders and discussion on local communities to determine the factors that influence the development strategy of the Watugong tourist village, mentioned in SWOT analyzed:

a. Strength, (1) Potential easy road access, (2) There are watersheds and water sources, (3) There is a heritage site of the Kanjuruhan Kingdom, (4) Production of handicrafts of local communities such as stoves, processing of plastic waste, and local culinary, (5) Availability of a gate that contain elements of history, (6) the surrounding area of education, (7) *guyub* (friendly) society and mutual cooperation, (8) the implementation of cultural activities, (9) the number of vacant land that can be used as a neighborhood park, (10) the attention of local residents against the historical heritage, (11) the high enthusiasm of the public to make the tourist village history.
b. Weakness, (1) Conditions pavilion grand (Pendopo Agung) less aesthetic, (2) the absence of identity of the region as an area that has historical value, (3) Not available cicerone (signage) towards the site Watugong, (4) A watershed yet managed properly
c. Opportunity, (1) The role of Government in synergy with the development of the tourist village history (*Kampung Tematik* Project), (2) The number of visitors from outside the city to an area of the site, (3) Close to educational facilities, (4) to be a lab or media research for institutions education, (5) Being a place of training on arts and culture, (6) Development of traditional culinary travel
d. Threats, (1) Transfer of land use, (2) The modernization current, (3) Nature of apathy towards the values of history and culture, (4) Decreased activities is cultural, (5) The density of residential areas, (6) Low the government's concern towards historic areas

3.1. Plan of Activities, Functions and Space Needs

Based on observations in planning location, there are some of results could be descriptively explained. 'Welcome area' located on two sites : the bridge area and Merjosari. The bridge area located in the north tend to be only passable by small vehicles because of a lot of the alley leading to the village, another case with a gate from the direction Merjosari (south) which has a wider space and connection with Joyogrand, Villa Bukit Tidar or Graha Dewata housing, so the circulation get in better direction from the south direction. Circulation is also heading towards the parking of vehicles. At some point in the township existing actually be eco-settlement in which there are many open spaces and planted with shading trees (mango), ornamental plants (hanjuang, labu kendi, rose, jasmine and gardenia), although some houses there immediately adjacent by road due to the limitations of the land owner. The location of settlements toward the site (pendopo) is not far away, and can be reached by foot. The total distance between the north and south entrance about 2 km. Another potential reason for this area is an owner of water source whose size is not too large, large discharge when the dry season. During this time only source of water used for bathing by some locals. Access to water sources is quite difficult, because it passes through rocks and slippery. The produced water to the river tends not clear anymore.

Pendopo area has an important point of the zone, which is a place to store historical objects, access is still limited. In addition, not commercialized, there is no zone that holds educational functions like a library. The public generally have access to the site after permission from a manager, when there is an event or cleaning the historical objects, called *ruwatan*. Another aspect from historical field, rural communities are also active in making local handicrafts such as stoves and plastic waste recycling, which products will be inviting tourists. People love to work together and cooperate in environmental hygiene, cultural activities and build the park. According to Mr. Aryadi as headman, this potential aspect can be developed in future village environmentally planning. As for the testimony of citizens who expect to their village planning for the tourism history area. It can improves the welfare of the community, strengthen the identity and towards a better life. By opening up the concept of the village as a tourist history area , local communities inviting domestic and international travelers to visit Watugong site.

As a village which refers to the concept of eco-tourism, the principles include:

a. The principle of conservation
b. The principle of community participation
c. The Economic principles
d. The principle of education

Step designing resumed on user activity plan as follows design user, the user consisting of villagers and tourists arrivals who have different needs. Travelers can be adults aged or children with visiting hours are 8:00 to 17:00. Circulation to welcome area with a path from the entrance to the south, which is where the road more accessible by vehicle. Main entrance from the north only accepts pedestrians and bicycles. The parking area is located on an open area which is quite wide and holds about 30 cars. Along the driveway, Watugong identity has been shown with various ornaments, carvings, street name signage, sclupture, and various types of vegetation as a shading, steering, and aesthetic elements. To reach the site, visitors can walk next to the side of the road which offered traditional hawker (street food) and souvenirs. Each of the entrance, there is a gate with unique ornaments Kanjuruhan kingdom. Pendopo, this building accommodates educational function by adding a library area, storefront Watugong and other historic objects. Visitors are also directed toward the source of the water and headed into the water next to river.

The space function is found based on the needs which contain functions and activities. They are classified based on a hierarchy of interests that consists of welcome area, a buffer area, a conservation zone and core zone. 'Welcome area' consists of the construction of the gate with the charge history, their welcome relief, slupture about Watugong and the addition of local vegetation. Buffer area consists of settlements, but has a special function to provide souvenirs, the development of local crafts, traditional culinary region, the addition of local vegetation and fence design each house with loads of history. The core zone, consisting of the restoration of the great pavilion, designed area of the village with historical ornaments, designing wall reliefs that discount the value of education and history. Planning the core zone, namely the restoration of the pavilion site, the design area of the village with historical ornaments. Conservation Area consists of the revitalization of the watersheds, reforestation area of the river, and the provision of arranging for fishing.

3.2. Basic Concept

The rationale thought for this Watugong landscape planning is the 'Conservation of History' which many elements like Kanjuruhan Kingdom form and ornament on site locations. This step illustrating the past in the present, but they appear in new materials and functions. Design applies to the lane road to the site and along the path of settlements, by adding a gate, post information, information boards, signage, sclupture, garden open spaces and selection of plants throughout the settlement.

IOP Conf. Series: Earth and Environmental Science **91** (2017) 012034 doi:10.1088/1755-1315/91/1/012034

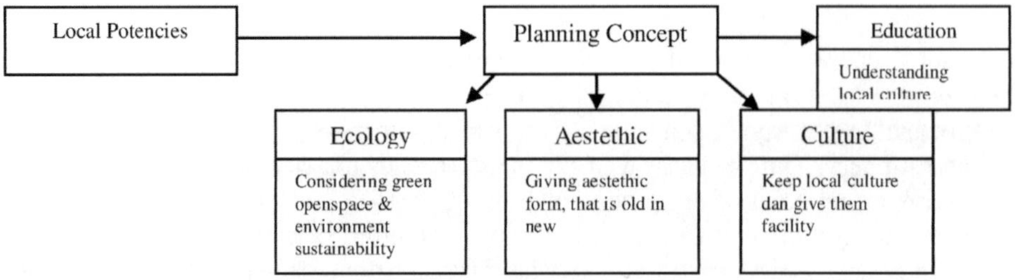

Figure 4. concept diagram

A. Mass Planning Concept

Mass structuring on Watugong tourist area is divided into four zones: the welcome area, a buffer area, conservation area, and the main zone. This structuring effect on the bearing elements of aesthetics at the site that is the gate, monument, post information, information boards and signage.

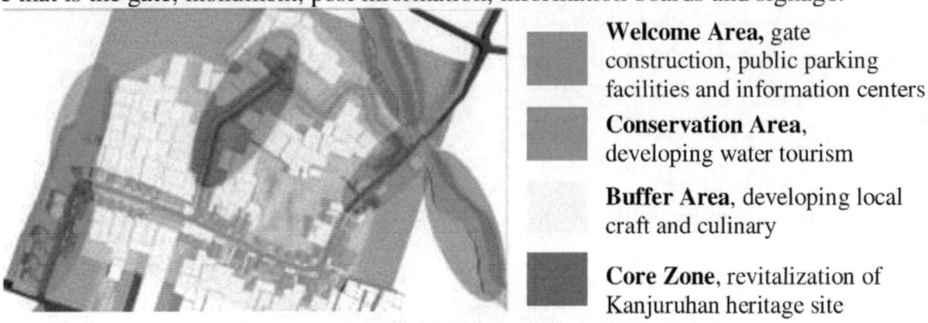

Figure 5. site zoning

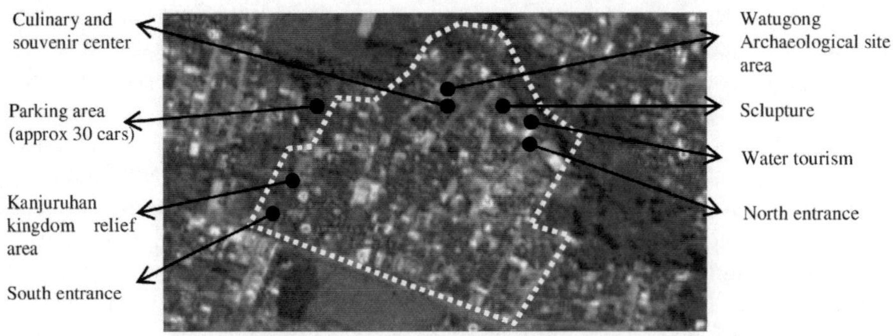

Figure 6. important location in site

B. Form Concept

Form follow the concept of history on the site, by using unique Kanjuruhan kingdom ornaments with jasmine buds characteristic or Kalamakara, the same form of the Candi Jago, Singosari and Candi Jago. Gong and the motives leaves the higher the conical worship Hyang Mring Waseso.

Figure 7. Form concept

C. Vegetation Concept
The concept of vegetation is achieved by the use of a typical plant Kanjuruhan kingdom like rose, jasmine, croton, gardenias, hanjuang, frangipani, and banyan. The arrangement of vegetation have been based functions such as ornamental plants placed in the garden or yard while the shade is located on a side street. Unisex uniform plant as a hallmark of the township is selected hanjuang.

As for the results obtained from the study of the concept of structuring the landscape of history in Kampung Watugong include (1) the concept of ecotourism is used for the design-based conservation, community participation, economic, and education, (2) The form is used to follow the concept of the history of the site, namely the Kingdom Kanjuruhan with characteristic of the buds of jasmine and shape Kalamakara, (3) Element outer designed limited to the arrangement of the fence, gate, posts information, information boards, signage, sclupture and parks, (4) Design of landscape history based on user requirements such as walking and enjoying the atmosphere around, see colorful flowers typical of Kanjuruhan, understanding the formation sclupture, monument and can be learned in the area of the site, (5) spatial zones are distinguished on 4 main welcome area, a buffer area, conservation area and the core zone. Structuring element adapted to function tread. The following illustration concept.

4. Conclusion
This study found the concept of landscape history is based on the potential regional or local wisdom by presenting an attractive view in village. Visitors are expected to enjoy the facilities presented in designs such as education zone, archaeological sites, and souvenirs and culinary. Landscape design in question is to determine patterns of structuring softscape and hardscape that support pedestrian as a space to walk towards the site. Ecotourism criteria also included the historical and cultural conservation, community participation, community economic development planning and improvement, and education.

Acknowledgement
Worship and praise to Allah SWT. who has given smoothness on this research activity. Neither the gratitude we say to the dean of the faculty of agriculture and whole Landscape Architecture lecturer in Tribhuwana Tunggadewi University especially Rizki Alfian who gave me help to look data from Tlogomas, also Wahidyanti who gave support to me, then Malang government, Tlogomas district office and their staff, because this article never exist without their program support, also along whole friends who participate in helping the completion of this study.

Reference
[1] I Setyabudi 2016 *Elemen dan Proses Desain Arsitektur Lanskap Taman Rumah Tinggal* (Book Style) Malang CV. Dream Litera Buana
[2] R Hakim 2012 *Komponen Perancangan Arsitektur Lansekap : Prinsip-Unsur dan Aplikasi Desain* (Book Style) Jakarta PT. Bumi Aksara

Landscape Design Process of Lakewood Nava Park BSD City Based on Smart Growth Concept

M Z Islami[1], R L Kaswanto[2]
[1]Research Scholar, Department of Landscape Architecture, Bogor Agricultural University, West Java, Indonesia.
[2]Lecturer, Department of Landscape Architecture, Bogor Agricultural University,

Email: zaenul2595@gmail.com

Abstract. A comfortable and green housing area in a city is a must for the people live in a city. The rapid development in a city caused greater need for land. This problem happens simultaneously with environmental problem globally such as growing number of people, pollution, excessive exploitation of resource, and decreasing in ethic of land uses. The design of Lakewood Nava Park BSD City prioritizes on pedestrian and walkable environment to apprehend those problems. Lakewood Nava Park is a landscape design project conducted by landscape consultant company, Sheils Flynn Asia. The concept of Smart Growth used as a recommendation for Lakewood Nava Park design. Smart Growth is a city planning and transportation theory which expand a city into a walkable city. The method used on this research is a comparison between landscape design process and Booth theory, also analyze ten principle concept of Smart Growth at the project. Generally, the comparison between design process and Booth theory resulted a slight difference in term and separate phase. The analysis result from Smart Growth concept is around 70% has been applied, and the rest 30% applied after the design has been built. By using Smart Growth principle, the purpose of Lakewood Nava Park design can be applied well.

1. Introduction

The development of cities in Indonesia is growing rapidly caused by infrastructure development, especially in property. The rapid rate of a city needs bigger land to build, meanwhile the area and land potential were static. This kind of development happen in time with environmental problem globally such as pollution, excessive use of resource and ethic decrease on land use [1]. The limitation of land and environmental problem will become serious problem for city development if every landscape architect didn't know about this.

The continuous development in a city caused a city can't afford more people anymore. The problem solving for these problem is a city must distract their development into suburban are, with serious concern in environmental issue. The uses of Smart Growth concept are a solution to optimize land use and environmental problem. Smart growth is a theory of city planning and transportation which expand a city into walkable city. The purpose of the theory is creating place uniqueness, expand transportation, business district, and housing network, preserve nature being, and public health [4].

The concept applied into Lakewood Nava Park design. Lakewood Nava Park is a residential area from Nava Park project in suburban area, BSD City. In design process, PT

Sheils Flynn Asia is involved on Lakewood Nava Park Project in BSD City for optimizing green expansion of land. PT SFA is an international landscape architecture consultant, branch from central office in Norfolk, United Kingdom. For environmentally friendly design, it need professional work for whole design process begin with preparation, analysis, concept, concept development, and build phase. The purpose of the research is analyzing the design process on Lakewood Nava Park conducted by PT SFA, and make recommendation on Lakewood Nava Park based on Smart Growth.

2. Methodology

2.1. Location

PT Sheils Flynn Asia located on Ir. H. Juanda No 13, Bogor National Park. The Lakewood Nava Park located on BSD Boulevard, Nava Park BSD City (Bumi Serpong Damai), Pagedangan, Tangerang District, Banten Province. The map location shown in Figure 1.

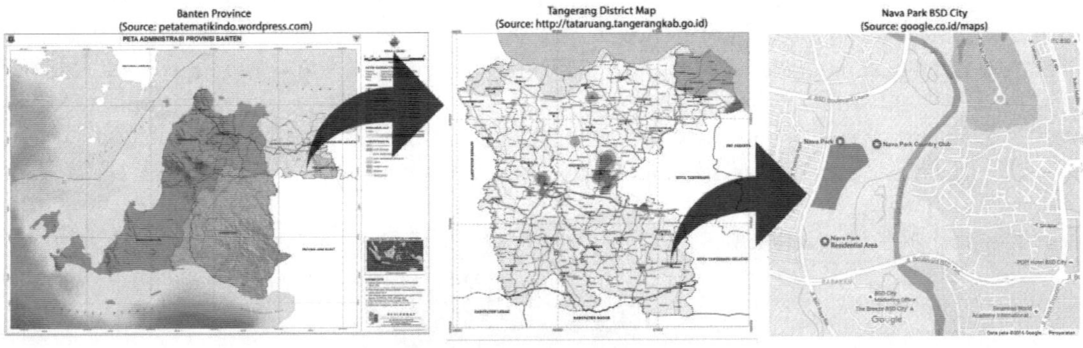

Figure 1. Project location

2.2. Method

The method used in this research is analyze design process in Lakewood Nava Park in BSD City conducted by PT SFA with Booth Theory, consist of:

- Project Acceptance.
- Research and Analysis.
- Design.
- Construction Drawings.

The method used on Smart Growth analysis is adjusting the ten Smart Growth principal which can be seen on smartgrowth.org website, consist of:

- Mix Land Uses.
- Take Advantage of Compact Building Design.
- Create a Range of Housing Opportunities and Choices.
- Create Walkable Neighbourhoods.
- Foster Distinctive, Attractive Communities with a Strong Sense of Place.
- Preserve Open Space, Farmland, Natural Beauty and Critical Environmental Areas.
- Strengthen and Direct Development Towards Existing Communities.
- Provide a Variety of Transportation Choices.
- Make Development Decisions Predictable, Fair and Cost Effective.
- Encourage Community and Stakeholder Collaboration in Development Decisions.

3. Results

3.1. Company Profile

PT Sheils Flynn Asia (SFA) is an International landscape architecture consultant. Established from 2001, SFA is a branch office from Sheils Flynn LTD (SF) located on Norfolk, United Kingdom (UK). SFA worked on various scale of project from landscape design and planning, landscape conservation, feasibility study, landscape restoration, and design and its implementation.

3.2. Lakewood Nava Park BSD City Design Project Description

Lakewood Nava Park is landscape design work in Nava Park, BSD City. Nava park is a new premium city area on 68 ha land in Serpong with lifestyle water community resort concept. Nava Park BSD City is a mixed-use project of a residential, apartment, and commercial area from trusted developer Sinarmas Land and Hong Kong Land from PT Bumi Parama Wisesa (BPW). Nava Park BSD City consist of several areas such as commercial, mixed use residential, low rise residential (Lancewood, Lakewood), Medium Rise Residential, High Rise Residential, Public Amenity, Internal Open Space, and Spring Water. The project worked by PT SFA is a landscape design in Low Rise Residential called Lakewood Nava Park. Master Plan Nava Park shown in Figure 2. Lakewood Nava Park consist of 110 house with 2 type (44 type 12 house, 86 type 10 house) with side park in each house, North Garden, Central Park, Linear Park, West Garden, and Connecting Park. The building of Lakewood Nava Park obtained by PT SFA as a result of landscape design proposal invitation by PT BPW for Nava Park. The project build started from January until July 2016.

Figure 2. Master Plan Lakewood Nava Park

3.3. Design Purpose

The design purpose from Lakewood Nava Park is to create functional and aesthetic landscape applied with various theme and organic design. This design creates comfortable landscape space. Design principle from this project is create walkable environment. The design principles applied through pedestrian priority connection, with a purpose to create integrated and connected Nava Park BSD City and comfortable environment for its users.

3.4. Lakewood Nava Park BSD City Landscape Process Design

The process conducted by PT SFA started from preparation, research and analysis, design concept, design development, and construction document.

- Inception. This phase consist of initial thought, is a preliminary presentation of proposal to client. The content of the proposal is general condition, analysis, design concept, site plan, and reference image as a support. Then contract phase conducted when both sides has an agreement between both sides about duration, amount of salary, reception ethic, and project phase. The next activity is survey to get accurate design and neglect error in design and build later.

- Research and Analysis. Consist of data collection of primary data from survey, and secondary data from client, then analyze it by adjusting with design purpose.
- Conceptual Design. This phase is an idea and concept creation which developed into site from basic concept to design concept. The concept uses by PT SFA preferably using a functional concept structurally and visually. This phase resulting conceptual plan and preliminary design.
- Design Development. Expandable design process focused on element design from its softscape, and hardscape. The result of this phase is master plan.
- Construction Documentation. This phase is a further detailing of drawing work and construction from hardscape and softscape element. This phase consists of planting plant (plant specification, amount and type of plant used), material plan (material specification and type used), construction detail and surfacing. The result of this phase used by contractor to build the project.

3.5. Lakewood Nava Park Analysis Process

3.5.1. Design Analysis Process.

Design process in SFA has certain standard in designing landscape project. The design process structured well until resulting a product. Generally, the design process is same with Booth design process. Started from proposal until implementation, there are some difference in uses of term and phase. The phase can be seen in the analysis process.

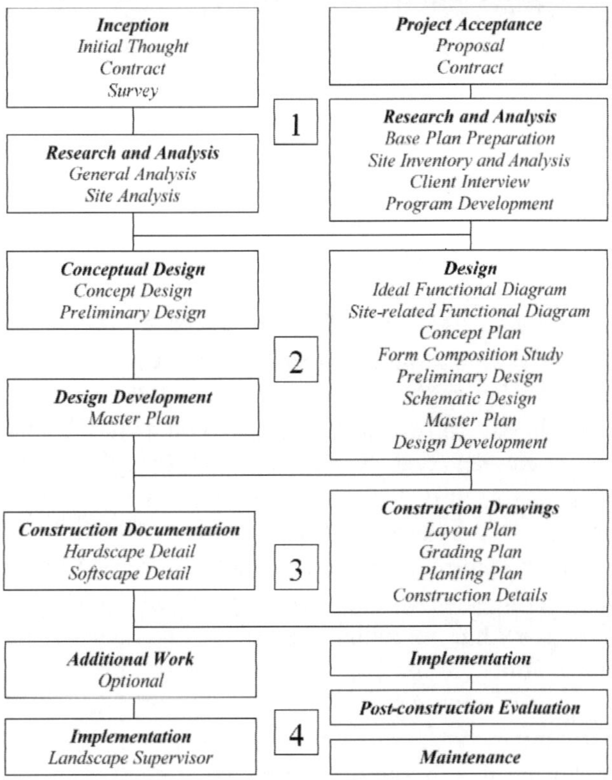

3.5.2. Smart Growth Analysis in Lakewood Nava Park Design Project.

There are a ten Smart Growth concept principle based on smartgrowth.org, which is.

- Mix Land Uses. Smart growth supports mixed land uses as a critical component of achieving better places to live. By putting residential, commercial and recreational uses in close proximity to one another, alternatives to driving, such as walking or biking, become viable. Mixed land uses also provide a more diverse and sizable population and commercial base for supporting viable public transit. Mixed land uses can contribute economic benefits. For example, siting commercial areas close to residential areas can raise property values, helping increase local tax receipts [5]. Nava Park is an area of mixed use, so that this principle can support the application of the concept of Smart Growth. Lakewood Nava Park is one part of a mixed-use area Nava Park as a residential area. The design of the Lakewood prioritizes areas of pedestrian in order to connect between areas on Nava Park.
- Take Advantage of Compact Building Design. Smart growth provides a means for communities to incorporate more-compact building design as an alternative to conventional, land-consumptive development. Compact building design suggests that communities be laid out in a way that preserves more open space, and that individual buildings make more efficient use of land and resources. For example, by encouraging buildings to grow vertically rather than horizontally, and by incorporating structured rather than surface parking, communities can reduce the footprint of new construction, and preserve more greenspace. This not only uses land efficiently, but it also protects more open land to absorb and filter rain water, reduce flooding and storm water drainage needs, and lower the amount of pollution washing into our streams, rivers and lakes. Increasing numbers of developments are successfully integrating compact design into community building efforts [5]. This principle has been used by Nava Park, where Lakewood as part of the design of the area. Its application is dominated by development performed vertically (house of three floors), as well as maximizing the use of open space. Comparison of percentage of green open spaces than spaces is awakened: 74%: 26% on the area of Lakewood Nava Park.
- Create a Range of Housing Opportunities and Choices. Providing quality housing for people of all income levels is an integral component in any smart growth strategy. Housing is a critical part of the way communities grow, because it constitutes a significant share of new construction and development. By using smart growth approaches to create a wider range of housing choices, communities can mitigate the environmental costs of auto-dependent development, use their infrastructure resources more efficiently, ensure a better jobs-housing balance, and generate astrong foundation of support for neighbourhood transit stops, commercial centers, and other services. No single type of housing can serve the varied needs of today's diverse households. Smart growth represents an opportunity for local communities to increase housing choice not only by modifying land-use patterns on newly developed land, but also by increasing housing supply in existing neighbourhoods and on land served by existing infrastructure. Most importantly, providing a range of housing choices allows all households to find their niche in a smart growth community – whether it is a garden apartment, a rowhouse, or a traditional single-family home [5]. Nava Park provides a wide choice of places to stay, ranging from apartments, hotels, and homes with different types of buildings. Lakewood Nava Park has two types of homes, namely house type 10 and type 12. Based on the analysis in terms of economics, these choices are reserved for among middle

income and above. Analysis in terms of the community grow, Lakewood area provides landscape facilities for a range of ages and facilities of home grown for occupants.

- Create Walkable Neighbourhoods. Walkable communities that are desirable places to live, work, learn, worship and play are a key component of smart growth. Their desirability comes from two factors. First, goods (such as housing, offices, and retail) and services (such as transportation, schools, libraries) are located within an easy and safe walk. Second, walkable communities make pedestrian activity possible, thus expanding transportation options, and creating a streetscape for a range of users – pedestrians, bicyclists, transit riders, and drivers. To foster walkability, communities must mix land uses and build compactly, as well as ensure safe and inviting pedestrian corridors. Land use and community design play a pivotal role in encouraging pedestrian environments. By building places with multiple destinations within close proximity, where the streets and sidewalks balance multiple forms of transportation, communities have the basic framework for walkability [5]. This principle is in accordance with the purpose of designing Lakewood Nava park, which is to create an environment that is walkable with pedestrian priority connection. Pedestrian facilities that are equipped with ramp for users of disabilities with a minimum width of 1.5 meters.
- Foster Distinctive, Attractive Communities with a Strong Sense of Place. Smart growth encourages communities to craft a vision and set standards for development that respect community values of architectural beauty and distinctiveness, as well as expand choices in housing and transportation. Smart growth seeks to create interesting, unique communities that reflect the values and cultures of the people who reside there, and foster physical environments that support a more cohesive community fabric. Smart growth ensures that the value of infill and greenfield development is determined as much by its accessibility (by car or other means) as its physical orientation to, and relationship with, other buildings and open space. By creating high-quality communities with architectural and natural elements that reflect the interests of all residents, there is a greater likelihood that buildings (and therefore entire neighbourhoods) will retain their economic vitality and value over time [5]. Lakewood Nava Park made to create a distinctive atmosphere through design with organic forms and water games in the open space. Selection of ornamental plants and forms typical embankment in Lakewood Nava Park. The design is accommodated so that the occupants can interact socially and share their culture.
- Preserve Open Space, Farmland, Natural Beauty and Critical Environmental Areas. "Open space" refers to natural areas that provide important community space, habitat for plants and animals, and recreational opportunities, as well as farm and ranch land (working lands), places of natural beauty, and critical environmental areas (e.g. wetlands). Open space preservation supports smart growth goals by bolstering local economies, preserving critical environmental areas, improving community quality of life, and guiding new growth into existing communities. Open space also provides significant environmental quality and health benefits. Open space protects animal and plant habitat, places of natural beauty, and working lands by removing development pressure and redirecting new growth to existing communities. Additionally, preservation of open space benefits the environment by combating air pollution, attenuatingnoise, controlling wind, providing erosion control, and moderating temperatures. Open space also protects surface- and ground-water resources by filtering trash, debris, and chemical pollutants before they enter a water system [5]. Preservation efforts conducted through Lakewood Nava Park design. Garden design created by minimizing pavement. At the park in the

IOP Conf. Series: Earth and Environmental Science **91** (2017) 012035 doi:10.1088/1755-1315/91/1/012035

North Garden and the West Garden electoral plant is functioning to invite insects and birds to the area. As well, the plant used for damping and noise pollution due to the location of the park adjacent to the main entrance Nava Park and adjacent to the Way BSD Boulevard.

- Strengthen and Direct Development Towards Existing Communities. Smart growth directs development towards existing communities already served by infrastructure, seeking to utilize the resources that existing neighbourhoods offer, and conserve open space and irreplaceable natural resources on the urban fringe. Development in existing neighbourhoods also represents an approach to growth that can be more cost-effective, and improves quality of life. Nevertheless, developers and communities are recognizing the opportunities presented by infill development, as suggested not only by demographic shifts, but also a growing awareness of the fiscal, environmental, and social costs of urban fringe development [5]. This principle has not been applied to the design of Lakewood Nava Park because there are no people in the Nava Park area.
- Provide a Variety of Transportation Choices. Providing people with more choices in housing, shopping, communities, and transportation is a key aim of smart growth. Communities are seeking a wider range of transportation options in an effort to improve beleaguered current systems. In response, communities are beginning to implement new approaches to transportation planning, such as better coordinating land use and transportation; increasing the availability of high-quality transit service; creating redundancy, resiliency and connectivity within their road networks; and ensuring connectivity between pedestrian, bike, transit, and road facilities. In short, they are coupling a multi-modal approach to transportation with supportive development patterns, to create a variety of transportation options [5]. Variations of transportation options have not been found in the Nava Park area, a mode of transportation that allows is by walk or drive by private vehicle.
- Make Development Decisions Predictable, Fair and Cost Effective. For a community to be successful in implementing smart growth, the concept must be embraced by the private sector. Only private capital markets can supply the large amounts of money needed to meet the growing demand for smart growth developments. If investors, bankers, developers, builders and others do not earn a profit, few smart growth projects will be built. Fortunately, government can help make smart growth more profitable for private investors and developers. Despite regulatory and financial barriers, developers have created successful examples of smart growth. In many cases, doing so has required them to spend time and money getting variances to the codes. y creating a supportive environment for development of innovative, pedestrian-oriented, mixed-use projects, government can provide smart growth leadership for the private sector [5]. This principle has been applied to the Nava Park, owner of the Nava Park area is renowned developers from the private sector, namely Sinarmas and Hong Kong Land. Effective development has been done through the master plan Nava Park.
- Encourage Community and Stakeholder Collaboration in Development Decisions. Growth can create great places to live, work and play—if it responds to a community's own sense of how and where it wants to grow. Communities have different needs and will emphasize some smart growth principles over others: those with robust economic growth may need to improve housing choices; others that have suffered from disinvestment may emphasize infill development; newer communities with separated

uses may be looking for the sense of place provided by mixed-use town centers; and still others with poor air quality may seek relief by offering transportation choices. Key actions in encouraging collaboration include developing an inclusionary process and a common understanding among diverse stakeholders, using effective and appropriate communication techniques, and working with local authorities [5]. The absence of people in the Nava Park area make this principle cannot yet applied to the site.

3.6. Design Recommendation in Lakewood Nava Park BSD City based on Smart Growth.

3.6.1. Design Process Recommendation.

Design process conducted by SFA is well structured, but to improve the effectiveness and efficiency, it is recommended to SFA to uses ideal functional diagram and site related functional diagram from Booth design process in conceptual design phase. The ideal function diagram is tools to look into design whether it is function or not based on the purpose. Site-related functional diagram used to learn and maximize the site potency which designed by connecting the main design and space connection to resulting best design solution. SFA process unintentionally also applied this tools, but this research conducted in same mainframe when design concept being analyze on preliminary design. It is recommended to create own phase to learn about ideal function diagram and site-related functional diagram before starting in any kind of design process. It is recommended to minimize error and the have bad design result.

3.6.2. Smart Growth Concept Recommendation on Design Process.

The application of the rest principle from smart growth concept, which is strengthening and direct-development into society, provide variation between transportation, and pushing stakeholder and people to collaborate and involved in development decision. The principle can be applied after the design has been built. It needs some effort to build a character in the people which applied by adjusting with the facility there. The second principle can be applied through the planning and designing of transportation in Nava park area to become more integrated with various area around it like BSD, Tangerang, DKI Jakarta. The other option for more environmentally friendly transportation are by using bicycle or walking for short distance. It is already built with a cycle way and pedestrian. The involvement of society and stakeholder can be applied when the society has been formed in there. This principal is the key for the other nine principles, because the developer and users need to collaborate in further expansion.

4. Conclusion

The phase uses in design process used by SFA generally is parallel with the design theory proposed by Booth starting from the proposal until the implementation. The difference was just term it uses and separated phase. The SFA phase are inception, research and analysis, conceptual design, design development, construction documentation, additional work and implementation. Meanwhile according to Booth, the design process are project acceptance, research and analysis, design, construction drawings, implementation, post-construction evaluation, and maintenance. The purpose of Lakewood Nava Park design is prioritizing the pedestrian and create walkable environment which is combined with Smart growth concept. There are ten smart growth principals, in Lakewood Nava park design these principals were 70% applied while the rest 30% still can't be achieved. The uses of smart growth principle can support the purpose of building Lakewood Nava Park.

5. References

[1] Kaswanto, Utami FNH. 2016. The Disparity of Watershed Development between Northern and Southern Region of Java Island. Procedia Environmental Sciences 33:21-26.

[2] Booth NK. Basic Elements of Landscape Architecture Design. New York: Ohio State University. 1983.

[3] Nava Park BSD City. Master Plan Nava Park BSD Hongkong Land & Sinarmas. http://www.navaparkbsdcity.com/2015/01/master-plan-nava-park-bsd-hongkong-land.html (accessed April 2016).

[4] Sari DP, Roychansyah MS. Konsep Smart Growth, Compact City dan "Retrofiiting" sebagai Solusi Urban Sprawl di Kota-kota Besar di Indonesia. Jurnal Arsitektur dan Desain. 1 (1):7. 2014.

[5] Smart Growth Online. Smart Growth Principles. http://smartgrowth.org/smart-growth-principles/ (accessed June 2016).

Redesign of Denggung Park as Sleman Urban Park based on Local Wisdom in Yogyakarta

[1]Sanjaya I and **[2]Fatimah IS**

[1]Student of Landscape Department of Architecture, Faculty of Agriculture, Bogor Agricultural University
[2] Lecture in Department of Landscape Architecture, Faculty of Agriculture, Bogor Agricultural University

Email : irsyadsanjaya@gmail.com

Abstract. Sleman Regency is one of the administrative area in Special Region of Yogyakarta Province which has increased the pace of infrastructure development activities that undertaken by the central government affects another surrounding area. The pace of infrastructure development impacts such problems in Sleman Regency such as, increasingly limited public spaces and changes in understanding the value of local wisdom. Sleman Regency has a park located in central government which is Denggung Park. This park has low visitors and less of aesthetic value which require re-design to improve the quality as public space for cultural identity space. The base concept of Urban Park adopted the philosophy that connects to four components in Javanese mythology. The four components in Javanese mythology symbolize the journey of human life in the Javanese cosmological theory, there are Mount Merapi, Keraton, Krapyak Stage, and South seas. The design concept inspired from pattern of Yogyakarta traditional clothing namely, Batik Kawung which describe of Philosophy "Four of Brotherhood and Five of Central itself" by means synergize four items creating world nature and human as life catalyzer. This study uses descriptive and spatial analysis method. The result of this research is expected to be a design recommendation for Sleman Regency governance in the urban park development.

1. Introduction

Sleman regency is one of the administrative area in the Special Region of Yogyakarta province. Sleman has an area of approximately 57.482 hectares with the central government in the Sleman district. Sleman district is the center of economic, social and cultural of Sleman Regency that marked by the development of city infrastructure. City infrastructure development can't be separated from modernization and influence lifestyle of surrounding people in Sleman Regency.

The pace of infrastructure development in Sleman regency characterized by increase of modern shopping center, settlement area, and industrial area in the suburban land. It is impacting to the city space planning, and environmental aspect in Sleman Regency which has been embraced the traditional development philosophy that called "Hamemayu Hayuning Bawono". This concept means that human should to preserve the environment of the universe created by God.

IOP Conf. Series: Earth and Environmental Science **91** (2017) 012036 doi:10.1088/1755-1315/91/1/012036

The impact of infrastructure development for city space planning are reduced public space. Therefore, the Sleman Regency government create a space planning strategy program as an effort to improve the quality and reconstructing of public spaces, especially downtown of Sleman Regency. Furthermore, the behavior of local people that increasingly modern causes the reduced understanding people of the local wisdom values which has been privileges character of Yogyakarta. The one of the public spaces in Sleman Regency that being a part of the Sleman Regency government program is Denggung Park.

Denggung Park is the urban park located in the central government area of Sleman Regency which bordered by national road access. However, the physical condition of this park has low value in aesthetics as the face of Sleman Regency urban park. Therefore, the result of this research expected to be a design recommendation to the Sleman Regency.

2. Objectives

- Identify and analyze the potential and the constraints for redesign the Denggung Park based on local wisdom
- Determine the basic concept, design concept and development concept to redesign of Denggung Park.

3. Methodology

This research was conducted in Denggung Park, in Sleman district, Sleman Regency,Yogyakarta Province. This research began in January 2016 until October 2016.

3.1. Research Method

- Commissioning
- Inventory
- Analysis
- Synthesis
- Concept (for Planning and Design)
- Construction and Operation

4. Results

4.1. General Condition

Denggung Park is an urban park located in central government area of Sleman Regency which used by local people for recreational area and venue for incidental events such as national independence ceremony, cultural performances, art and sport festivals. This Park bordered by traditional market on the northern side, KRT Pringgodiningrat street on the southern side, Magelang highway street on the western side, and government offices area on the eastern side. Denggung Park is managed by landscape division in environmental agencies of Sleman Regency.

4.2. Analysis and Synthesis

4.2.1. Accessibility and Circulation

Denggung Park located nearby to national road access which is Magelang highway street. There are two types of circulation at the park, Primary circulation and Secondary Circulation. Primary circulation intended for pedestrians people, then Secondary circulation intended for vehicles.

4.2.2. Topography

Denggung Park has a flat topography with a slope about 0-2 %. The flat topography can be a potential to developed a variety of activities and addition of landscape elements to attract the visitors/user.

4.2.3. Vegetation and Animals

The existing vegetation types that found in Denggung Park are dominated by tree species such as, *Pterocarpus indicus*, *Ficus benjamina*, and *Roystonea regia*. The big trees species have a potential as a shade in the daytime.

4.2.4. Visual
There are some potential good view in Denggung Park, in the southern side of the park there is an elephant sculpture as a focal point on the Denggung Park. Furthermore, also in the southern side there is signage of Denggung Park and electronic video (videotron) that looks glorious.
4.2.5. Facilities
The facilities that already available at the Denggung Park such as, bench, trash bin and toilet. However, the condition has not proper to accommodate the visitors. In addition, the visitors want to restructuring and relocation to the street traders who build up semi-permanent stall.
4.2.6. Users
Denggung Park fairly crowded in the afternoon by people, especially children and adolescents. When a specific event, the intensity of visits increase compared with the weekdays. Denggung Park used as a venue for events such as national independence ceremony, cultural exhibition and art festival.
4.2.7. Local Wisdom Values
Based on the research, the potential value of local wisdom philosophy that developed in the park is space planning aspect refers to Javanese local wisdom philosophy adapted from Yogyakarta kingdom which is "Keraton". Furthermore, traditional architecture that applied through the traditional construction, local plants, traditional rituals and arts.

4.3. Concept
4.3.1 Basic Concept
The concept that will be applied to Denggung Park is the historical axis of Yogyakarta, which is the implementation of traditional development philosophy called "*Hamemayu Hayuning Bawono*". The historical axis of Yogyakarta is connecting the four components include, Mount Merapi, Keraton Yogyakarta, Krapyak Stage, and South Seas. This concept symbolizes the cycle of human life that starts from born to the world and live a life as a child, then entered a transition period that called adolescence. Then after going through this period, human will enters the adulthood and the old ages period before returning to God.
4.3.2 Design Concept
The design concept adopted a motif of Yogyakarta traditional clothing, called *Batik Kawung*. This motif contains a philosophy "Four brotherhood and five of central itself" that means a harmony between four elements of the universe and human as a symbol of earthly life. Batik Kawung contains two geometrical pattern, circle and square. The circle pattern on the motif is adopted as a planting pattern. The square pattern adopted as a space and hardscape patterns.

4.4. Development Concept
4.4.1 Space Concept
The space concept of redesign Denggung Park implementing the cycle of human life philosophy. It is categorized according to age levels including, children, adolescents, adults and elderly. According to cycle of human life philosophy generated four classifications area including, welcome area, actualization area, service area and contemplation area.
4.4.2 Circulation Concept
Circulation type that used on the research site is controlled form. The controlled form has characteristics strict and formal. (Booth 1983). Therefore, this type is suitable to be applied as a circulation form pattern. Circulation development concept of

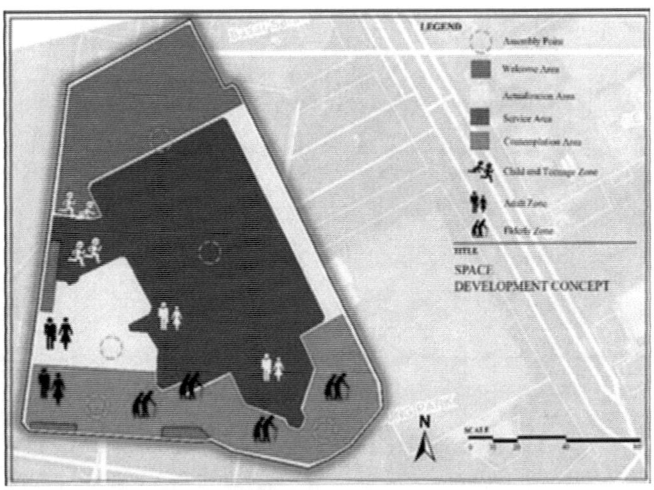

Figure 1. Space Concept

Denggung Park categorized into two types, primary circulation and secondary circulation. The primary circulation is the main circulating inside the park that reserved for pedestrians. Secondary circulation reserved for vehicles outside of the park.

4.4.3 Vegetation Concept

Vegetation development concept of Denggung Park is categorized into five classification functions, ground cover, shades, boundary, corridors and display. The development concept using a composition and the traditional philosophy of the local species plants.

4.5. Development Design

Development design generates an overall plan, called site plan. Welcome area is located in the northern side of the park which designed more attractive as a place to receive the visitors and become the entrance gate.

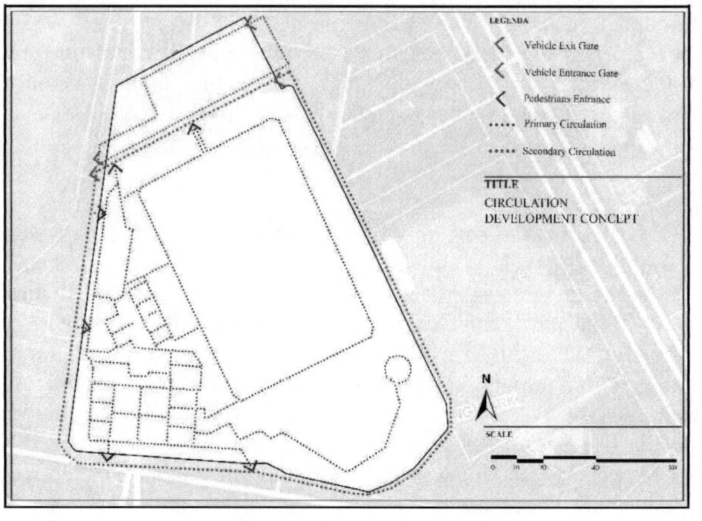

Figure 2. Circulation Concept

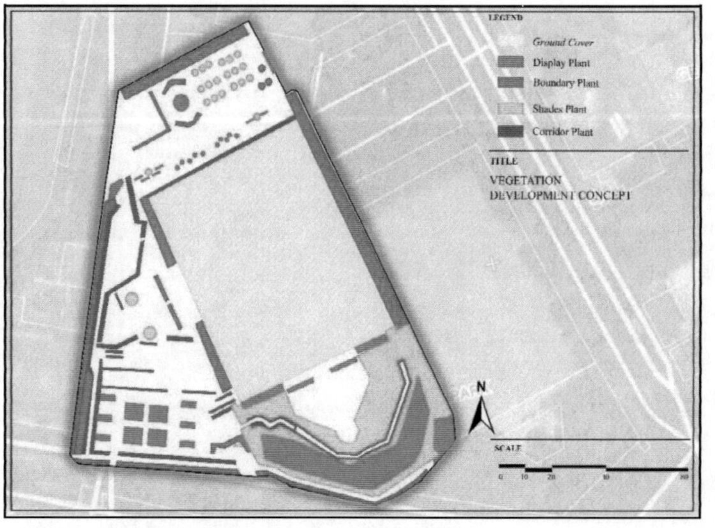

Figure 3. Vegetation Concept

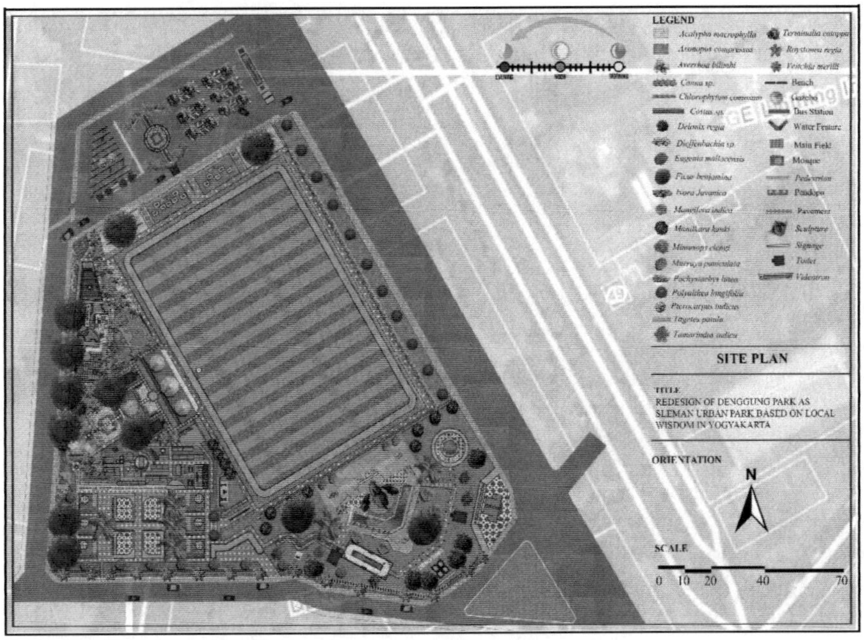

Figure 4. Site Plan

Furthermore, Actualization area also located in the northern side of the park that intended to accommodate the teen community. The main facilities in this area include, main field, outdoor gym, stage for arts and cultural performance, mini skate park, and jogging track surrounding the main field. The third is service area in the southern side that reserved for adult visitors. The main facilities include, a mosque, wireless fidelity (*Wi-fi*) zone, and trades stall centre. The fourth is contemplation area in the southern side. It is reserved to elderly and disable people. The facilities in this area bring a relaxing atmosphere for the visitors. Reflexology path in this area provided for medical therapy, then Plaza and Water Feature that presents the water attractions to improve the visual character in this area.

5. Conclusion

Denggung Park is an urban park that located in the Sleman Regency government centre and bordered by national access road. Based on analysis and synthesis, Denggung Park has a potential to be developed. The basic concept that adopted from historical axis of Yogyakarta which is the implementation of traditional development philosophy, Hamemayu Hayuning Bawono

Historical axis of Yogyakarta connecting the four components from the north side to south side there is, Mount Merapi, Keraton Yogyakarta, Krapyak Stage, and South Seas. This concept also represent the cycle of human life from childhood, adolescence, adulthood and elderly. The design concept inspired from pattern of Yogyakarta traditional clothing namely, Batik kawung which describe of Philosophy "Sedulur Papat Limo Pancer" that means four of brotherhood and five of central itself. Local wisdom symbolized by spatial plan aspect, traditional architecture, vegetation and traditional rituals. This research is expected to be a design recommendation for Sleman Regency government in the urban park development program.

References

[1] Anshoriy HMN, Sudarsono. 2007. *Kearifan Lingkungan dalam Perspektif Budaya Jawa*. Jakarta (ID): Obor Indonesia Foundation

[2] Booth NK. 1983. *Basic Elements of Landscape Architectural Design*. Illinois (US): Waveland Press.

[3] Simonds, JO. 1983. Landscape Architecture: A Manual of Site Planning and Design. New York: McGraw-Hill Companies, Inc

Green Campus Study by using 10 UNEP's Green University Toolkit Criteria in IPB Dramaga Campus

Saraswati Sisriany[1] and Indung Sitti Fatimah[2]

[1]Fresh Graduate, Department of Landscape Architecture, IPB
[2]Lecture, Department of Landscape Architecture, IPB

saraswatisisriany@gmail.com

Abstract. Campus landscape is an important part of campus life, because it is regarded as a physical manifestation of the value of a college. Green campus is a concept to build sustainable living practices that are environmentally friendly in educational institutions around the world, including in IPB Dramaga Campus. The main objective of this study is to identified and analyze IPB Dramaga Campus sustainability used green campus criteria from UNEP (United Nations Environment Programme). The methods stages are data collection, analysis and assessment, and recommendation as the synthesis. All the data analyzed with gap analysis, then it assess with Likert Scale scoring. The results showed that green level of IPB Dramaga Campus is classified as Moderate, with total score 32. The result from each criterias are, Energy, Carbon and Climate Change is Moderate; Water is Not Good; Waste is Moderate; Biodiversity and Ecosystem Services is Very Good; Planning Design & Development is Good; Procurement is Moderate; Green Office is Very Not Good; Green Lab is Moderate; Green IT is Good; and Transport is Good. The Green Level of IPB Dramaga Campus will reach Very Good if these recommendation of strategies applied. The strategies are Green Office, Green Campus Audit, Green Champion, Green Financial Strategies, Water Treatment, Green Lab dan Off Campus Transportation.

1. Introduction

Campus landscape is an important part of campus life, because it is regarded as a physical manifestation of the value of a college. Green campus is a concept to build sustainable living practices that are environmentally friendly in educational institutions around the world including IPB Dramaga Campus. IPB is one of campuses in Indonesia who have started implementing the environmental concept by having IPB Green Campus in 2020 as it vision. There are several references for green campus, two of them are UIGreenmetric and UNEP's Greening University Toolkit. The first one, UIGreenmetric is a green campus rating which been applied in IPB Dramaga Campus and own six criteria of green campus. The second one, UNEP's Greening University Toolkit is a green campus toolkit which has not applied before in IPB Dramaga Campus and own ten criteria of green campus. Therefore, this study use UNEP's green campus criteria to re-examine the green campus implementation in IPB Dramaga Campus as an additional reference in order to actualize IPB Green Campus 2020. The purposes of this study are to identified and analyse green campus implementation in IPB Dramaga Campus, with the final output is a recommendation of green campus strategies that could be applied in IPB Dramaga Campus. This research is expected to be a reference to the relevant parties in designing green campus and to help achieve IPB Green Campus 2020.

2. Methods

This study take place in IPB Dramaga Kampus, West Java, or geographically in 6°33'38.3"S 106°43'33.2"E. It started from April 2015 until September 2016. The research consist of several stages, they are

2.1 Preparation and Data Inventories

This study started with desk study for green campus references from article, paper, thesis, or journal. Data inventories for primary data is by interview the key responden, and the secondary data is from literature or direct data from the relevant parties. Data inventories are based upon UNEP's Greening University Toolkit Criteria.

2.2 Analysis and Assessment

All the data then identified through 10 UNEP's green campus criteria and analyse using gap analysis by comparing the actual condition and the ideal condition of green campus in IPB Dramaga Campus. The assessment use Likert scoring methods which divided into five level of scoring. Score 1: No Policy, 2: Policy Preparation, 3: Policy in initial implementation, 4: Policy Implemented with some problems, 5: Policy in full implementation. This assessment is consist of several stages from sub criteria, to each criteria, and finally the overall assessment of green campus.

2.3 Synthesis

Final stage is to make a recommendation green campus strategies that could be implemented in IPB Dramaga Campus.

3. Results and Discussions

3.1. General Conditions

IPB Green Campus first announce in early 2015, with UIGreenmetric as it concept based. The purpose of IPB Green Campus are develop the behavior of academicians environmentally friendly, realizing IPB as an environmental campus: zero waste, zero emission, rich in bio-diversity and cultural, creating the conditions campuses that are safe, comfortable and conducive to successful learning systems, as well as to support agro-eco-edu-tourism program. There are several green campus element that are plan to be implemented they are Green Transportatation 2015, Green Movement 2016, Green Energy 2018, Green Building 2019 to finally reach IPB Green Campus 2020.

3.2 IPB Dramaga Campus in UNEP's Green Campus Criteria

3.2.1 Energy, Carbon and Climate Change

Energy, Carbon and Climate Change is one of the basic issues of sustainability. Although IPB Green Energy is still in it preparation but some of the indicators are already implemented or in its initial stage. The result shows Energy Conservation is Not Good, Energy Efficiency is Moderate, and Renewable and Alternative Energy is Moderate. Overall results for Criteria of Energy, Carbon and Climate Change is Moderate (detail in Table 1).

Tabel 1. Energy, Carbon and Climate Change

Indicators		Score
1. Energy Conservation	**Not Good (2)**	**22**
Employment of Energy Manager		5
Energy efficiency standards for new construction and refurbishments		2
Energy efficiency purchasing standards		2
Staff energy conservation training		3
Improved space utilisation to avoid new construction or heating/cooling of underutilised space		2
Thermal comfort policy (e.g. widening heating/cooling temperature settings)		2
Financial straegies to assign energy cost incurred - and savings achieved - to the responsible cost centres		1
Energy / climate change awareness programs		4
Establishment of "energy champions" network across campus building		1
2. Energy efficiency	**Moderate (3)**	**19**
Detailed energy audit to identify prority areas		1
Periodic recommissioning and building tuning to optimise energy efficiency		2
Building retrofitting		4
Lighting		2
Heating, ventilation and air-conditioning (HVAC)		4
Laboratory ventilation and fume hoods		4
Installation of building management and control systems (BMCS) and sub-metering for major building energy uses, energy use displays.		2
3. Renewable and alternative energy	**Moderate (3)**	**16**
Purchase of certified "green power".		5
Installation of photovoltaic, wind, biomass, etc. systems.		3
Installation of cogeneration and trigeneration.		1
Fuel switching		4
University managed revegetation program to offset greenhouse emissions.		3
Total Score Criteria of Energy, Carbon and Climate Change	**Moderate (8)**	

IOP Conf. Series: Earth and Environmental Science **91** (2017) 012037 doi:10.1088/1755-1315/91/1/012037

3.2.2 Water

IPB Dramaga Campus doesn't have element for Water, both in UIGreenmetric and this study shows that water practices in this campus is not good. The only good part is they use treatment water for daily consumes, as an alternative of soil water. The result shows Water Conservation is Not Good, Water Efficiency is Moderate, and Water Reuse and Recycling is Not Good. Overall result for Water is Not Good (detail in Table 2).

Tabel 2. Water

Indicators		**Score**
1. Water conservation	**Not Good (2)**	**17**
Employment of Water Manager (can be combined Energy / Water Manager position).		5
Water efficiency standards for new construction and refurbishments.		2
Water efficiency purchasing standards.		2
Staff water conservation training (can combine with energy conservation training).		3
Financial strategies to assign water costs incurred – and savings achieved – to the responsible cost centres.		1
Water conservation awareness programs – posters, stickers, events and competitions, websites, awards and incentives.		3
Extension of "energy champions" network to incorporate water conservation.		1
2. Water efficiency	**Moderate (3)**	**23**
Detailed water audit and campus water balance to identify priority areas.		1
Active maintenance program of early detection and repair of faulty plant, equipment and fixtures.		4
Retrofitting of water saving devices		2
Underground pipework leak detection and repair.		4
Use of pervious paving.		4
Specification of low water use species for campus grounds.		3
Laboratory water use		4
Installation of building management and control systems (BMCS) and sub-metering for major building water uses, water use displays.		1
3. Water reuse and recycling	**Not Good (2)**	**14**
Capture and reuse of rainwater from roofs and other hard surfaces for non-potable uses (irrigation, laboratories, toilet flushing, cooling towers, construction works, swimming pools, etc.)		1
Substitution of borewater for non-potable uses, when combined with managed aquifer recharge to ensure more water is returned to the aquifer than extracted.		5
Installation of greywater recycling system for treatment of kitchen, laundry and shower water		1
Composting toilets and urine recovery for fertiliser.		1
Installation of blackwater recycling system to treat sewage for non-potable uses.		5
Recovery and reuse of fire system test water, vehicle washdown water, etc.		1
Total Score Criteria of Water	**Not Good (7)**	

3.2.3 Waste

Waste practices in IPB Dramaga Campus is in its initials stage, and some of the indicators has been fulfilled even has some problems with them. The result shows Policy and Behaviour Change is Moderate, Waste Management is Moderate, and Cloosing the Loop is Moderate. Overall result for Waste Criteria is Moderate (detail in Table 3).

Tabel 3. Waste

Indicators		**Score**
1. Policy and behaviour change	**Moderate (3)**	**15**
Employment of Waste Manager.		5
Sustainable procurement standards which address longevity, durability, repairability, recyclability and recycled content.		2
Financial strategies to assign waste costs incurred – and savings achieved – to the responsible cost centres.		1
Waste management awareness programs		3
Programs targeting teaching and research to minimise generation of hazardous wastes.		4
2. Waste management	**Moderate (3)**	**18**
Waste characterisation study to identify waste stream components and prioritise response.		3
Individual staged and prioritised programs for waste minimisation which address each component of the university waste stream according to environmental impact.		2
Performance-based waste management contracts to specify resource recovery targets.		1
In-house collection of recyclables (e.g. paper / cardboard) where practicable, to support local job creation.		4
Provision of adequate storage spaces for waste and recyclables.		4

Secure storage spaces for hazardous wastes to minimise risk of spillage / leakage.		4
3. Closing the loop	**Moderate (3)**	**8**
Campus based exchange and reuse programs		2
On-site composting of food and garden organics for reuse on campus grounds.		4
Campus based programs to process collected recyclables		2
Total Score Waste Criteria	**Moderate (9)**	

3.2.4 Biodiversity and Ecosystem Service

This criteria is the best attained from IPB Dramaga Campus. In April 2016 IPB officially announced as Biodiversity Campus, and no wonder this criteria reach maximum stage. The result shows Policy, Design and Development is Very Good, and Managememt and Maintenance is also Very Good. Overall result for Biodiversity and Ecosystem Service Criteria is Very Good (detail in Table 4).

Table 4. Biodiversity and Ecosystem Service

Indicators		**Score**
1. Policy, design and development	**Very Good (5)**	**35**
Survey and evaluation of campus biodiversity and ecosystem services.		5
Extension of campus green space (consolidation / intensification of campus buildings over time, installation of green roofs / walls).		5
Increase density of campus vegetation, e.g. through additional tree planting.		3
Enhance diversity of campus vegetation.		3
Green infrastructure / ecological engineering projects (green roofs / walls, designed wetlands for wastewater treatment, phytoremediation of contaminated land, indoor landscapes for biofiltration / indoor environmental quality).		4
Development of productive landscape systems (permaculture, aquaponics) to provide food / fibre / timber.		5
Restorative and enabling landscapes for contemplation, recreation and wellbeing.		5
Campus grounds and green infrastructure used in teaching and research.		5
2. Management and maintenance	**Very Good (5)**	**20**
Specify local native species		5
Preserve significant vegetation during building works		5
Avoid monocultures		5
Avoid environmental weeds		5
Total Score Biodiversity and Ecosystem Service	**Very Good (10)**	

3.2.5 Planning, Design and Development

This criteria is in preparation, and planned to be implement in 2019 as IPB Green Building, but some of the indicators is already fulfilled by the current campus building and landscape. The result shows Campus Planning is Very Good, Campus Building Design is Not Good, and Campus Construction Management is Good. Overall result for Planning, Design and Development is Good (detail in Table 5).

Table 5. Planning, Design and Development

Indicators		**Score**
1. Campus Planning	**Very Good (5)**	**18**
Campus-specific sustainability objectives included in all campus planning instruments		5
Space planning at campus, precinct and building scale to optimise flexibility, adaptability, diversity and multifunctionality of spaces.		5
Investigation of non-building solutions to accommodate university growth.		5
Physical accessibility of the campus to the external community, different age groups and people with a disability.		3
2. Campus Building Design	**Not Good (2)**	**4**
Design to the appropriate green building rating system as the minimum starting point.		2
Each new building / major refurbishment to incorporate at least one innovative sustainability feature beyond the requirements of the green building rating system.		2
3. Campus Construction Management	**Good (4)**	**12**
Construction contractors certified to ISO 14001.		5
Contractor staff inducted to the university's sustainability management system.		5
Management of campus construction/demolition to minimise on- and off-site impacts.		2
Total Score Planning, Design and Development	**Good (11)**	

3.2.6 Procurement

IPB Dramaga Campus while on its way to IPB Green Campus 2020 is also improving their administration in procurement even if it still in their initial stages. The result shows that Developing Specification is Moderate, Tender Evaluation is Moderate, and Contract Management is Moderate. Overall result of Procurement is Moderate (detail in Table 6).

Table 6. Procurement

Indicators		Score
1. Developing Specification	**Moderate (3)**	**9**
Evaluation of university contracts for procurement of goods and services on the basis of cost, complexity and actual/potential sustainability impacts to determine priorities.		3
Staged development of sustainable procurement standards / specifications based on identified priorities.		3
Inclusion of sustainability criteria in tender specifications for procurement of goods and services.		3
2. Tender Evaluation	**Moderate (3)**	**3**
Inclusion of sustainability criteria in tender evaluation procedures.		3
3. Contract Management	**Moderate (3)**	**6**
Inclusion of sustainability objectives and targets in contract management documentation, and regular monitoring of progress.		3
"Second party" audits of providers to drive continual improvement through the supply chain.		3
Total Score Procurement	**Moderate (9)**	

3.2.7 Green Office

IPB Dramaga Campus is also doesn't have element to manage the Green Office, therefore this criteria result is the worst. The result shows Policy and Behaviour Change is Very Not Good, and Office Practices is also Very Not Good. Overall result for Green Office is Very Not Good (detail in Table 7).

Table 7. Green Office

Indicators		Score
1. Policy and behaviour change	**Very Not Good (1)**	**6**
Employment of Green Office Manager.		1
Sustainable procurement standards for office equipment and consumables.		1
Education, training and awareness programs		3
Establishment of "Green Office champions" network across campus buildings as the vehicle for the energy and water conservation network		1
2. Office Practices	**Very Not Good (1)**	**2**
Campus- wide audit of office practices disaggregated to department level		1
Establishment of department-specific targets to drive continual improvement		1
Total Score Green Office	**Very Not Good (2)**	

3.2.8 Green Lab

Green Lab practice is also doesn't applied in IPB Green Campus policy, but some of the indicators fulfilled by standard lab of IPB Campus. The result shows Policy and Behaviour Change is Very Not Good, Laboratory Practice is Not Good, and Maintenance and Capital Works is Moderate. Overall result for Green Lab is Not Good (detail in Table 8).

Table 8. Green Lab

Indicators		Score
1. Policy and behaviour change	**Very Not Good (1)**	**7**
Employment of a Green Lab manager.		1
Development of a "green chemistry" program.		1
Sustainable procurement standards for lab equipment and consumables.		4
Green Lab online and face-to-face training.		1
2. Laboratory Practices	**Not Good (2)**	**10**
Campus wide audit of university laboratories – energy, water, input and output of chemicals, hazardous waste management.		1
Establishment of lab-specific prioritised targets for improvement.		5
Development of online tracking system for chemical management (inputs, processes and outputs).		3
Establish lab equipment / consumables exchange program to minimise waste.		1
Maintenance and capital works	**Moderate (3)**	**13**

IOP Conf. Series: Earth and Environmental Science **91** (2017) 012037 doi:10.1088/1755-1315/91/1/012037

Development of green laboratory design standards, e.g. referencing Labs21.		1
Laboratory ventilation and fume hoods		4
Laboratory water use		4
Secure storage spaces for hazardous wastes to minimise risk of spillage / leakage.		4
Total Score Green Lab	**Not Good (6)**	

3.2.9 Green IT

Even if, Green IT is not applied in one of IPB Green Campus element but some of the indicators fulfilled by the IT standards of IPB Campus. The result shows IT Policy and behaviour change is Moderate, and IT Management and Capital Works is Good. Overall result for Green IT is Good (detail in Table 9).

Table 9. Green IT

Indicators		**Score**
1. IT Policy and behaviour change	**Moderate (3)**	**9**
Adoption and implementation of IT purchasing standards (e.g. IEEE, EPEAT, etc.).		5
"Switch off when not in use" awareness programs		3
Standard operating environments (hardware and software).		1
2. IT Management and Capital Works	**Good (4)**	**17**
Reduce frequency of computer replacement programs – substitute software upgrades for hardware upgrades where possible.		5
Centralised / dedicated server space(s) to avoid dispersing server heat loads across multiple buildings.		5
Computer reuse program, e.g. donation to community groups / schools.		2
E-waste program.		4
Ensure energy saving features are enabled.		1
Total Score Green IT	**Good (7)**	

3.2.10 Transport

Transport is the first green campus practices that been applied in IPB Dramaga Campus. Most of the indicators is already done but not at its maximum effort because of some problems like it's not acceptable yet in IPB Dramaga Campus environment. The result shows Transport in General is Very Good, Commuter Transport is Moderate, and Travel on University Business is Good. Overall result for Transport is Good (detail in Table 10).

Table 10. Transport

Indicators		**Score**
1. General	**Very Good (5)**	**9**
Employment of Transport Manager.		5
Development of university transport policy.		4
2. Commuter Transport	**Moderate (3)**	**32**
Student housing and services on or close to campus.		5
Awareness and promotion of alternatives to private transport – posters, stickers, events and competitions, websites, awards and incentives.		3
Regular liaison with public transport providers to optimise services to the campus.		1
Incentives for staff committing to forego use of private commuter transport.		1
Secure, undercover bike racks, and shower facilities, lockers and bike repair workshop for cyclists.		4
Car pooling programs.		4
Reduction of car parking spaces and provision of dedicated spaces for car pool vehicles and electric vehicles (and also charging points).		4
Establishment of shuttle bus service where the university has multiple campuses.		5
Acknowledgement that for reasons of social equity, disability, etc. some staff and students will still need to use private vehicles to access the campus.		1
Pedestrian-friendly campus to minimise internal motor vehicle trips.		4
3. Travel on University Bussiness	**Good (4)**	**12**
Acquisition and promotion of video conferencing technology to staff and students.		3
University managed revegetation program to offset emissions for air travel, and/or commitment to "third party" carbon credit / carbon offset program.		1
Purchase of fuel efficient vehicles for university fleet.		3
Regular maintenance to optimise motor vehicle fleet fuel efficiency.		5
Total Score Transport	**Good (12)**	

3.3 Green Level of IPB Dramaga Campus
The result shows that green level of IPB Dramaga Campus is Moderate with total score 31 (Table 11)

Table 11. Overall Result of IPB Dramaga Campus

No	Criteria	Score	Nilai
1	Energy, Carbon, and Climate Change	3	Moderate [a]
2	Water	2	Not Good[a]
3	Waste	3	Moderate[a]
4	Biodiversity and Ecosystem Service	5	Very Good[a]
5	Planning, Design and Development	4	Good[a]
6	Procurement	3	Moderate[a]
7	Green Office	1	Very Not Good[a]
8	Green Lab	2	Not Good[a]
9	Green IT	4	Good[a]
10	Transport	4	Good[a]
	Total	**31**	**Moderate[b]**

[a] Score (1= Very Not Good; 2 = Not Good; 3 = Moderate**;** 4 = Good; 5 = Very Good)
[b] Total Score (10 – 17 = Very Not Good; 18 – 25 = Not Good; **26 – 33 = Moderate;** 34 – 41 = Good; 42 – 50 = Very Good)

3.4 Recommendation Green Campus Strategies for IPB Dramaga Campus
Based on the results of the analysis, some of strategies that need to be added to IPB Green Campus are the Green Office, Green Campus Audit, Green Champion, Green Financial Strategies, Water Treatment, Green Lab and Off Campus Transportation. All policies could be made simultaneously, but the implementation can be done gradually in order to be sustainable, particularly for Green Office, Green Campus Audit, Green Champion, and Green Financial Strategies and Green Financial Strategies, whose policies are closely interconnected. Green Office, Green Campus Audit, Green Champion, Green Financial Strategies and Green Lab will be based upon the indicators of UNEP's Greening University Toolkit. Meanwhile the Water Treatment strategies will be implemented as Waste Water Garden to treat greywater waste from IPB Dramaga Campus. Off-Campus Transportation is an additional strategies to IPB Dramaga Campus so they should not only concern about in-campus matter but also surround the campus. If those strategies applied in IPB Dramaga Campus, the green level will increase from Moderate to Good or even Very Good depended of the implementation itself.

4. Conclusion
The study results that IPB Dramaga Campus green level through 10 UNEP's Green Campus is Moderate with total score 31. That green level will be increase from Moderate to Very Good with a proper implementation of Green Office, Green Campus Audit, Green Champion, Green Financial Strategies, Water Treatment, Green Lab and Off Campus Transportation.

References
[1] [UNEP] United Nations Environment Programme. 2013. Greening University Toolkit. China : UNEP - Tongji Institute of Environment and Sustainable Development.
[2] [UI] Universitas Indonesia. 2014. Guidelines of UIGreenmetric World University Ranking. Depok : Universitas Indonesia.
[3] ________.1996. Campus Landscape Master Plan of University of California Riverside. http://pdc.ucr.edu/does2.toc.ucr.htm
[4] Merlita R. 2014. Desain Waste Water Garden Untuk Instalasi Pengolahan Air Limbah (IPAL) Domestik Komunal Di Perumahan Nusantara Permai Sukabumi Bandar Lampung [skripsi]. Lampung : Universitas Malahayati Lampung.
[5] Simpson W. 2009. Cool Campus! A How-To Guide for College and University Climate Action Planning. AASHE.
[6] Wiedmann, T. and Minx, J. (2008). A Definition of 'Carbon Footprint'. In: C. C. Pertsova, Ecological Economics Research Trends: Chapter 1, pp. 1-11, Nova Science Publishers, Hauppauge NY, USA.

IOP Conf. Series: Earth and Environmental Science **91** (2017) 012038 doi:10.1088/1755-1315/91/1/012038

International tourist preference of Lodok Rice Field natural elements, the cultural rice field from Manggarai – Indonesia

Ray March Syahadat[1], Priambudi Trie Putra[1], Nuraini[2], Balqis Nailufar[2], Desy Fatmala Makhmud[1]

[1]National Institute of Science and Technology, Jakarta, Indonesia
[2]Tribhuwana Tunggadewi University, Malang, Indonesia

E-mail: ray.arl@istn.ac.id

Abstract. Lodok Rice Field or usually known as spiderweb rice field is a system of land division. It cultural rice field only found on Manggarai, Province of East Nusa Tenggara, Indonesia. The landscape of Lodok Rice Field was aesthetic and it has big potential for tourism development. The aim of this study was to know the perception of natural elements of Lodok Rice Field landscape that could influence international tourist to visited Lodok Rice Field. If we know the elements that could influenced the international tourist, we could used the landscape image for tourism media promotion. The methods of this study used scenic beauty estimation (SBE) by 85 respondents from 34 countries and Kruskal Wallis H test. The countries grouped by five continents (Asia, America, Europe, Africa, and Oceania). The result showed that the Asian respondents liked the elements of sky, mountain, and the rice field. Then, the other respondent from another continent liked the elements of sunshine, mountain, and the rice field. Although the Asian had different perception about landscape elements of rice field's good view, it's not differ significantly by Kruskal Wallis H test.

1. The first section in your paper

Indonesia has 21 million farmers, the land area of 11.5 million hectares, and the rice become a staple food for about 95% Indonesian people [1]. Rice farming influenced by cultural conditions and also local knowledge. [2] [3]. Local knowledge is custom or habit that has become a tradition, conducted by a group of people from generation to generation which is still maintained by the customary law community in a region, and has significant value.

Lodok rice field from Manggarai, Province of East Nusa Tenggara is one of cultural rice field in Indonesia. This Rice Field has pattern look like spiderweb. Therefore this rice field also known as spiderweb rice field. Manggarai community called land sharing tradition as *lingko*. *Lingko* allegedly created since the people in Flores Island know agricultural system until they formed a village that called *beo*. Each *lingko* has a leader that called *tua golo*. *Tua golo* has power to regulate the *lingko*, *beo*, and all costum [4]. The system of regulate a land area (*lingko*) called *lodok*. This system concentrated in the big house that called *mbaru gendang* that has philosophy as a symbol of indigenous power.

Land sharing of Lodok Rice Field depending on the number of recipient, relationship with the landlord, and the social status in the *beo*. *Tua golo* usually get the biggest part because has many responsibilities. In the center of Lodok rice field, there is a *teno*, i.e. wood become the focal point to make a line. Physically, Lodok Rice Field has outstanding attraction as well as the philosophy and also the landscape. Unfortunately the potential of Lodok Rice Field has not been optimized primarily as a tourism attraction. From the last research, Lodok Rice Field favoured by the tourists (domestic and international) and they interest to visit it [5]. The aim of this study was to know the perception of natural elements of Lodok Rice Field landscape that could influence international tourist to visited Lodok Rice Field. If we know the elements that could influenced the international tourist, we could used the landscape image for tourism media promotion.

2. Methods

The method that used of this study was survey to 85 respondents from outside Indonesia. The countries grouped by five continents which are Asia, America, Europe, Africa, and Oceania. The data analyzed by descriptive statistic, scenic beauty estimation (SBE), and Kruskal Wallis H test. SBE used to know where the landscape that has good view. We showed four different Lodok Rice Field's landscapes and the respondents give a score from 1 (bad view) to 10 (good view). The fourth landscapes described different situations, angle, view points, scenic routes, and also elements (Figure 1).

Figure 1. The fourth landscapes of Lodok Rice Field

The image of landscape 1 just focuses on the rice field as the main element. The landscape 2 has elements of sky, mountain, and rice field. Next, the image of landscape 3 has elements of sunshine, mountain, and rice field. Then, the landscape 4 has elements of forest, mountain, and rice field. All images used landscape layout orientation except landscape 2 that used portrait layout orientation. Kruskal Wallis H test used to ensure there or not differences in SBE result by respondents in each continent.

3. Result and discussion

Tourism and the images of landscape have connectivity because landscape is motivating factor for tourists [6] [7] [8]. It has influence for the tourist to make decision for their destination [9] [10]. Thus, the images of landscape for the tourism promotion should choose carefully.

The result showed that almost all images of landscape had good view because it score there are in interval -20 to 20 (Figure 2). Asian respondents liked the elements of sky, mountain, and the rice field (Landscape 2). They like the portrait layout orientation. We assumed it because the Asian culture. Most of Asian people lived in East Asia and as we know, the East Asian people have culture to written down (vertically). Much old paint in Asia also used the same orientation which is portrait layout orientation.

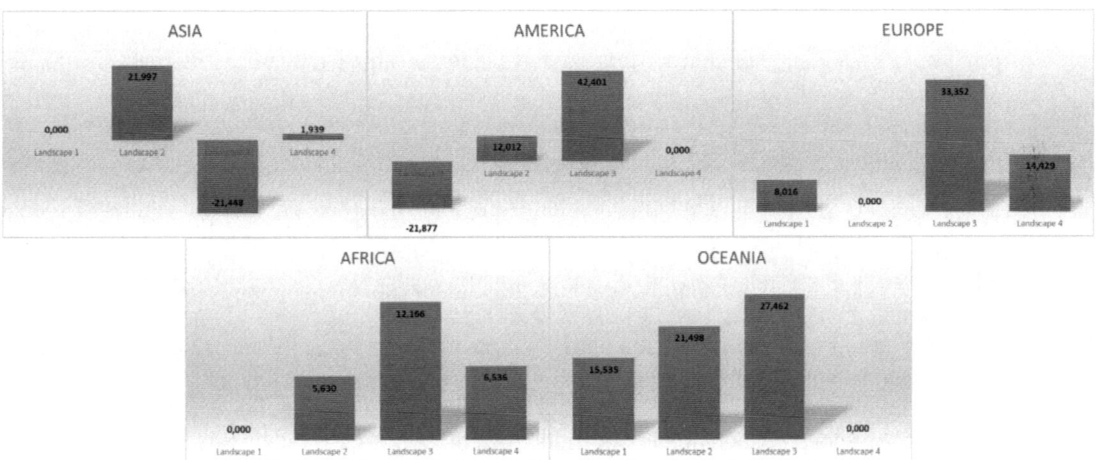

Figure 2. SBE result

The respondent from the other continents had different perception with Asian respondents. They liked the elements of sunshine, mountain, and the rice field (Landscape 3). The American respondents had contrast perception with Asian respondent. They think the portrait image is worst image of all. Most of respondent think the landscape image with landscape layout orientation had good view than portrait layout orientation. Bodnár ever reported the differences perception influenced by viewpoints and scenic routes [11]. Our research proved it, the different angle and layout orientation affected the result of SBE.

Because of the variety result, we used Kruskal Wallis H test to ensure it. The result showed that although the Asian had different perception about landscape elements of rice field's good view, it's did not differ significantly by Kruskal Wallis H test (Table 1). It means the landscape that had good view and could be used for tourism media promotion of Lodok Rice Field by people around the world is landscape 3 with elements of sunshine, mountain, and the rice field.

Table 1. Kruskal Wallis H test result

	SBE
Chi-Square	4,224
df	3,000
Asymp. Sig.	0,238

But, if the government want to promote the Lodok Rice Field to specific continent for example Asia, they can be used the landscape 2 with elements of sky, mountain, and the rice field with portrait layout orientation. This result also informs us although the main element for the tourism object is the rice field, the surrounding natural elements can be considered to increase aesthetic value of the object.

4. Conclusion

The Asian respondents liked the elements of sky, mountain, and the rice field. Then, the other respondent from another continent liked the elements of sunshine, mountain, and the rice field. Although the Asian had different perception about landscape elements of rice field's good view, it's did not differ significantly by Kruskal Wallis H test. It means the landscape that had good view and could be used for tourism media promotion of Lodok Rice Field by people around the world is landscape with elements of sunshine, mountain, and the rice field

References

[1] Swastika D K S, Wargiono J, Soejitno and Hasanuddin A 2007 Analisa kebijakan peningkatan produksi padi melalui efisiensi pemanfaatan lahan sawah di Indonesia *Analisis Kebijakan Pertanian* 5(1) p 36-52

[2] Sumodiningrat G 2001 *Menuju swasembada pangan. revolusi hijau II: introduksi manajemen dalam pertanian* (Jakarta: RBI)

[3] Uker D and Fanany R 2011 The traditional decision-making process of Berkaul in Tanjung Emas, West Sumatra: its nature and significance. *SOJOURN: Journal of Social Issues in Southeast Asia* 26(1) p 1-15

[4] Jemahat L 2011 Pola relasi sosial elit tradisional: kasus Desa Rana Mbeling, Kabupaten Manggarai Timur, NTT *Jurnal Demokrasi* 10(1) p 67-84

[5] Nailufar B, Nuraini, Syahadat R M and Nurisjah S 2015 *Tourist preference of Lodok Rice Field, the spiderweb rice field from Manggarai Indonesia In: Proceeding ACLA International Symposium Agricultural Landscapes of Asia: Learning, Preserving, and Redifining* ed Asian Cultural Landscape Association (Bali: Indonesia) p 1-5

[6] Molina M A R, Jamilena D M F and Garcia C 2015 The contribution of website design to the generation of tourist destination image: The moderating effect of involvement *Tourism Management* 47(2015) p 303-317

[7] Tasci A & Garther W 2007 Destination image and its functional relationships *Journal of Travel Research* 45 (2007) p 413-425

[8] Bodnár R 2010 *Landscape as a motivating factor for tourists. In: Recreation, Tourism and Nature in Changing World, Proceedings of The Fifth International Conference on Monitoring and Management of Visitor Flows in Recreational and Protected Areas* ed Gooseen M, Elands B, Marwick, R van May 30-June 3 2010 (Wageningen: Netherlands) pp 144-146

[9] Nadeau J, Heslop L, O'Reilly N and Luk P 2008 Destination in a country image context *Annals of Tourism Research* 35(1) p 84-106

[10] Chen C and Tsai D 2007 How destination image and evaluative factors affect behavioural intentions *Tourism Management* 28(2007) p 1115-1122

[11] Bodnár R K 2011 Tourist aspects of assessing landscape change *GeoJournal of Tourism and Geosites* 7(1) p 39-50